Development of g-C$_3$N$_4$-Based Photocatalysts: Environmental Purification and Energy Conversion

Development of g-C_3N_4-Based Photocatalysts: Environmental Purification and Energy Conversion

Editors

Weilong Shi
Feng Guo
Xue Lin
Yuanzhi Hong

Basel • Beijing • Wuhan • Barcelona • Belgrade • Novi Sad • Cluj • Manchester

Editors

Weilong Shi
School of Material Science
and Engineering
Jiangsu University of Science
and Technology
Zhenjiang
China

Feng Guo
School of Energy and Power
Jiangsu University of Science
and Technology
Zhenjiang
China

Xue Lin
School of Material Science
and Engineering
Beihua University
Jilin
China

Yuanzhi Hong
School of Material Science
and Engineering
Beihua University
Jilin
China

Editorial Office
MDPI
St. Alban-Anlage 66
4052 Basel, Switzerland

This is a reprint of articles from the Special Issue published online in the open access journal *Catalysts* (ISSN 2073-4344) (available at: https://www.mdpi.com/journal/catalysts/special_issues/ 06846G0EH5).

For citation purposes, cite each article independently as indicated on the article page online and as indicated below:

Lastname, A.A.; Lastname, B.B. Article Title. *Journal Name* **Year**, *Volume Number*, Page Range.

ISBN 978-3-0365-9661-7 (Hbk)
ISBN 978-3-0365-9660-0 (PDF)
doi.org/10.3390/books978-3-0365-9660-0

© 2023 by the authors. Articles in this book are Open Access and distributed under the Creative Commons Attribution (CC BY) license. The book as a whole is distributed by MDPI under the terms and conditions of the Creative Commons Attribution-NonCommercial-NoDerivs (CC BY-NC-ND) license.

Contents

Loic Jiresse Nguetsa Kuate, Zhouze Chen, Jialin Lu, Huabing Wen, Feng Guo and Weilong Shi
Photothermal-Assisted Photocatalytic Degradation of Tetracycline in Seawater Based on the Black g-C_3N_4 Nanosheets with Cyano Group Defects
Reprinted from: *Catalysts* 2023, 13, 1147, doi:10.3390/catal13071147 1

Tejaswi Tanaji Salunkhe, Thirumala Rao Gurugubelli, Bathula Babu and Kisoo Yoo
Recent Innovative Progress of Metal Oxide Quantum-Dot- Integrated g-C_3N_4 (0D-2D) Synergistic Nanocomposites for Photocatalytic Applications
Reprinted from: *Catalysts* 2023, 13, 1414, doi:10.3390/catal13111414 17

Junxiang Pei, Haofeng Li, Songlin Zhuang, Dawei Zhang and Dechao Yu
Recent Advances in g-C_3N_4 Photocatalysts: A Review of Reaction Parameters, Structure Design and Exfoliation Methods
Reprinted from: *Catalysts* 2023, 13, 1402, doi:10.3390/catal13111402 37

Taoming Yu, Doudou Wang, Lili Li, Wenjing Song, Xuan Pang and Ce Liang
A Novel Organic/Inorganic Dual Z-Scheme Photocatalyst with Visible-Light Response for Organic Pollutants Degradation
Reprinted from: *Catalysts* 2023, 13, 1391, doi:10.3390/catal13111391 55

Weili Fang and Liang Wang
S-Scheme Heterojunction Photocatalyst for Photocatalytic H_2O_2 Production: A Review
Reprinted from: *Catalysts* 2023, 13, 1325, doi:10.3390/catal13101325 75

Subhadeep Biswas and Anjali Pal
A Brief Review on the Latest Developments on Pharmaceutical Compound Degradation Using g-C_3N_4-Based Composite Catalysts
Reprinted from: *Catalysts* 2023, 13, 925, doi:10.3390/catal13060925 93

Shuai Zhang, Enhui Jiang, Ji Wu, Zhonghuan Liu, Yan Yan, Pengwei Huo and Yongsheng Yan
Visible-Light-Driven GO/Rh-$SrTiO_3$ Photocatalyst for Efficient Overall Water Splitting
Reprinted from: *Catalysts* 2023, 13, 851, doi:10.3390/catal13050851 125

Shiyun Li, Yuqiong Guo, Lina Liu, Jiangang Wang, Luxi Zhang, Weilong Shi, et al.
Fabrication of FeTCPP@CNNS for Efficient Photocatalytic Performance of p-Nitrophenol under Visible Light
Reprinted from: *Catalysts* 2023, 13, 732, doi:10.3390/catal13040732 139

Yawei Xiao, Zhezhe Wang, Bo Yao, Yunhua Chen, Ting Chen and Yude Wang
Hollow g-C_3N_4@$Cu_{0.5}In_{0.5}S$ Core-Shell S-Scheme Heterojunction Photothermal Nanoreactors with Broad-Spectrum Response and Enhanced Photocatalytic Performance
Reprinted from: *Catalysts* 2023, 13, 723, doi:10.3390/catal13040723 157

Jiajia Wei, Xing Chen, Xitong Ren, Shufang Tian and Feng Bai
Facile Construction of Intramolecular g-CN-PTCDA Donor-Acceptor System for Efficient CO_2 Photoreduction
Reprinted from: *Catalysts* 2023, 13, 600, doi:10.3390/catal13030600 173

Faisal Al Marzouqi and Rengaraj Selvaraj
Surface Plasmon Resonance Induced Photocatalysis in 2D/2D Graphene/g-C$_3$N$_4$ Heterostructure for Enhanced Degradation of Amine-Based Pharmaceuticals under Solar Light Illumination
Reprinted from: *Catalysts* **2023**, *13*, 560, doi:10.3390/catal13030560 . . . . **187**

Article

Photothermal-Assisted Photocatalytic Degradation of Tetracycline in Seawater Based on the Black g-C₃N₄ Nanosheets with Cyano Group Defects

Loic Jiresse Nguetsa Kuate [1,†], Zhouze Chen [2,†], Jialin Lu [1], Huabing Wen [1], Feng Guo [1,*] and Weilong Shi [2,*]

1. School of Energy and Power, Jiangsu University of Science and Technology, Zhenjiang 212003, China
2. School of Material Science and Engineering, Jiangsu University of Science and Technology, Zhenjiang 212003, China
* Correspondence: gfeng0105@126.com (F.G.); shiwl@just.edu.cn (W.S.)
† These authors contributed equally to this work.

Abstract: As a broad-spectrum antibiotic, tetracycline (TC) has been continually detected in soil and seawater environments, which poses a great threat to the ecological environment and human health. Herein, a black graphitic carbon nitride (CN-B) photocatalyst was synthesized by the one-step calcination method of urea and phloxine B for the degradation of tetracycline TC in seawater under visible light irradiation. The experimental results showed that the photocatalytic degradation rate of optimal CN-B-0.1 for TC degradation was 92% at room temperature within 2 h, which was 1.3 times that of pure CN (69%). This excellent photocatalytic degradation performance stems from the following factors: (i) ultrathin nanosheet thickness reduces the charge transfer distance; (ii) the cyanogen defect promotes photogenerated carriers' separation; (iii) and the photothermal effect of CN-B increases the reaction temperature and enhances the photocatalytic activity. This study provides new insight into the design of photocatalysts for the photothermal-assisted photocatalytic degradation of antibiotic pollutants.

Keywords: photothermal-assisted; black g-C₃N₄; photocatalytic; degradation; cyano group defects

1. Introduction

In the past half a century, the rapid development of the mariculture industry has led to an increase in human demand for seafood. Many mariculture sites in coastal areas have been used to breed all kinds of seafood, and the bacteria in seawater are prone to the outbreak of infectious diseases in the aquaculture water bodies, endangering social health [1–4]. Tetracycline (TC), a commonly used broad-spectrum antibiotic to prevent infectious diseases and treat bacterial infections, can be widely used to eliminate pathogenic bacteria and viruses from seawater [5–7]. However, due to the strong toxicity of TC molecules, it is easy to discharge living organisms through urine or feces, polluting water bodies, causing great harm to the seawater environment and organisms, and causing many adverse effects such as bacterial drug resistance and biological toxicity [8,9].

In recent years, researchers have used Fenton oxidation, biological treatment, adsorption, and membrane separation to remove antibiotics from seawater [10–12]. Unfortunately, the above methods have high energy consumption, low removal efficiency, and high cost, which limits their practical application [13,14]. In contrast, photocatalysis is an advanced oxidation technology with low consumption, strong oxidation capacity, and high mineralization rate, which can directly use photocatalyst to absorb solar energy and produce active substances to effectively remove various kinds of low-concentration pollutants in seawater [15,16]. Among the many photocatalytic materials, graphitic carbon nitride (g-C₃N₄) is a two-dimensional (2D) layered nanosheets composed of carbon and nitrogen atoms, which has attracted wide attention due to its unique electronic band structure, excellent

optical and electronic properties, easy preparation, and high stability [17–20]. Nevertheless, the photocatalytic activity of g-C_3N_4 is limited by its inherent defects, including poor visible light absorption (λ < 450 nm), high electron–hole pair recombination rate, and slow charge migration rate [21–23]. In order to address these problems, various strategies have been developed to enhance the photocatalytic activity of the original g-C_3N_4, such as heterojunction construction, element doping, morphological control, etc. [24–26]. Among them, the introduction of cyano group defects in g-C_3N_4 can promote visible light absorption, and photo-generation electron–hole separation is considered an effective method to improve photocatalytic activity [27]. For example, Hu et al. prepared a BaCN-C_3N_4 photocatalyst that promotes visible light absorption and higher photocatalytic activity due to the presence of Ba^{2+}, which promotes partial ring-opening through heptazine rings leading to cyano group defects production [28]. However, due to the low ambient temperature and photon utilization rate, the photocatalytic degradation efficiency of photocatalysts is still low. Fortunately, the construction of the photothermal-assisted photocatalytic degradation system can increase the temperature in the reaction system through the photothermal effect of the photothermal material itself, thus accelerating the chemical reaction kinetics on the surface of the photocatalyst, promoting the transfer of photogenerated charge, and improving the photocatalytic degradation activity [29]. For instance, Yang et al. synthesized core–shell $CoTiO_3$@MnO_2 photocatalyst photothermal catalyst, which can show excellent TC degradation performance under light irradiation due to the broad-spectrum absorption and photothermal effect of MnO_2 that can promote the charge separation [30]. Similarly, Wang et al. reported the development of a photothermal-assisted photocatalytic system using graphene oxide as a photothermal substrate and g-C_3N_4 as a photocatalyst, which exhibited excellent stability and reusability for the degradation of antibiotics in wastewater treatment [31]. However, most of the above-mentioned researches on photothermal-assisted photocatalysis is realized by combining photothermal materials, which often has the disadvantages of a complex preparation process, poor stability, and high cost [32,33]. Considering this, it is of great significance to design an ultrathin g-C_3N_4 nanosheet with its own photothermal effect for the photothermal-assisted photocatalytic degradation of antibiotics.

Herein, the black carbon nitride (CN-B) photocatalysts were constructed by adding Phloxine B to the process of conventional urea preparation carbon nitride as well as using a one-step calcination method to achieve photothermal-assisted photocatalytic degradation of TC in simulated seawater under visible light irradiation. Our results exhibit that the ultrathin nanosheet thickness of CN-B reduces the charge transfer distance, and the introduction of Phloxine B leads to cyano group defects to promote the photogenerated charge transfer and make it have a strong thermal effect to improve the temperature of the reaction system and the photocatalytic activity.

2. Results and Discussion

Scanning electron microscopy (SEM) was used to observe the morphology of the as-prepared photocatalysts. As given in Figure 1a,b, the SEM images of pure CN and CN-B-0.1 photocatalysts showed small morphological differences, both of which are stacked nanosheet structures. Furthermore, the microstructure and elemental composition of CN and CN-B-0.1 were analyzed by transmission electron microscopy (TEM). As can be seen from Figure 1c,d, the thickness of CN-B-0.1 nanosheet is about 12–18 nm, indicating that the introduction of phloxine B during the process of synthesis can not only maintain the nanosheet structure of pure CN but also the ultrathin nanosheet thickness can shorten the charge transfer distance, which is more conducive to improve the photocatalytic activity. In addition, in order to further explore the elements of the as-prepared CN-B photocatalyst, the element mapping images (Figure 1e), energy dispersive X-ray spectra (EDX, Figure S1), and element composition analysis based on X-ray photoelectron spectroscopy (XPS) measurements (Table S1) presented the existence of the corresponding elements of C, N, O, and Na.

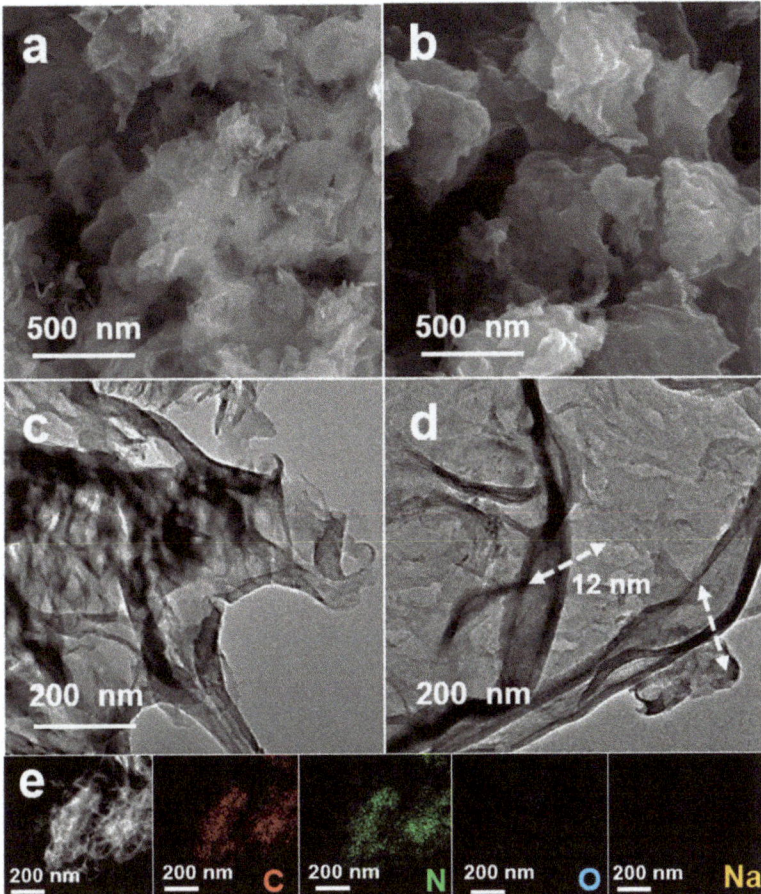

Figure 1. (a,b) SEM images of CN and CN-B-0.1. (c,d) TEM images of CN and CN-B-0.1. (e) Elemental mapping images of CN-B-0.1.

The X-ray diffraction (XRD) patterns of as-prepared CN and CN-B samples are given in Figure 2a. For the original CN sample, two distinct characteristic diffraction peaks were observed, and the first peak located at 13.2° corresponds to the (001) plane structure, specifically representing the arrangement of tris-triazine units (with the lattice of 6.76 Å) within the network of g-C_3N_4 [34,35], while the second peak situated at 27.5° is associated with the (002) plane, indicating the periodic accumulation of carbon nitride nanosheets along the c-axis [36,37]. Furthermore, compared with the XRD pattern of pure CN, the (001) peak intensity of the CN-B photocatalyst gradually decreased with the increase in the phloxine B content in the precursor, indicating a periodic disruption of the planar structure [38]. In addition, the (002) crystal plane diffraction peaks for CN-B photocatalysts exhibit slight blue shifts due to the strengthening of the interface interaction between CN-B nanosheets, decreasing the layer spacing [39,40]. The functional groups of the synthesized samples were further determined using Fourier transform infrared spectroscopy (FT-IR). Figure 2b presents the FT-IR spectra of CN-B photocatalysts are consistent with the typical vibration modes of CN, indicating that the functional group structure of CN was not significantly changed after addition of phloxine B in the precursor. Typical absorption bands located at 1200–1700 cm^{-1} represent the stretching mode of unique aromatic CN heterocycles, while the broadband in the 3000–3500 cm^{-1} range may interact with stretched vibrations of the O-H and N-H bonds adsorbed on the CN surface [41,42]. The peak signal at 810 cm^{-1} is thought to be caused by the characteristic stretching vibration peak of the triazine unit in CN [43,44]. Another interesting finding was that a new peak was observed at 2170 cm^{-1} of the CN-B photocatalyst, caused by the tensile vibration of the cyano groups (C≡N) generated by the catalytic pyrolysis on the CN surface during the calcination of urea and phloxine B [45,46].

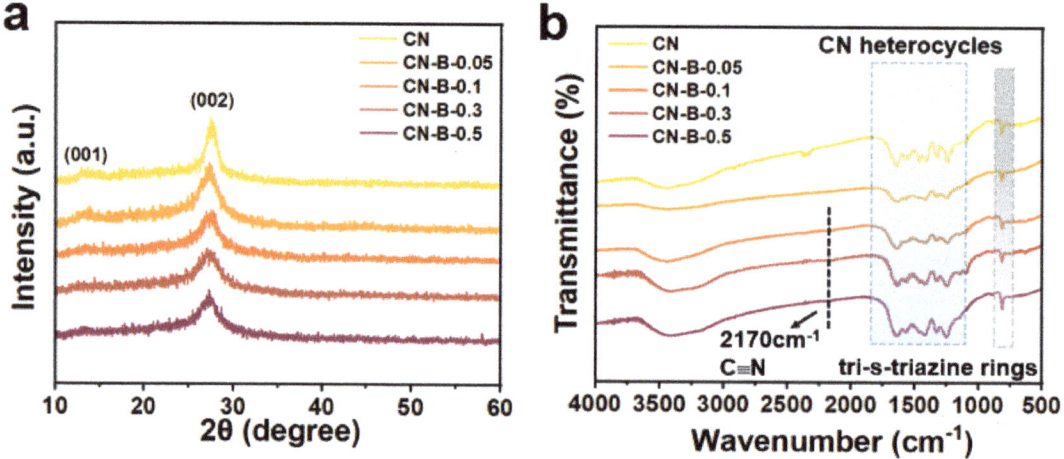

Figure 2. (a) XRD patterns and (b) FT-IR spectra of CN and CN-B photocatalysts.

The chemical composition and state of as-prepared CN and CN-B samples were revealed by XPS analysis. From the XPS survey spectra in Figure 3a, it can be observed that the CN-B-0.1 photocatalyst not only presents the C 1s, N 1s, and O 1s elements of the original CN but also detects additional Na 1s elements, which perfectly matches the results of EDX spectrum and element mapping images. Moreover, a small amount of the O element can be attributed to the adsorbed water on the CN surface, while the trace amount of the Na element may be caused by the low content of phloxine B in the photocatalyst. Compared with pure CN, CN-B-0.1 possesses a higher C atom ratio, which may be due to the increase in C content caused by the addition of phloxine B during the preparation process, resulting in an increase in the C/N atomic ratio (Table S1). As disclosed in Figure 3b, the high-resolution C1s spectrum of the CN-B-0.1 photocatalyst can be fitted to the three prominent peaks at 284.8, 285.9, and 288.2 eV, which correspond to C-C, C-N-H, and C-N=C, respectively [47,48]. The three peaks of the high-resolution N 1s spectrum of CN-B-0.1 in Figure 3c at 398.8, 399.6, and 400.8 eV are attributed to the triazine ring (C-N=C, N$_{2c}$), nitrogen N-(C)$_3$ group, and N-H bond, respectively [49,50]. For the high-resolution O 1s spectrum, CN-B-0.1 has two peaks at binding energies 532.0 and 533.4 eV (Figure 3d), which can be attributed to the C=O and C-O-H bonds [51,52]. Additionally, as presented in Figure S2, compared with pure CN, the peak of CN-B-0.1 photocatalyst at 1071.4 eV corresponds to Na 1s element [53,54]. In summary, it is evident that the positions of the C1s, N 1s, and O 1s peaks on the CN-B photocatalyst have undergone a slight shift compared to CN, indicating that the physical environment of the internal structure of CN could be changed, which affects electron transfer in the CN network to a certain extent [55,56].

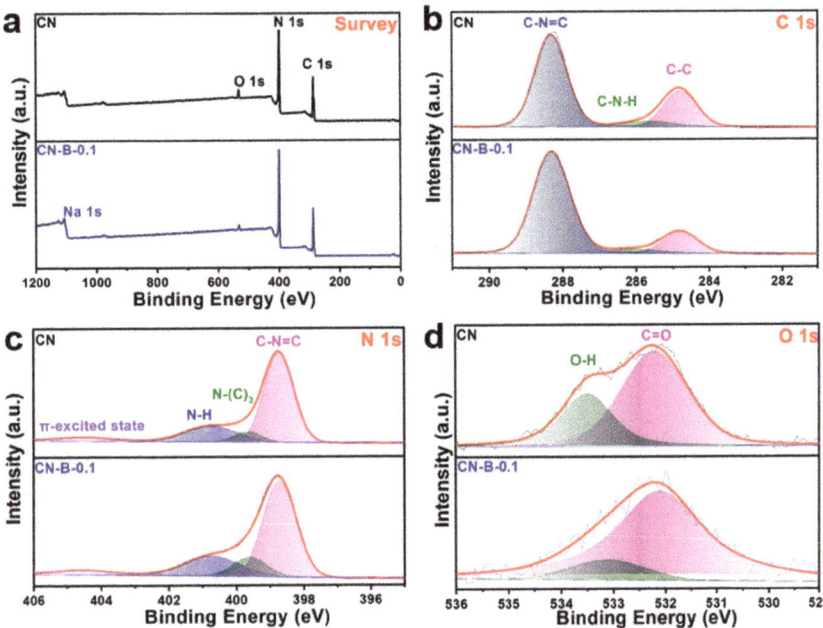

Figure 3. (**a**) XPS survey spectra and high-resolution XPS spectra of (**b**) C 1s, (**c**) N 1s, and (**d**) O 1s for CN and CN-B-0.1 photocatalysts.

The optical properties of the synthesized samples were characterized by UV-vis diffuse reflection spectra (DRS) at the absorption wavelength range of 300–800 nm. As presented in Figure 4a, pure CN exhibits the characteristic absorption pattern common in organic semiconductors, with a gap absorption of about 450 nm near the absorption edge [7,57,58]. For the CN-B photocatalysts, the absorption edge exhibits a significant red shift, increasing from 450 to 600 nm, indicating the reduced band gap and increase in visible light absorption in the visible region compared with CN-B photocatalysts. Figure S3 exhibits a digital photo of the as-prepared samples; the pure CN is pale yellow, while as the precursor increases the contents of phloxine B, the color of the CN-B photocatalyst gradually changes black, thus enhancing the absorption of visible light. In addition, the band gap values of as-prepared CN and CN-B samples were studied by using the Tauc function $(\alpha h\nu)^2 = A(h\nu - E_g)$ [59], and the corresponding bandgaps (E_g) of pure CN and different proportions of CN-B photocatalysts are given in Figure 4b. In order to further insights into the band structures of CN and CN-B, the flat band potentials (E_{fb}) of CN and CN-B-0.1 were analyzed by Mott–Schottky (M-S) measurement at tested the frequencies of 800, 1000, and 1200 Hz, respectively. As can be seen from Figure 4c, the E_{fb} of CN and CN-B-0.1 are −0.72 and −0.62 V vs. reversible hydrogen electrode (RHE), respectively, indicating that the CN-B exhibits the more negative conduction band (CB) position, thereby enhancing the photocatalytic redox ability [60,61]. Based on the position of flat band potentials and the equation of $E_{VB} = E_{CB} + E_g$, the valence band (VB) of CN and CN-B-0.1 photocatalyst is calculated to be 2.07 and 2.01 eV, respectively, and the corresponding band structure diagrams were summarized in Figure 4d.

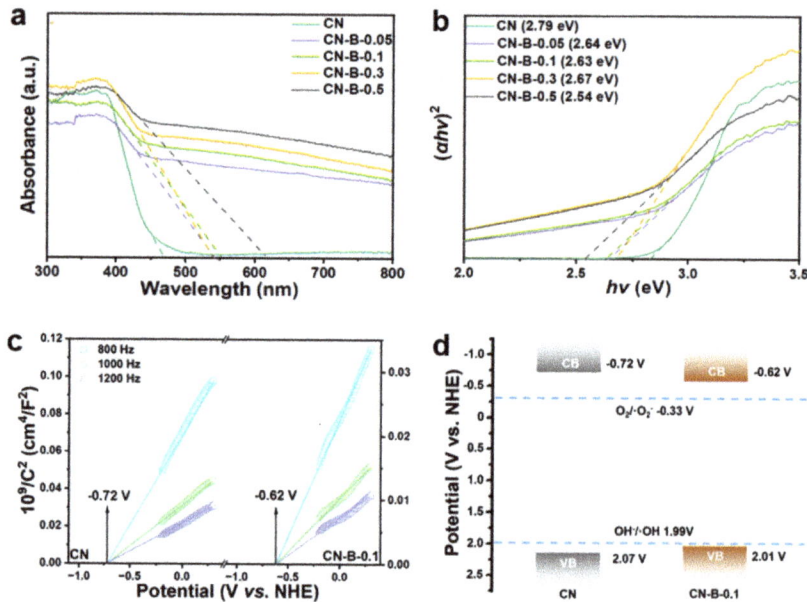

Figure 4. (a) UV–vis DRS of as-prepared photocatalysts. (b) Measured band gap values of pure CN and CN-B samples. (c) Mott–Schottky plots. (d) Energy diagrams of CN and CN-B-0.1 photocatalysts.

In order to evaluate the photocatalytic degradation efficiency of the as-prepared photocatalyst, the photocatalytic degradation experiment was carried out in simulated seawater with TC (30 mg/L) as the target pollutant under visible light irradiation. Before the photocatalytic reaction, the mixture of TC and photocatalysts was vigorously stirred without light for 30 min to ensure the adsorption-desorption equilibrium. As presented in Figure 5a, pure CN degraded only 69% of TC within 2 h under visible light radiation in simulated seawater, while the degradation rates of CN-B-0.05, CN-B-0.1, CN-B-0.3, and CN-B-0.5 were 89.4%, 91.7%, 90.9%, and 90.7%, respectively. Based on the above results, a certain amount of cyano group defect can promote photogenerated electron migration and improve the photocatalytic degradation activity of CN, while the excessive cyano group defect may lead to new photogenerated carrier recombination centers, reduce the production of active substances and thus CN-B-0.1 shows the highest photocatalytic degradation efficiency [40]. In addition, Table S2 and Figure S4 compare the performance of different materials for TC degradation reported in the literature, revealing that CN-B-0.1 photocatalyst has excellent photocatalytic degradation activity. The pseudo-first-order kinetic model was also used to fit the TC photocatalytic degradation kinetic curve and corresponding kinetic constants of the as-prepared photocatalysts (Figure 5b,c). The apparent rate constant (k) of CN-B-0.1 is 0.0242 min^{-1}, which is 2.44 times that of pure CN (0.0099 min^{-1}). Additionally, considering that the stability of the photocatalyst is one of the key factors in practical application, four continuous experiments of CN-B-0.1 photocatalytic TC degradation in simulated seawater were carried out. It is worth mentioning that CN-B-0.1 photocatalyst still maintained 90% TC degradation efficiency after four cycles of TC degradation experiment (Figure 5d), indicating that the CN-B-0.1 photocatalyst possesses superior stability. The O 1s XPS spectrum of the CN-B-0.1 photocatalyst did not remarkably changed after the photocatalysis, demonstrating that the structural stability of the CN-B-0.1 photocatalyst is well maintained (Figure S5).

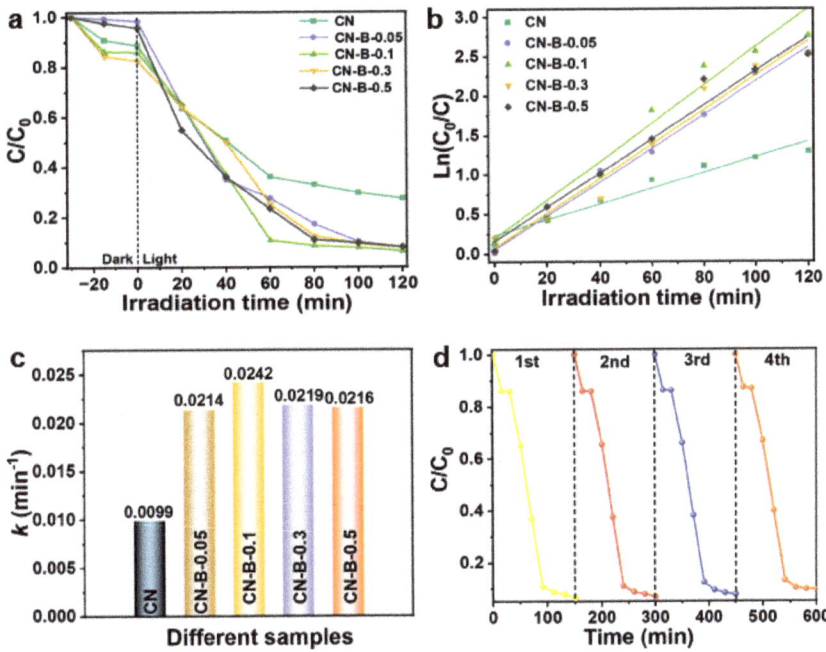

Figure 5. (a) Photocatalytic TC degradation activity of as-prepared samples in simulated seawater. (b,c) The pseudo-first-order degradation kinetic curves and corresponding degradation rate constants of as-prepared samples. (d) Four cycles of experiments for photocatalysis degradation of TC over CN-B-0.1 photocatalyst.

In order to reveal the contribution of the introduction of phloxine B precursor for the enhanced photocatalytic degradation activity, CN and CN-B-0.1 photocatalyst powders were irradiated for 150 s using a 300 W Xenon lamp, while surface temperature measurements were recorded at 30 s intervals using an infrared thermal camera. The temperature changes in CN and CN-B-0.1 photocatalysts are given in Figure 6a,b, where the temperature of CN-B-0.1 rapidly increases from 27.7 °C to 71.6 °C and remains stable, which is higher than that of pure CN (44.1 °C). This is because the introduction of phloxine B precursor deepens the CN color, prompting the photothermal effect of CN-B. Under the action of the photothermal effect, low-energy visible photons can be effectively converted into heat energy to achieve efficient photocatalytic degradation of TC, and thus the photothermal conversion efficiency of CN-B-0.1 photocatalyst (Figure 6c) was further investigated. When the entire system reaches equilibrium, the calculated time constant of CN-B-0.1 photocatalyst is 279.47 S, and the corresponding photothermal conversion efficiency (η) is 88.06%, further proving that CN-B-0.1 photocatalyst can quickly and efficiently convert light into heat energy, further accelerating the charge separation and transfer, and enhancing the photocatalytic activity. Based on the above analysis, the photocatalytic degradation activity of CN-B-1 photocatalyst was tested at different temperatures (5 °C, 10 °C, and room temperature (RT)) by controlling the reaction temperature through a circulating condensate device system. As presented in Figure 6d, pristine CN has a certain temperature sensitivity, the degradation rate decreases with the decrease in temperature, and the degradation rate reaches 69% at RT condition. In addition, the photocatalytic degradation rate of CN-B-0.1 was positively correlated with temperature, and the photocatalytic degradation rate reached 92% at the same time, indicating that the introduction of phloxine B precursor can enhance the photocatalytic degradation activity of CN by synergistic reaction of photothermal effect. Under the action of photothermal-assisted photocatalytic degradation, the degradation kinetic curves and kinetic constants of pure CN and CN-B-0.1 photocatalysts conform to the pseudo-first-order rate equation (Figure 6e,f). The apparent rate constant k (min^{-1}) value of CN-B-0.1 photocatalyst at different temperatures (5 °C, 10 °C, and RT) is 0.005, 0.007, and 0.024 min^{-1}, which is 2.5, 1.75, and 2.4 times that of pure CN (0.002, 0.004, and 0.010 min^{-1}), respectively, indicating that increased temperature can accelerate the photocatalytic degradation rate constant and improve the photocatalytic activity. Correspondingly, photoelectrochemical characterizations at different temperatures were tested to investigate the role of photothermal effects on the catalysts' carrier dynamics. As presented in Figure 6g, at the same temperature, CN-B-0.1 RT exhibited a lower PL signal compared with pure CN due to the higher photon-generated carrier separation rate of CN-B enabled by the introduction of cyano group defects. The PL signal of CN-B-0.1 decreased significantly with increasing temperature, indicating that increasing temperature can further improve the charge separation efficiency [62–64]. In Figure 6h, CN-B-0.1 exhibited a high photocurrent response intensity, which proves the strong electron transfer capability of the modified photocatalyst [65,66]. As the ambient temperature increases, the photocurrent response intensity of CN-B-0.1 increases, which further indicates that the higher temperature further promotes electron migration and transfer, thus improving the photocatalytic production activity. For the electrochemical impedance spectrum (EIS) plots in Figure 6i, the arc radius of the CN-B-0.1 sample is less than CN and decreases as the ambient temperature increases, which again demonstrates the promotion of the cyanogen defect and the photothermal effect on electron transfer [67–69].

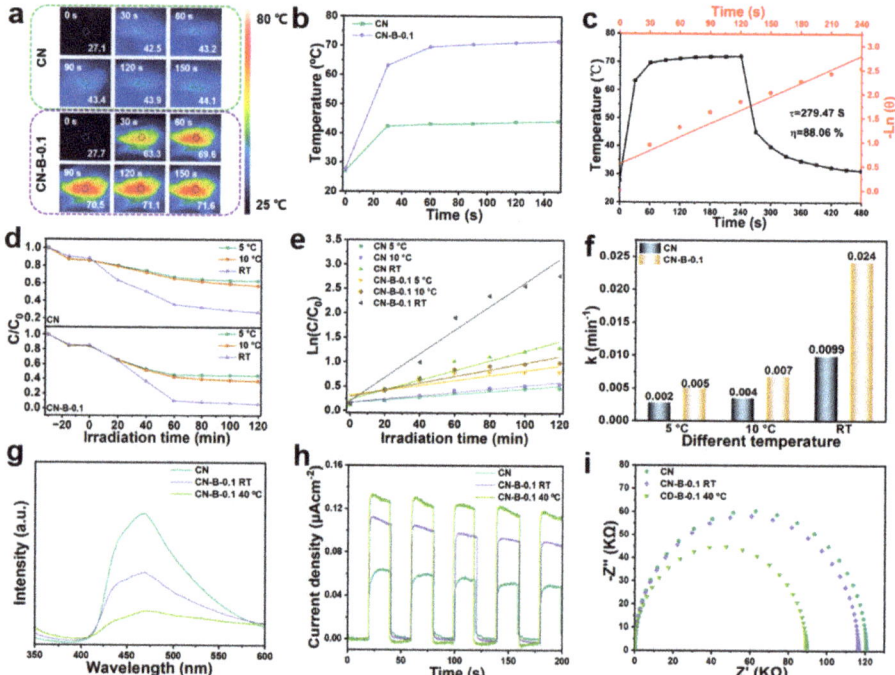

Figure 6. (**a**) Photothermal infrared thermal images and (**b**) corresponding temperature curves of CN and CN-B-0.1 in simulated seawater. (**c**) Temperature and photothermal conversion efficiency of CN-B-0.1 photocatalyst. (**d**) Photocatalytic TC degradation activity of CN and CN-B-0.1 in simulated seawater at different temperatures. (**e**,**f**) The corresponding pseudo-first-order degradation kinetic curves and kinetic constants. (**g**) PL spectra. (**h**) Transient photocurrent response curves. (**i**) EIS plots of pristine CN and CN-B-0.1 at RT and 40 °C.

In order to determine the main reactive radical species responsible for the photocatalytic degradation TC for the CN-B system, corresponding free radical scavenging experiments were performed [70,71]. The scavengers of 1,4-Benzoquinone (BQ), triethanolamine (TEOA), and isopropanol (IPA) were used to quench superoxide radicals ($\bullet O_2^-$), holes (h^+), and hydroxy radicals ($\bullet OH$), respectively. As presented in Figure 7a,b, after adding BQ, the final degradation efficiency of TC was effectively inhibited, indicating that $\bullet O_2^-$ radicals were the main free active radicals in the CN-B system. It is worth noting that when TEOA was added, the degradation efficiency of TC was reduced, indicating that h^+ played a secondary role in the degradation process.

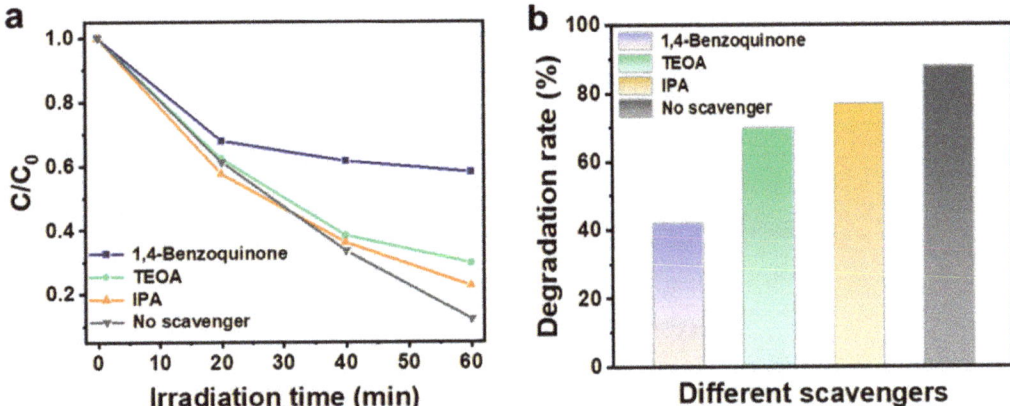

Figure 7. (**a**) Degradation curves and (**b**) corresponding degradation rate of CN-B-0.1 as the photocatalyst by adding different free radical trapping agents.

Based on the above experimental results, a possible mechanism of photothermal-assisted photocatalytic degradation TC by the CN-B photocatalyst under visible irradiation was proposed, as provided in Figure 8. The CN-B photocatalyst produces electrons and holes under visible light irradiation, while the high separation efficiency of photon-generated carriers is achieved due to the introduction of cyano group defects. More importantly, the introduction of the phloxine B precursor in CN-B can make the sample color darker to enhance the photocatalytic reaction temperature through the photothermal effect, thus enhancing the photocatalytic degradation activity. Specifically, according to the band structure of the photocatalyst, the electrons in the CB position of CN-B react with dissolved oxygen in the water to form a superoxide radical ($\bullet O_2^-$) (E_0 ($O_2/\bullet O_2^-$) = -0.33 eV vs. NHE), while the h^+ in the VB of CN-B convert water molecules into $\bullet OH$ to degrade pollutants. Finally, $\bullet O_2^-$, h^+ and $\bullet OH$ degrade the TC into CO_2, H_2O, etc. In conclusion, in the CN-B photocatalytic system, active free radicals can be generated in the following ways:

$$CN\text{-}B + h\nu \rightarrow CN\text{-}B\ (e^- + h^+) \quad (1)$$

$$CN\text{-}B\ (e^-) + O_2 \rightarrow \bullet O_2^- \quad (2)$$

$$CN\text{-}B\ (h^+) + H_2O + OH^- \rightarrow \bullet OH \quad (3)$$

$$\bullet O_2^-/h^+/\bullet OH + \text{pollutions} \rightarrow CO_2 + H_2O \quad (4)$$

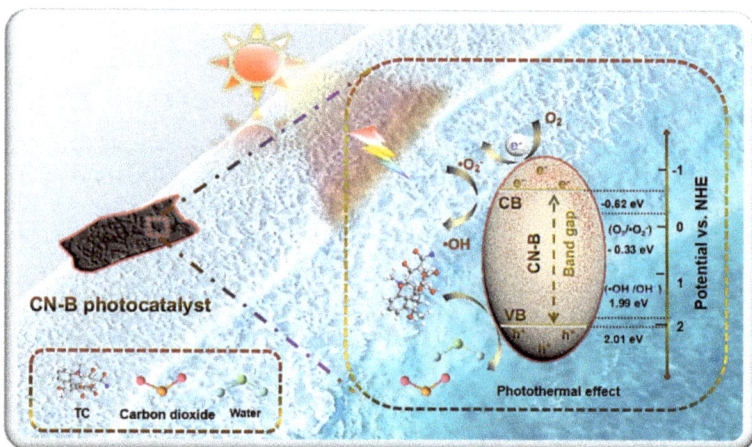

Figure 8. Possible mechanism of photothermal-assisted photocatalytic degradation TC in seawater by CN-B photocatalyst under visible light irradiation.

3. Experimental Section

3.1. Materials

The urea ($CO(NH_2)_2$) (A.R. $\geq$ 98%) and phloxine B ($C_{20}H_2Br_4Cl_4Na_2O_5$) (A.R. $\geq$ 80%) were purchased from Sinopharm Chemical Reagent Co., Ltd. (Shanghai, China), and were analytical grade without further purification.

3.2. Preparation of CN and CN-B Materials

The black g-C_3N_4 (CN-B) materials were synthesized by a one-step calcination method with controlling the additional amount of phloxine B powder (Scheme 1). Firstly, X mg (X = 5, 10, 20, and 30) of phloxine B was ground with 10 g of urea in a mortar. Next, after sufficient grinding, the mixture in the mortar was placed in a crucible and heated to 520 °C at a heating rate of 5 °C/min under an air atmosphere and maintained for 2 h. Finally, after natural cooling, the resulting powder was washed and dried to obtain CN-B photocatalysts. According to the quality of the phloxine B placed in the precursor before the calcination, the resulting samples were labeled as CN-B-0.05, CN-B-0.1, CN-B-0.2, and CN-B-0.3, respectively. Additionally, the pure CN sample was synthesized under the same reaction conditions except for the addition of phloxine B.

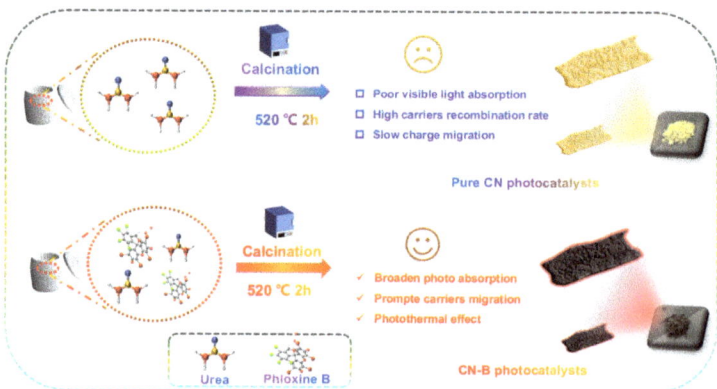

Scheme 1. Schematic diagram of the synthesis process of the pure CN and CN-B nanosheets.

3.3. Photocatalytic Degradation Experiments

The photocatalytic degradation performances of as-prepared samples were evaluated by using a parallel photochemical reaction instrument (CEL-LAB200E7, Beijing Zhongyang Jinyuan Technology Co., LTD., Beijing, China)) and 30W LED lamp as vis light source (λ: 410–760 nm). Specifically, a certain amount of photocatalyst powder and TC aqueous solution (50 mL, 30 mg/L) were dispersed into seawater (3.5 wt% NaCl solution) and stirred in the dark for 30 min to achieve adsorption–desorption equilibrium. After the adsorption equilibrium was reached, the LED lamp was opened for 120 min to evaluate the degradation efficiency of CN and CN-B photocatalysts under visible light irradiation. The photocatalytic degradation of TC was detected by UV-Vis spectrophotometer (UV-2450, Shanghai, China) at a maximum wavelength of 357 nm with 20 min intervals; withdrawn samples were extracted from the suspension and centrifuged. In order to test the cyclic degradation stability of the synthesized sample, the photocatalyst, after degradation, was collected by centrifuge and washed and dried for repeated testing. In addition, the solution temperature of the reactor can be controlled by a condensate circulation device to allow photocatalytic degradation experiments to be performed at different temperatures (such as 5 and 10 °C).

The specific experimental procedures of characterizations, photoelectrochemical properties measurements, and photothermal performance measurements were listed in the Supporting Information.

4. Conclusions

In summary, a stable photothermal-assisted photocatalytic degradation TC system was designed, and a black carbon nitride (CN-B) photocatalyst was constructed by one-step calcination of urea and phloxine B. The results showed that the photocatalytic degradation TC in simulated seawater for the optimal sample CN-B-0.1 was 92% within 2 h at room temperature under visible light irradiation. The excellent photocatalytic degradation performance is mainly attributed to the following reasons: (i) the charge transfer distance reduced by the thickness of ultra-thin nanosheets, (ii) the introduction of cyanogen defects promotes photogenerated carrier separation and migration, and (iii) the photothermal effect of CN-B increases the temperature of the reaction system and further improves the photocatalytic degradation performance. This work provides a promising strategy and systematic method for developing g-C_3N_4-based photocatalysts for photothermal-assisted photocatalytic degradation of antibiotics.

Supplementary Materials: The following supporting information can be downloaded at: https://www.mdpi.com/article/10.3390/catal13071147/s1, Figure S1. Energy dispersive X-ray spectra (EDX) spectrum of CN-B-0.1 sample. Figure S2. High-resolution XPS spectra of Na 1s for CN and CN-B-0.1 photocatalysts. Figure S3. Digital photos of (a) CN, (b) CN-B-0.05, (c) CN-B-0.1, (d) CN-B-0.3 and (e) CN-B-0.5. Figure S4. Photocatalytic Tc degradation performance of CN-B-0.1 photocatalyst compared with the previously reported of the different materials. Figure S5. High-resolution XPS spectra of O 1s of CN-B-0.1 photocatalyst before and after photocatalysis. Table S1. Surface relative element content of CN and CN-B-0.1 from XPS characterizes. Table S2. Photocatalytic Tc degradation performance of CN-B-0.1 photocatalyst compared with the previously reported of the different materials [72–75].

Author Contributions: Data curation, L.J.N.K. and J.L.; Writing—original draft, L.J.N.K. and Z.C.; Writing—review & editing, H.W., F.G. and W.S. All authors have read and agreed to the published version of the manuscript.

Funding: The authors would like to acknowledge the funding support from the National Natural Science Foundation of China (No. 22006057 and 21908115) and "Doctor of Mass entrepreneurship and innovation" Project in Jiangsu Province.

Data Availability Statement: Samples of the compounds are available or not available from the authors.

Conflicts of Interest: The authors declare no conflict of interest.

References

1. Zheng, D.; Chang, Q.; Gao, M.; She, Z.; Jin, C.; Guo, L.; Zhao, Y.; Wang, S.; Wang, X. Performance evaluation and microbial community of a sequencing batch biofilm reactor (SBBR) treating mariculture wastewater at different chlortetracycline concentrations. *J. Environ. Manag.* **2016**, *182*, 496–504. [CrossRef] [PubMed]
2. Liu, X.; Steele, J.C.; Meng, X.Z. Usage, residue, and human health risk of antibiotics in Chinese aquaculture: A review. *Environ. Pollut.* **2017**, *223*, 161–169. [CrossRef] [PubMed]
3. Hu, J.; Sun, C.; Wu, L.-X.; Zhao, G.-Q.; Liu, H.-Y.; Jiao, F.-P. Halogen doped g-C_3N_4/ZnAl-LDH hybrid as a Z-scheme photocatalyst for efficient degradation for tetracycline in seawater. *Sep. Purif. Technol.* **2023**, *309*, 123047. [CrossRef]
4. Zhu, D.; Cai, L.; Sun, Z.; Zhang, A.; Heroux, P.; Kim, H.; Yu, W.; Liu, Y. Efficient degradation of tetracycline by RGO@black titanium dioxide nanofluid via enhanced catalysis and photothermal conversion. *Sci. Total Environ.* **2021**, *787*, 147536. [CrossRef] [PubMed]
5. Lu, Z.; Yu, Z.; Dong, J.; Song, M.; Liu, Y.; Liu, X.; Ma, Z.; Su, H.; Yan, Y.; Huo, P. Facile microwave synthesis of a Z-scheme imprinted $ZnFe_2O_4$/Ag/PEDOT with the specific recognition ability towards improving photocatalytic activity and selectivity for tetracycline. *Chem. Eng. J.* **2018**, *337*, 228–241. [CrossRef]
6. Lou, J.; Xu, X.; Gao, Y.; Zheng, D.; Wang, J.; Li, Z. Preparation of magnetic activated carbon from waste rice husk for the determination of tetracycline antibiotics in water samples. *RSC Adv.* **2016**, *6*, 112166–112174. [CrossRef]
7. Sun, H.; Guo, F.; Pan, J.; Huang, W.; Wang, K.; Shi, W. One-pot thermal polymerization route to prepare N-deficient modified g-C_3N_4 for the degradation of tetracycline by the synergistic effect of photocatalysis and persulfate-based advanced oxidation process. *Chem. Eng. J.* **2021**, *406*, 126844. [CrossRef]
8. Davison, J. Genetic exchange between bacteria in the environment. *Plasmid* **1999**, *42*, 73–91. [CrossRef]
9. Niu, J.; Lin, H.-Z.; Jiang, S.-G.; Chen, X.; Wu, K.-C.; Liu, Y.-J.; Wang, S.; Tian, L.-X. Comparison of effect of chitin, chitosan, chitosan oligosaccharide and N-acetyl-D-glucosamine on growth performance, antioxidant defenses and oxidative stress status of Penaeus monodon. *Aquaculture* **2013**, *372–375*, 1–8. [CrossRef]
10. Jeong, W.G.; Kim, J.G.; Baek, K. Removal of 1,2-dichloroethane in groundwater using Fenton oxidation. *J. Hazard. Mater.* **2022**, *428*, 128253. [CrossRef]
11. Liu, Z.; Lompe, K.M.; Mohseni, M.; Berube, P.R.; Sauve, S.; Barbeau, B. Biological ion exchange as an alternative to biological activated carbon for drinking water treatment. *Water Res.* **2020**, *168*, 115148. [CrossRef]
12. Liu, Y.; Yu, X. Carbon dioxide adsorption properties and adsorption/desorption kinetics of amine-functionalized KIT-6. *Appl. Energy* **2018**, *211*, 1080–1088. [CrossRef]
13. Guo, F.; Shi, W.; Li, M.; Shi, Y.; Wen, H. 2D/2D Z-scheme heterojunction of $CuInS_2$/g-C_3N_4 for enhanced visible-light-driven photocatalytic activity towards the degradation of tetracycline. *Sep. Purif. Technol.* **2019**, *210*, 608–615. [CrossRef]
14. Zhang, T.; Liu, M.; Meng, Y.; Huang, B.; Pu, X.; Shao, X. A novel method for the synthesis of Ag_3VO_4/$Ag_2VO_2PO_4$ heterojunction photocatalysts with improved visible-light photocatalytic properties. *Sep. Purif. Technol.* **2018**, *206*, 149–157. [CrossRef]
15. Lu, C.; Guo, F.; Yan, Q.; Zhang, Z.; Li, D.; Wang, L.; Zhou, Y. Hydrothermal synthesis of type II $ZnIn_2S_4$/$BiPO_4$ heterojunction photocatalyst with dandelion-like microflower structure for enhanced photocatalytic degradation of tetracycline under simulated solar light. *J. Alloys Compd.* **2019**, *811*, 151976. [CrossRef]
16. Zu, M.; Zhou, X.; Zhang, S.; Qian, S.; Li, D.-S.; Liu, X.; Zhang, S. Sustainable engineering of TiO_2-based advanced oxidation technologies: From photocatalyst to application devices. *J. Mater. Sci. Technol.* **2021**, *78*, 202–222. [CrossRef]
17. Qiao, X.; Wang, C.; Niu, Y. N-Benzyl HMTA induced self-assembly of organic-inorganic hybrid materials for efficient photocatalytic degradation of tetracycline. *J. Hazard. Mater.* **2020**, *391*, 122121. [CrossRef]
18. Reddy, K.R.; Reddy, C.V.; Nadagouda, M.N.; Shetti, N.P.; Jaesool, S.; Aminabhavi, T.M. Polymeric graphitic carbon nitride (g-C_3N_4)-based semiconducting nanostructured materials: Synthesis methods, properties and photocatalytic applications. *J. Environ. Manag.* **2019**, *238*, 25–40. [CrossRef]
19. Shi, W.; Li, M.; Huang, X.; Ren, H.; Yan, C.; Guo, F. Facile synthesis of 2D/2D $Co_3(PO_4)_2$/g-C_3N_4 heterojunction for highly photocatalytic overall water splitting under visible light. *Chem. Eng. J.* **2020**, *382*, 122960. [CrossRef]
20. Shi, W.; Ren, H.; Huang, X.; Li, M.; Tang, Y.; Guo, F. Low cost red mud modified graphitic carbon nitride for the removal of organic pollutants in wastewater by the synergistic effect of adsorption and photocatalysis. *Sep. Purif. Technol.* **2020**, *237*, 116477. [CrossRef]
21. Mishra, A.; Mehta, A.; Basu, S.; Shetti, N.P.; Reddy, K.R.; Aminabhavi, T.M. Graphitic carbon nitride (g–C_3N_4)–based metal-free photocatalysts for water splitting: A review. *Carbon* **2019**, *149*, 693–721. [CrossRef]
22. Yu, W.; Shan, X.; Zhao, Z. Unique nitrogen-deficient carbon nitride homojunction prepared by a facile inserting-removing strategy as an efficient photocatalyst for visible light-driven hydrogen evolution. *Appl. Catal. B Environ.* **2020**, *269*, 118778. [CrossRef]
23. Shi, W.; Shu, K.; Huang, X.; Ren, H.; Li, M.; Chen, F.; Guo, F. Enhancement of visible-light photocatalytic degradation performance over nitrogen-deficient g-C_3N_4/$KNbO_3$ heterojunction photocatalyst. *J. Chem. Technol. Biotechnol.* **2020**, *95*, 1476–1486. [CrossRef]
24. Feng, C.; Lu, Z.; Zhang, Y.; Liang, Q.; Zhou, M.; Li, X.; Yao, C.; Li, Z.; Xu, S. A magnetically recyclable dual Z-scheme GCNQDs-$CoTiO_3$/$CoFe_2O_4$ composite photocatalyst for efficient photocatalytic degradation of oxytetracycline. *Chem. Eng. J.* **2022**, *435*, 134833. [CrossRef]
25. Wu, M.; He, X.; Jing, B.; Wang, T.; Wang, C.; Qin, Y.; Ao, Z.; Wang, S.; An, T. Novel carbon and defects co-modified g-C_3N_4 for highly efficient photocatalytic degradation of bisphenol A under visible light. *J. Hazard. Mater.* **2020**, *384*, 121323. [CrossRef]

26. Li, Y.; Fang, Y.; Cao, Z.; Li, N.; Chen, D.; Xu, Q.; Lu, J. Construction of g-C$_3$N$_4$/PDI@MOF heterojunctions for the highly efficient visible light-driven degradation of pharmaceutical and phenolic micropollutants. *Appl. Catal. B Environ.* **2019**, *250*, 150–162. [CrossRef]
27. Liu, G.; Dong, G.; Zeng, Y.; Wang, C. The photocatalytic performance and active sites of g-C$_3$N$_4$ effected by the coordination doping of Fe(III). *Chin. J. Catal.* **2020**, *41*, 1564–1572. [CrossRef]
28. Hu, X.; Lu, P.; Pan, R.; Li, Y.; Bai, J.; He, Y.; Zhang, C.; Jia, F.; Fu, M. Metal-ion-assisted construction of cyano group defects in g-C$_3$N$_4$ to simultaneously degrade wastewater and produce hydrogen. *Chem. Eng. J.* **2021**, *423*, 130278. [CrossRef]
29. Lu, Y.; Zhang, H.; Fan, D.; Chen, Z.; Yang, X. Coupling solar-driven photothermal effect into photocatalysis for sustainable water treatment. *J. Hazard. Mater.* **2022**, *423*, 127128. [CrossRef]
30. Yang, X.; Wei, S.; Ma, X.; Gao, Z.; Huang, W.; Wang, D.; Liu, Z.; Wang, J. Core–shell CoTiO$_3$@MnO$_2$ heterostructure for the photothermal degradation of tetracycline. *J. Mater. Sci.* **2023**, *58*, 3551–3567. [CrossRef]
31. Wang, T.; Bai, Z.; Wei, W.; Hou, F.; Guo, W.; Wei, A. beta-Cyclodextrin-Derivative-Functionalized Graphene Oxide/Graphitic Carbon Nitride Composites with a Synergistic Effect for Rapid and Efficient Sterilization. *ACS Appl. Mater. Interfaces* **2022**, *14*, 474–483. [CrossRef]
32. Zhang, X.; Ma, Y.; Zhang, X.; Pang, X.; Yang, Z. Bio-inspired self-assembled bacteriochlorin nanoparticles for superior visualization and photothermal ablation of tumors. *Biomed. Pharmacother.* **2023**, *165*, 115014. [CrossRef] [PubMed]
33. Hou, B.; Shi, Z.; Kong, D.; Chen, Z.; Yang, K.; Ming, X.; Wang, X. Scalable porous Al foil/reduced graphene oxide/Mn$_3$O$_4$ composites for efficient fresh water generation. *Mater. Today Energy* **2020**, *15*, 100371. [CrossRef]
34. Wang, Y.; Yang, W.; Chen, X.; Wang, J.; Zhu, Y. Photocatalytic activity enhancement of core-shell structure g-C$_3$N$_4$@TiO$_2$ via controlled ultrathin g-C$_3$N$_4$ layer. *Appl. Catal. B Environ.* **2018**, *220*, 337–347. [CrossRef]
35. Guo, F.; Sun, H.; Huang, X.; Shi, W.; Yan, C. Fabrication of TiO$_2$/high-crystalline g-C$_3$N$_4$ composite with enhanced visible-light photocatalytic performance for tetracycline degradation. *J. Chem. Technol. Biotechnol.* **2020**, *95*, 2684–2693. [CrossRef]
36. Tan, X.; Jiang, K.; Zhai, S.; Zhou, J.; Wang, J.; Cadien, K.; Li, Z. X-Ray Spectromicroscopy Investigation of Heterogeneous Sodiation in Hard Carbon Nanosheets with Vertically Oriented (002) Planes. *Small* **2021**, *17*, e2102109. [CrossRef]
37. Guo, F.; Wang, L.; Sun, H.; Li, M.; Shi, W. High-efficiency photocatalytic water splitting by a N-doped porous g-C$_3$N$_4$ nanosheet polymer photocatalyst derived from urea and N,N-dimethylformamide. *Inorg. Chem. Front.* **2020**, *7*, 1770–1779. [CrossRef]
38. Jin, W.; Ji, Y.; Larsen, D.H.; Huang, Y.; Heuvelink, E.; Marcelis, L.F.M. Gradually increasing light intensity during the growth period increases dry weight production compared to constant or gradually decreasing light intensity in lettuce. *Sci. Hortic.* **2023**, *311*, 111807. [CrossRef]
39. Shi, Y.; Li, L.; Sun, H.; Xu, Z.; Cai, Y.; Shi, W.; Guo, F.; Du, X. Engineering ultrathin oxygen-doped g-C$_3$N$_4$ nanosheet for boosted photoredox catalytic activity based on a facile thermal gas-shocking exfoliation effect. *Sep. Purif. Technol.* **2022**, *292*, 121038. [CrossRef]
40. Shi, W.; Shu, K.; Sun, H.; Ren, H.; Li, M.; Chen, F.; Guo, F. Dual enhancement of capturing photogenerated electrons by loading CoP nanoparticles on N-deficient graphitic carbon nitride for efficient photocatalytic degradation of tetracycline under visible light. *Sep. Purif. Technol.* **2020**, *246*, 116930. [CrossRef]
41. Wang, J.; Huang, J.; Xie, H.; Qu, A. Synthesis of g-C$_3$N$_4$/TiO$_2$ with enhanced photocatalytic activity for H$_2$ evolution by a simple method. *Int. J. Hydrogen Energy* **2014**, *39*, 6354–6363. [CrossRef]
42. Guo, F.; Huang, X.; Chen, Z.; Sun, H.; Chen, L. Prominent co-catalytic effect of CoP nanoparticles anchored on high-crystalline g-C$_3$N$_4$ nanosheets for enhanced visible-light photocatalytic degradation of tetracycline in wastewater. *Chem. Eng. J.* **2020**, *395*, 125118. [CrossRef]
43. Guo, F.; Li, M.; Ren, H.; Huang, X.; Shu, K.; Shi, W.; Lu, C. Facile bottom-up preparation of Cl-doped porous g-C$_3$N$_4$ nanosheets for enhanced photocatalytic degradation of tetracycline under visible light. *Sep. Purif. Technol.* **2019**, *228*, 115770. [CrossRef]
44. Guo, F.; Wang, L.; Sun, H.; Li, M.; Shi, W.; Lin, X. A one-pot sealed ammonia self-etching strategy to synthesis of N-defective g-C$_3$N$_4$ for enhanced visible-light photocatalytic hydrogen. *Int. J. Hydrogen Energy* **2020**, *45*, 30521–30532. [CrossRef]
45. Zeng, D.; Ong, W.-J.; Chen, Y.; Tee, S.Y.; Chua, C.S.; Peng, D.-L.; Han, M.-Y. Co$_2$P Nanorods as an Efficient Cocatalyst Decorated Porous g-C$_3$N$_4$ Nanosheets for Photocatalytic Hydrogen Production under Visible Light Irradiation. *Part. Part. Syst. Charact.* **2018**, *35*, 1700251. [CrossRef]
46. Shi, W.; Yang, S.; Sun, H.; Wang, J.; Lin, X.; Guo, F.; Shi, J. Carbon dots anchored high-crystalline g-C$_3$N$_4$ as a metal-free composite photocatalyst for boosted photocatalytic degradation of tetracycline under visible light. *J. Mater. Sci.* **2020**, *56*, 2226–2240. [CrossRef]
47. Guo, S.; Deng, Z.; Li, M.; Jiang, B.; Tian, C.; Pan, Q.; Fu, H. Phosphorus-Doped Carbon Nitride Tubes with a Layered Micro-nanostructure for Enhanced Visible-Light Photocatalytic Hydrogen Evolution. *Angew. Chem. Int. Ed. Engl.* **2016**, *55*, 1830–1834. [CrossRef] [PubMed]
48. Zhu, X.; Guo, F.; Pan, J.; Sun, H.; Gao, L.; Deng, J.; Zhu, X.; Shi, W. Fabrication of visible-light-response face-contact ZnSnO$_3$@g-C$_3$N$_4$ core–shell heterojunction for highly efficient photocatalytic degradation of tetracycline contaminant and mechanism insight. *J. Mater. Sci.* **2020**, *56*, 4366–4379. [CrossRef]
49. Qin, J.; Chen, J.; Salisbury, J.B. Photon transferred TL signals from potassium feldspars and their effects on post-IR IRSL measurements. *J. Lumin.* **2015**, *160*, 1–8. [CrossRef]

50. Guo, F.; Huang, X.; Chen, Z.; Cao, L.; Cheng, X.; Chen, L.; Shi, W. Construction of Cu_3P-$ZnSnO_3$-g-C_3N_4 p-n-n heterojunction with multiple built-in electric fields for effectively boosting visible-light photocatalytic degradation of broad-spectrum antibiotics. *Sep. Purif. Technol.* **2021**, *265*, 118477. [CrossRef]
51. Fu, J.; Zhu, B.; Jiang, C.; Cheng, B.; You, W.; Yu, J. Hierarchical Porous O-Doped g-C_3N_4 with Enhanced Photocatalytic CO_2 Reduction Activity. *Small* **2017**, *13*, 1603938. [CrossRef] [PubMed]
52. Zhang, W.; Shi, W.; Sun, H.; Shi, Y.; Luo, H.; Jing, S.; Fan, Y.; Guo, F.; Lu, C. Fabrication of ternary CoO/g-C_3N_4/Co_3O_4 nanocomposite with p-n-p type heterojunction for boosted visible-light photocatalytic performance. *J. Chem. Technol. Biotechnol.* **2021**, *96*, 1854–1863. [CrossRef]
53. Liu, G.; Yan, S.; Shi, L.; Yao, L. The Improvement of Photocatalysis H_2 Evolution Over g-C_3N_4 With Na and Cyano-Group Co-modification. *Front. Chem.* **2019**, *7*, 639. [CrossRef]
54. Guo, F.; Chen, Z.; Huang, X.; Cao, L.; Cheng, X.; Shi, W.; Chen, L. Cu_3P nanoparticles decorated hollow tubular carbon nitride as a superior photocatalyst for photodegradation of tetracycline under visible light. *Sep. Purif. Technol.* **2021**, *275*, 119223. [CrossRef]
55. Chen, H.; Yu, Y.; Yu, Y.; Ye, J.; Zhang, S.; Chen, J. Exogenous electron transfer mediator enhancing gaseous toluene degradation in a microbial fuel cell: Performance and electron transfer mechanism. *Chemosphere* **2021**, *282*, 131028. [CrossRef]
56. Shi, Y.; Li, L.; Xu, Z.; Sun, H.; Guo, F.; Shi, W. One-step simple green method to prepare carbon-doped graphitic carbon nitride nanosheets for boosting visible-light photocatalytic degradation of tetracycline. *J. Chem. Technol. Biotechnol.* **2021**, *96*, 3122–3133. [CrossRef]
57. Guo, F.; Chen, Z.; Huang, X.; Cao, L.; Cheng, X.; Shi, W.; Chen, L. Ternary Ni2P/Bi2MoO6/g-C_3N_4 composite with Z-scheme electron transfer path for enhanced removal broad-spectrum antibiotics by the synergistic effect of adsorption and photocatalysis. *Chin. J. Chem. Eng.* **2022**, *44*, 157–168. [CrossRef]
58. Xu, Z.; Shi, Y.; Li, L.; Sun, H.; Amin, M.D.S.; Guo, F.; Wen, H.; Shi, W. Fabrication of 2D/2D Z-scheme highly crystalline carbon nitride/δ-Bi_2O_3 heterojunction photocatalyst with enhanced photocatalytic degradation of tetracycline. *J. Alloys Compd.* **2022**, *895*, 162667. [CrossRef]
59. Liu, Y.; Yang, Z.-H.; Song, P.-P.; Xu, R.; Wang, H. Facile synthesis of Bi_2MoO_6/$ZnSnO_3$ heterojunction with enhanced visible light photocatalytic degradation of methylene blue. *Appl. Surf. Sci.* **2018**, *430*, 561–570. [CrossRef]
60. Cheng, Q.; Yang, W.; Chen, Q.; Zhu, J.; Li, D.; Fu, L.; Zhou, L. Fe-doped zirconia nanoparticles with highly negative conduction band potential for enhancing visible light photocatalytic performance. *Appl. Surf. Sci.* **2020**, *530*, 147291. [CrossRef]
61. Sun, H.; Wang, L.; Guo, F.; Shi, Y.; Li, L.; Xu, Z.; Yan, X.; Shi, W. Fe-doped g-C_3N_4 derived from biowaste material with Fe-N bonds for enhanced synergistic effect between photocatalysis and Fenton degradation activity in a broad pH range. *J. Alloys Compd.* **2022**, *900*, 163410. [CrossRef]
62. Guo, F.; Chen, Z.; Shi, Y.; Cao, L.; Cheng, X.; Shi, W.; Chen, L.; Lin, X. A ragged porous hollow tubular carbon nitride towards boosting visible-light photocatalytic hydrogen production in water and seawater. *Renew. Energy* **2022**, *188*, 1–10. [CrossRef]
63. Shi, Y.; Li, L.; Xu, Z.; Sun, H.; Amin, S.; Guo, F.; Shi, W.; Li, Y. Engineering of 2D/3D architectures type II heterojunction with high-crystalline g-C_3N_4 nanosheets on yolk-shell $ZnFe_2O_4$ for enhanced photocatalytic tetracycline degradation. *Mater. Res. Bull.* **2022**, *150*, 111789. [CrossRef]
64. Sun, H.; Shi, Y.; Shi, W.; Guo, F. High-crystalline/amorphous g-C_3N_4 S-scheme homojunction for boosted photocatalytic H_2 production in water/simulated seawater: Interfacial charge transfer and mechanism insight. *Appl. Surf. Sci.* **2022**, *593*, 153281. [CrossRef]
65. Sun, X.; Shi, Y.; Lu, J.; Shi, W.; Guo, F. Template-free self-assembly of three-dimensional porous graphitic carbon nitride nanovesicles with size-dependent photocatalytic activity for hydrogen evolution. *Appl. Surf. Sci.* **2022**, *606*, 154841. [CrossRef]
66. Shi, W.; Cao, L.; Shi, Y.; Chen, Z.; Cai, Y.; Guo, F.; Du, X. Environmentally friendly supermolecule self-assembly preparation of S-doped hollow porous tubular g-C_3N_4 for boosted photocatalytic H_2 production. *Ceram. Int.* **2023**, *49*, 11989–11998. [CrossRef]
67. Shi, Y.; Li, L.; Xu, Z.; Guo, F.; Li, Y.; Shi, W. Synergistic coupling of piezoelectric and plasmonic effects regulates the Schottky barrier in Ag nanoparticles/ultrathin g-C_3N_4 nanosheets heterostructure to enhance the photocatalytic activity. *Appl. Surf. Sci.* **2023**, *616*, 156466. [CrossRef]
68. Guo, F.; Li, L.; Shi, Y.; Shi, W.; Yang, X. Synthesis of N-deficient g-C_3N_4/epoxy composite coating for enhanced photocatalytic corrosion resistance and water purification. *J. Mater. Sci.* **2023**, *58*, 4223–4239. [CrossRef]
69. Guo, F.; Li, L.; Shi, Y.; Shi, W.; Yang, X.; Li, H. Achieving superior anticorrosion and antibiofouling performance of polyaniline/graphitic carbon nitride composite coating. *Progress. Org. Coat.* **2023**, *179*, 107512. [CrossRef]
70. Li, L.; Zhang, Y.; Shi, Y.; Guo, F.; Yang, X.; Shi, W. A hydrophobic high-crystalline g-C_3N_4/epoxy resin composite coating with excellent durability and stability for long-term corrosion resistance. *Mater. Today Commun.* **2023**, *35*, 105692. [CrossRef]
71. Yuan, H.; Sun, H.; Shi, Y.; Wang, J.; Bian, A.; Hu, Y.; Guo, F.; Shi, W.; Du, X.; Kang, Z. Cooperation of carbon doping and carbon loading boosts photocatalytic activity by the optimum photo-induced electron trapping and interfacial charge transfer. *Chem. Eng. J.* **2023**, *472*, 144654. [CrossRef]
72. Xu, W.; Lai, S.; Pillai, S.C.; Chu, W.; Hu, Y.; Jiang, X.; Fu, M.; Wu, X.; Li, F.; Wang, H. Visible light photocatalytic degradation of tetracycline with porous Ag/graphite carbon nitride plasmonic composite: Degradation pathways and mechanism. *J. Colloid Interface Sci.* **2020**, *574*, 110–121. [CrossRef] [PubMed]

73. Hernandez-Uresti, D.B.; Vazquez, A.; Sanchez-Martinez, D.; Obregon, S. Performance of the polymeric g-C_3N_4 photocatalyst through the degradation of pharmaceutical pollutants under UV-vis irradiation. *J. Photo Chem. Photo Biol. A* **2016**, *324*, 47–52. [CrossRef]
74. Hong, Y.; Li, C.; Zhang, G.; Meng, Y.; Yin, B.; Zhao, Y.; Shi, W. Efficient and stable Nb_2O_5 modified g-C_3N_4 photocatalyst for removal of antibiotic pollutant. *Chem. Eng. J.* **2016**, *299*, 74–84. [CrossRef]
75. Chen, D.; Wu, S.; Fang, J.; Lu, S.; Zhou, G.; Feng, W.; Yang, F.; Chen, Y.; Fang, Z. A nanosheet-like α-Bi_2O_3/g-C_3N_4 heterostructure modified by plasmonic metallic Bi and oxygen vacancies with high photodegradation activity of organic pollutants. *Sep. Purif. Technol.* **2018**, *193*, 232–241. [CrossRef]

Disclaimer/Publisher's Note: The statements, opinions and data contained in all publications are solely those of the individual author(s) and contributor(s) and not of MDPI and/or the editor(s). MDPI and/or the editor(s) disclaim responsibility for any injury to people or property resulting from any ideas, methods, instructions or products referred to in the content.

Review

Recent Innovative Progress of Metal Oxide Quantum-Dot-Integrated g-C$_3$N$_4$ (0D-2D) Synergistic Nanocomposites for Photocatalytic Applications

Tejaswi Tanaji Salunkhe [1,†], Thirumala Rao Gurugubelli [2,†], Bathula Babu [3,*] and Kisoo Yoo [3,*]

1. Department of Chemical and Biological Engineering, Gachon University, Seongnam-si 13120, Republic of Korea; tejaswisalunkhe235@gmail.com
2. Department of Physics, School of Sciences, SR University, Warangal 506 371, India; thirumala.phy@gmail.com
3. School of Mechanical Engineering, Yeungnam University, Gyeongsan 38541, Republic of Korea
* Correspondence: babuphysicist@ynu.ac.kr (B.B.); kisooyoo@yu.ac.kr (K.Y.)
† These authors contributed equally to this work.

Abstract: Modern industrialization has unleashed unprecedented environmental challenges, primarily in the form of pollution. In response to these pressing issues, the quest for innovative and sustainable solutions has intensified. Photocatalysis, with its unique capabilities, has emerged as a potent technology to combat the adverse effects of industrialization on the environment. This review highlights recent advances in harnessing photocatalysis to address environmental pollution. Photocatalysis offers a multifaceted approach, utilizing solar energy for catalytic reactions and enabling efficient pollutant removal. Quantum dots and graphitic carbon nitride (g-C$_3$N$_4$) are essential elements in this science. In contrast to quantum dots, which have enormous potential due to their size-dependent bandgap tunability and effective charge carrier production, g-C$_3$N$_4$ has properties like chemical stability and a configurable bandgap that make it a versatile material for photocatalysis. In this review, we explore recent achievements in integrating metal oxide quantum dots with g-C$_3$N$_4$, forming nanocomposites with superior photocatalytic activity. These nanocomposites exhibit extended light absorption ranges and enhanced charge separation efficiency, positioning them at the forefront of diverse photocatalytic applications. In conclusion, this comprehensive review underscores the critical role of photocatalysis as a potent tool to counteract the adverse environmental effects of modern industrialization. By emphasizing recent advancements in g-C$_3$N$_4$ and quantum dots and highlighting the advantages of metal oxide quantum dots decorated/integrated with g-C$_3$N$_4$ nanocomposites, this work contributes to the evolving landscape of sustainable solutions for environmental remediation and pollution control. These innovations hold promise for a cleaner and more sustainable future.

Keywords: quantum dots; g-C$_3$N$_4$; photocatalytic; nanocomposite

1. Introduction

The ongoing march of progress, for eons, has been marked by humanity's relentless pursuit of industrialization. From the spinning jenny to the state-of-the-art factories dotting our landscapes, industrial processes have been the harbingers of prosperity, growth, and the advancement of our species [1]. Yet, this coin possesses a tarnished flip side. Our industrial accomplishments, while monumental, have brought with them undeniable environmental degradation. However, with its plethora of benefits, industrialization inadvertently ushered in a myriad of ecological challenges. Rapid urbanization and unchecked manufacturing processes spawned large-scale environmental pollution [2,3]. As industries mushroomed, so did the emissions, leading to the degradation of both air and water quality, affecting the very tenets of human health and environmental sustainability [4,5]. Among the myriad solutions that have been tabled to combat environmental degradation, one that stands out for its potential and innovation is photocatalysis [6]. Rooted in the confluence of physics and chemistry, photocatalysis offers an avenue where pollutants are degraded under the influence of light [7]. This process is not only environment-friendly but also sustainable. The advantages of photocatalysis are manifold. Aside from its ability to degrade organic pollutants, it exhibits the potential to harness solar energy efficiently, making it an eco-friendly solution to some of the pressing challenges of our times [8].

Historically, the world of photocatalysis has witnessed the introduction and application of numerous catalysts, each bringing its own set of benefits and challenges. From the pioneering work on TiO_2 to the utilization of complex organic polymers, the field has never ceased to evolve [9,10]. Among these myriad materials, the realm of 2D and 3D materials has garnered significant interest [11,12]. Their unique morphologies, structural attributes, and ease of manipulation have made them frontrunners in the race to find the most efficient photocatalyst. In recent years, a material that has captured the imagination of researchers and scientists alike is graphitic carbon nitride, or $g-C_3N_4$ [13]. This two-dimensional material, with its layered structure reminiscent of graphene, has shown exceptional promise as a photocatalyst. The journey of $g-C_3N_4$ in the photocatalytic domain has been both evolutionary and revolutionary. Various precursors, ranging from urea to dicyandiamide, have been utilized in its synthesis, leading to variations in its properties and, consequently, its photocatalytic efficiency [14]. The synthesis of $g-C_3N_4$ is as intriguing as its properties. Various precursors, including melamine, dicyandiamide, and others, have been deployed to extract this material, with each method yielding slightly varied material properties, thereby influencing its overall photocatalytic performance [15].

Yet, even the most promising of materials present challenges. Issues with $g-C_3N_4$ include quick photogenerated electron–hole pair recombination, a narrow light absorption spectrum, and certain stability concerns, which have the potential to modestly shade the material's otherwise brilliant prospects. An innovative approach to overcoming these challenges is the hybridization of $g-C_3N_4$ with other materials, especially quantum dots, culminating in the creation of superior nanocomposites [16]. Quantum dots, particularly metal oxide quantum dots, have properties–such as size-tunable band gaps and a high surface-to-volume ratio–that are extraordinarily beneficial for photocatalysis. Their amalgamation with $g-C_3N_4$ brings forth synergistic effects, where the strengths of one complement the weaknesses of the other [17]. To combat $g-C_3N_4$'s limitations and further enhance its capabilities, researchers looked towards nanotechnology, specifically quantum dots. These nanosized semiconducting particles, notable for their quantum mechanical properties, brought several benefits [18]. Their size-dependent tunable band gaps, high surface reactivity, and ability to be easily integrated with other materials made them ideal partners for $g-C_3N_4$ [19]. Several quantum dots such as metal oxides, metal sulfides, carbon quantum dots, graphene quantum dots, etc., are integrated with $g-C_3N_4$ for superior photocatalytic performance [20–23].

Quantum size effects become prominent when the size of the semiconductor particles is reduced to the nanometer scale, approaching the exciton Bohr radius, resulting in quantum confinement. This phenomenon significantly alters the materials' electronic and optical properties, thereby influencing their photocatalytic behavior. For instance, as the particle size of these metal oxides decreases to the quantum scale, the bandgap can broaden due to the increased energy difference between the valence and conduction bands. This bandgap modification enhances light absorption efficiency and, subsequently, the photocatalytic performance under visible light, a feature not often observed in bulk counterparts. Specifically, in the realm of photocatalysis, nanosized TiO_2 and ZnO have demonstrated improved charge carrier generation due to their quantum-dot-like behavior. For instance, studies have shown that TiO_2 nanoparticles, with sizes reduced to the quantum realm (below 10 nm), exhibit a shift in their optical absorption toward the visible region, a direct consequence of the quantum size effect. Similarly, ZnO quantum dots (QDs) have been observed to display a higher photocatalytic activity compared to their bulk equivalents due to their enlarged bandgap and more efficient charge separation, critical for processes like degradation of pollutants.

Within the quantum dot realm, metal oxide (TiO_2, SnO_2, CuO, ZnO) and metal sulfide (CdS, SnS_2, MoS_2) quantum dots garnered significant attention [24–26]. Their inherent stability, coupled with favorable electronic properties, made them especially suited for photocatalytic applications [18,27,28]. Intense study has been conducted on the combination of g-C_3N_4 with quantum dots, also known as 2D-0D nanocomposites [20]. Significant milestones have been achieved, demonstrating enhanced photocatalytic activities, stability, and a broader range of light absorption [29]. This narrative, rich in scientific endeavors, merits a thorough review. It is imperative to collate, analyze, and critique the vast body of work that exists on this subject. For researchers delving deeper into this domain, there exists a myriad of opportunities [30]. The optimization of synthesis methods, exploring new combinations of quantum dots, and tuning the interfaces of these nanocomposites are just a few of the avenues that hold promise [31]. As this exciting chapter in photocatalysis continues to unfold, it remains a beacon of hope for a cleaner, more sustainable future. While this review intends to offer a comprehensive overview, it also serves as a clarion call to researchers worldwide. The field of g-C_3N_4 and quantum dot nanocomposites, though richly explored, is brimming with possibilities. Fresh perspectives, interdisciplinary collaborations, and novel methodologies can unearth nuances previously overlooked.

Impact of particle size on charge separation and catalytic kinetics:

Quantum Confinement Effect: as particle size reduces to the nanoscale, approaching the exciton Bohr radius, quantum confinement becomes prominent. This phenomenon significantly impacts the electronic properties of semiconductors, including the bandgap's widening. For photocatalysts, this can enhance absorption in the visible light range, crucial for solar-driven applications.

Increased Surface Area: smaller particles imply a larger surface area relative to volume, increasing the number of active sites available for reactions. This also facilitates the separation of charge carriers, as electrons and holes have shorter distances to travel to reach the surface, reducing recombination rates and enhancing photocatalytic efficiency.

Enhanced Charge Carrier Dynamics: the high surface-to-volume ratio at the nanoscale influences the redox potential of the material surface, creating more favorable conditions for charge transfer to the reactants, further discouraging charge recombination.

Catalyst–Reactant Interaction: smaller particles allow for more intimate contact with reactants due to their increased surface area, enhancing interaction frequency and energy transfer efficacy, which are critical for reaction kinetics.

Diffusion and Reaction Rates: nanosized materials modify diffusion rates of reactants and products. The shortened diffusion paths in smaller particles accelerate reaction rates, making them more efficient catalysts.

Activation Energy: the quantum size effect can modify the activation energy required for certain reactions. Quantum dots, due to their discrete energy levels, can lower the activation energy barriers, thereby accelerating the reaction kinetics.

2. Synthesis Protocols of g-C_3N_4

g-C_3N_4, an emerging two-dimensional polymer, has captivated researchers' interest primarily due to its remarkable physicochemical properties, making it a potential candidate for various applications, including photocatalysis [13]. Its unique electronic structure, environmental compatibility, and abundant natural precursors make g-C_3N_4 an attractive and eco-friendly material. In this comprehensive review, we delve into the synthesis protocols, precursor variations, morphology alterations, and their corresponding implications on photocatalytic performance. The synthesis of g-C_3N_4 generally revolves around thermally induced polymerization of nitrogen-rich precursors. The following are notable synthesis routes: (a) Direct Thermal Polymerization—in this method, nitrogen-rich organic precursors like melamine or urea are directly heated to temperatures between 500 and 600 °C, leading to polymerization and subsequent formation of g-C_3N_4 [32]. (b) Solvent-assisted

Synthesis—by incorporating solvents, the crystallinity and porosity of g-C_3N_4 can be manipulated, enhancing its performance in specific applications. (c) Microwave-assisted Synthesis—this method harnesses the rapid heating capabilities of microwaves to achieve efficient and homogenous polymerization.

Several precursors for g-C_3N_4 synthesis, distinguished by their nitrogen content, can be employed for g-C_3N_4 synthesis: (a) Melamine—owing to its high nitrogen content, melamine is a widely preferred precursor. It affords good crystallinity and relatively more uncondensed amino groups, beneficial for certain applications [33]. (b) Urea—a more economical alternative, urea can be directly polymerized to yield g-C_3N_4. However, urea-derived g-C_3N_4 often displays inferior crystallinity [32]. (c) Dicyandiamide—it serves as an intermediate between melamine and urea in terms of the nitrogen content and resulting material properties [34]. The morphological attributes of g-C_3N_4 are significantly dictated by the thermal treatment it undergoes (Figure 1). Two parameters, the duration and ramping rate of the heating process, play pivotal roles: (a) Temperature Duration—extended heating times can refine the crystalline structure but might lead to an over-condensed framework, reducing activity in certain applications [32]. (b) Ramping Rate—a rapid ramping rate can cause abrupt polymerization, potentially resulting in non-uniform morphology. Gradual heating, on the other hand, allows a more ordered structure to be formed.

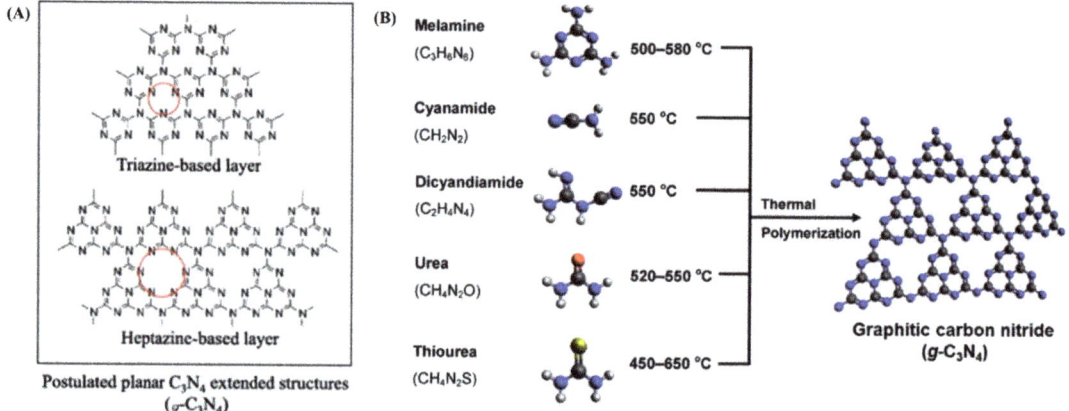

Figure 1. (**A**,**B**) Synthesis of g-C3N4 from different precursors and temperature conditions [35].

Over the years, researchers have reported a diverse array of g-C_3N_4 morphologies like nanosheets, nanorods, nanolayers, etc., due to their specific reasons [36]. Bulk g-C_3N_4 provides the primary structure obtained from direct thermal polymerization, characterized by its layered architecture. g-C_3N_4 nanosheets are thin, 2D planar structures that exhibit a higher surface area compared to the bulk counterpart. g-C_3N_4 nanorods are one-dimensional nanostructures offering directional pathways for charge transport. g-C_3N_4 nanolayers are ultrathin variants of nanosheets, further maximizing the surface area. g-C_3N_4 porous nanosheets introducing porosity can facilitate greater substance interaction and accessibility [37].

Several strategies exist to induce desired morphological changes in g-C_3N_4. (a) Exfoliation: mechanical or chemical means can be employed to delaminate bulk g-C_3N_4, producing nanosheets [38]. (b) Template Method: using sacrificial templates, g-C_3N_4 can be cast into specific morphologies, like rods or spheres, which are subsequently removed [39]. (c) Direct Synthesis: by manipulating synthesis conditions, such as precursor concentration or solvent choice, varied morphologies can be achieved [40]. The morphological modifications directly impact g-C_3N_4's performance in photocatalysis. Enhanced surface area, improved light absorption, efficient charge separation, and facile substance accessibility are some of the benefits brought by morphology optimization. From its synthesis from diverse precursors to the myriad morphologies it can adopt, g-C_3N_4 offers an exciting playground for materials scientists and engineers. As our understanding of its structure–performance relationship deepens, fine-tuned g-C_3N_4 materials, specifically designed for target applications, can be anticipated. By marrying the principles of green chemistry with advanced characterization and simulation techniques, the next chapter in the g-C_3N_4 saga is set to be even more promising.

Role of advanced spectroscopic techniques:

Researchers must know the importance of understanding the underlying mechanisms responsible for the high photocatalytic activity of g-C_3N_4-QD nanocomposites.

Absorption and Bandgap Analysis: enhanced absorption in the visible-light region is a key indicator of improved photocatalytic activity. When quantum dots (QDs) are integrated with g-C_3N_4, a noticeable shift in the absorption edge towards longer wavelengths can be observed. This suggests a reduced bandgap, allowing the composite to harness a greater portion of the solar spectrum. A reduced bandgap often leads to increased electron–hole pair generation, thus driving photocatalytic reactions more efficiently [26].

Photoluminescence (PL) Spectroscopy: PL studies offer invaluable insights into the recombination rates of photoinduced electron–hole pairs. For an effective photocatalyst, the suppression of this recombination is crucial. Post integration of QDs with g-C_3N_4, a significant quenching or decrease in the PL intensity can be observed, signifying reduced recombination rates. This reduced recombination, as evidenced by the PL studies, supports the notion of heightened photocatalytic performance of the composite [19].

X-ray Photoelectron Spectroscopy (XPS): XPS is instrumental in probing the surface electronic states of materials. Upon forming a heterojunction between g-C_3N_4 and QDs, shifts in the XPS peak positions can be noticed, indicating a change in the electronic environment and suggesting charge transfer between the constituents. The altered intensities can hint at the difference in elemental composition, showcasing the successful integration of QDs onto g-C_3N_4. Such charge transfer is paramount for separating the photogenerated electron–hole pairs, thereby enhancing the photocatalytic efficiency [21].

Other Spectroscopic Techniques: Electron Paramagnetic Resonance (EPR) can be employed to detect photoinduced radical species, directly supporting the photocatalytic activity of the material. Additionally, techniques like Fourier-transform infrared spectroscopy (FTIR) can offer insights into the surface functional groups, ensuring the stability and integrity of the composite during photocatalytic reactions [24].

3. Metal Oxide QD-g-C_3N_4 Nanocomposites

Advantages of quantum dots over nanoparticles:

Size and Quantum Confinement: the primary distinction lies in the size. Quantum dots are typically smaller than nanoparticles and are in the range of 1–10 nanometers (approximately 10–50 atoms in diameter). At this scale, quantum effects significantly influence the material's properties, leading to phenomena like quantum confinement in semiconductor QDs, which is not observed in larger nanoparticles. This results in unique optical, electronic, and catalytic properties for QDs.

Optical Properties: due to quantum confinement, QDs exhibit size-dependent tunable photoluminescence, allowing them to absorb and emit light over a wide spectrum. This property is crucial for various applications, including photocatalysis, and is not prominently observed in larger nanoparticles.

Surface Properties: the high surface-to-volume ratio of QDs leads to a significant proportion of atoms being at the surface, which profoundly impacts their chemical reactivity and catalytic activity. While nanoparticles also have a high surface-to-volume ratio, the quantum effects in QDs enhance these surface-related properties.

Energy Band Structures: the discrete energy levels in QDs, a consequence of quantum confinement, differ substantially from the continuous band structure of bulk materials or larger nanoparticles. This affects their interaction with light, charge carrier generation, and transfer—critical factors in photocatalytic processes.

3.1. Wide-Bandgap Metal Oxide QD-g-C$_3$N$_4$ Nanocomposites

In recent half-decade research, TiO$_2$ quantum dots (QDs) have solidified their position as stalwarts in the realm of nanotechnology, largely attributed to their exceptional photocatalytic proficiencies [41]. These quantum entities hold immense promise in efficiently absorbing solar energy, paving the way for their integration into an array of environmental and energy-focused applications. Properties intrinsic to TiO$_2$ QDs set them apart in the vast quantum landscape. The quantum confinement effect empowers them with a modifiable bandgap, a boon for diversifying photocatalytic ventures. Their magnified surface-to-volume ratio augments their inherent reactivity, and their commendable photostability ensures longevity in demanding applications [42]. As for their real-world implications, these QDs shine in water purification, adeptly obliterating organic contaminants. Their prowess extends to hydrogen production, where they serve as linchpins in photoelectrochemical water splitting. Furthermore, their capabilities in air purification, specifically in annihilating noxious air pollutants, have been documented. In essence, the ongoing research narrative accentuates the transformative potential of TiO$_2$ QDs, suggesting a luminous path ahead in environmental rejuvenation and sustainable energy paradigms. According to Wang et al., creating a p-TiO$_2$ QDs@g-C$_3$N$_4$ p-n junction results in better photocatalytic performance than using pure g-C$_3$N$_4$. The improved performance is a result of the p-n heterojunction, strong interface interaction, and quantum-size impact [43]. Wang (2021) synthesized an F-doped TiO$_2$ quantum dot/g-C$_3$N$_4$ nanosheet Z-scheme photocatalyst through chemical bonding, resulting in improved oxidizability, reducibility, and interfacial charge transfer ability [44]. Lee (2023) created a 0D/2D heterojunction nanocomposite with TiO$_2$ quantum dots anchored on g-C$_3$N$_4$ nanosheets (Figure 2) that demonstrated accelerated solar-driven photocatalysis [45]. The integration of anatase/rutile homojunction quantum dots onto g-C$_3$N$_4$ nanosheets, which is intended to target the breakdown of antibiotics in saltwater matrices, was documented by Hu and colleagues. Their study delves into the combined mechanism of adsorption and photocatalysis, shedding light on its underlying intricacies. They further evaluated the ternary heterojunctions formed between anatase/rutile quantum dots (QDs) and g-C$_3$N$_4$, emphasizing their effectiveness in the removal of Oxytetracycline (OTC). Additionally, the research gauges the toxicity levels of the resultant intermediates detected post-process [24].

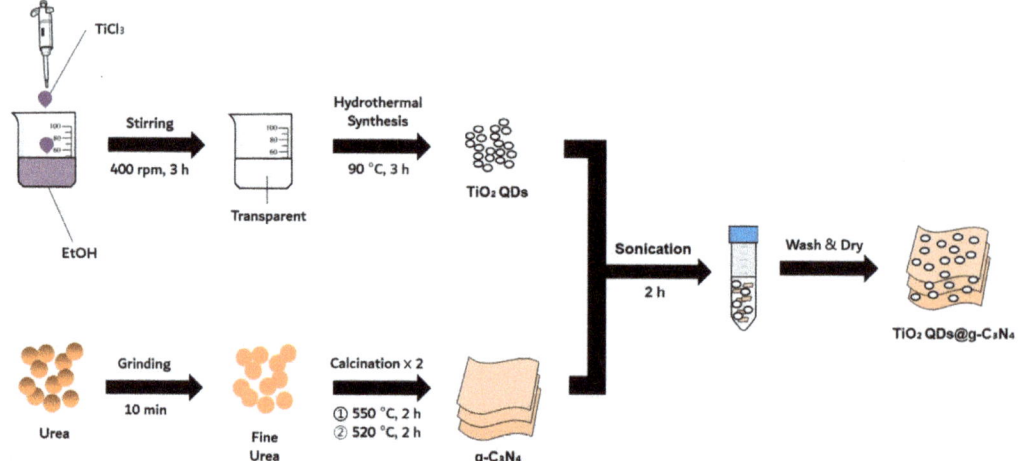

Figure 2. Synthetic process for fabricating TiO$_2$ QDs@g-C$_3$N$_4$ nanocomposite [45].

In recent years, the world of nanotechnology has seen an upswing in interest towards SnO$_2$ quantum dots (QDs), primarily owing to their potent capabilities in photocatalysis [46]. These quantum dots excel in efficiently harnessing light, thereby driving effective charge separation and curbing recombination—traits indispensable for successful photocatalysis. Delving into their synthesis, several cutting-edge methods have emerged over the past half-decade. The hydrothermal method, which revolves around reacting tin salts in water under specific temperature and pressure conditions, remains a favored choice [47]. However, the sol-gel approach, where a precursor solution transitions from a gel-like consistency to the desired quantum dots upon drying and calcination, is also prevalent [46]. Not to be overshadowed, the microwave-assisted synthesis leverages the power of uniform and rapid microwave heating, often resulting in SnO_2 QDs of superior crystallinity in a fraction of the conventional synthesis time [48]. What truly sets SnO_2 QDs apart are their intrinsic properties. The quantum confinement effect grants researchers the liberty to tweak their bandgap, ensuring adaptability for a range of light-driven reactions. Their nanoscale stature bestows upon them a vast surface area, ideal for fostering enhanced reactant interactions. Furthermore, they stand out in the quantum dot family for their remarkable chemical and thermal stability. On the application front, these QDs have been instrumental in several arenas, from water splitting, where they play a role in converting water into hydrogen fuel using sunlight, to the degradation of persistent organic pollutants in water [49]. Another noteworthy application is their potential in reducing CO_2, where they can transform atmospheric carbon dioxide into valuable fuels, presenting a promising avenue to combat escalating CO_2 levels [50]. Recent studies and trends hint that the true potential of SnO_2 QDs, especially when amalgamated with complementary materials, is yet to be fully unlocked, holding promises for advances in sustainable energy and environmental solutions.

SnO_2 quantum dots (QDs) with graphitic carbon nitride (g-C_3N_4) can improve photocatalytic activity. In 2018, Babu found that when exposed to sunlight, the mixture of SnO_2 QDs and g-C_3N_4 nanolayers displayed increased photocatalytic performance, effectively breaking down methyl orange. This increase in sunlight-driven photocatalytic activity is attributed to the cooperative interaction between the g-C_3N_4 nanolayers and SnO_2 quantum dots. These results highlight the potential of g-C_3N_4 nanolayers and SnO_2 QDs as powerful sunlight-responsive photocatalysts, particularly for the degradation of pollutants like methyl orange [27]. In 2019, Yousaf noted a marked increase in photocatalytic performance upon the embellishment of g-C_3N_4 with SnO_2 QDs, which led to the successful decomposition of Rhodamine B. The relative proportion of SnO_2 to g-C_3N_4 in these nanohybrids plays a pivotal role in determining their photocatalytic efficacy (Figure 3). Such findings highlight the potency of SnO_2/g-C_3N_4 nanocomposites, particularly in the domain of degrading contaminants like Rhodamine B (RhB) in solutions [51]. In 2017, Ji pioneered the synthesis of a composite photocatalyst combining SnO_2 with graphene-like g-C_3N_4. This composite exhibited superior visible-light-driven activities in degrading organic pollutants. Remarkably, its optimal photocatalytic efficiency under visible light exposure surpassed that of SnO_2 and g-C_3N_4 by almost 9 and 2.5 times, respectively [52]. The synergy between SnO_2 and graphene-like g-C_3N_4 is highlighted in this composite, which is represented as SnO_2/graphene-like g-C_3N_4, underlining its potential in photocatalytic processes. Its process in the degradation of Rhodamine B (RhB) with visible light in particular provides encouraging insights into its functional possibilities. All of these results point to the possibility that SnO_2 quantum dots and g-C_3N_4 work better together to accelerate photocatalytic breakdown of organic contaminants in water.

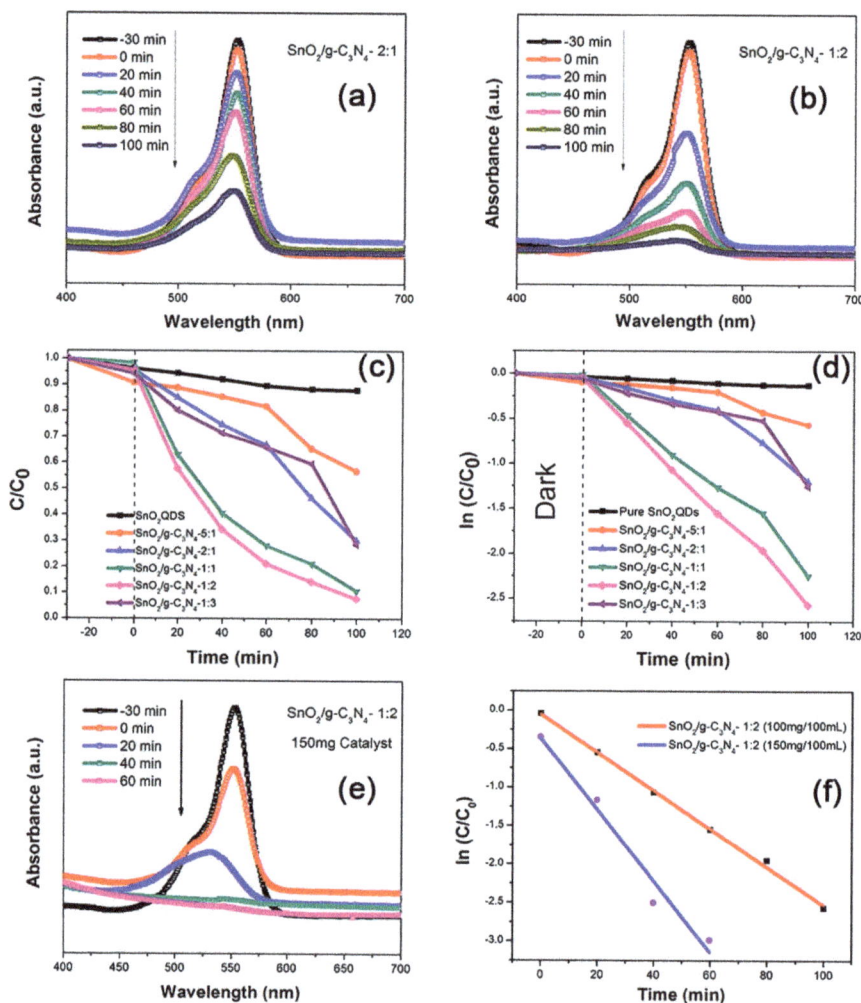

Figure 3. (a–f) Photocatalytic application of SnO$_2$ QD/g-C$_3$N$_4$ nanocomposite [51].

Zinc oxide (ZnO) quantum dots have shown significant promise for photocatalytic applications, driven by their unique physicochemical properties. ZnO quantum dots exhibit enhanced photocatalytic efficiency owing to their high surface area and quantum confinement effects [53]. Their ability to generate reactive oxygen species upon light irradiation makes them potent catalysts for degrading organic pollutants. Researchers have delved into surface modifications to improve the photocatalytic performance of ZnO quantum dots [54]. Techniques like doping, coating, or hybridizing with other materials have been shown to enhance their stability and photocatalytic activity. The advent of black ZnO quantum dots has opened up the possibility of utilizing visible light, significantly broadening the spectrum of light that can be used for photocatalytic applications. Innovations in the design of heterostructures with ZnO quantum dots have shown promise in promoting charge separation, which is crucial for efficient photocatalysis. ZnO quantum dots have found real-world applications in water treatment, air purification, and energy conversion, embodying the translation of academic research to practical solutions [55]. Studies have showcased the robustness of ZnO quantum dots in diverse environmental conditions, highlighting their potential for outdoor applications. The integration of ZnO quantum dots with other nanomaterials like graphitic carbon nitride has led to the creation of novel nanocomposites with superior photocatalytic properties [56]. Ren et al. created a composite by mixing graphitic carbon nitride (g-C_3N_4) and ZnO quantum dots (QDs) with the goal of enhancing the material's photocatalytic properties. It was impressive to see how well the composite degraded Rhodamine B when exposed to visible light. A 96.8% degradation rate of Rhodamine B was attained under visible light within a 40 min window, which was a startling achievement. This heightened photocatalytic efficacy is believed to emanate from the synergistic interplay between ZnO QDs, GO, and g-C_3N_4. The separation of photogenerated electron–hole pairs was accelerated with this combination [57]. In conclusion, the ZnO QD/GO/g-C_3N_4 composite emerges as a potent contender for the remediation of organic pollutants in wastewater, presenting a workable solution for real-world wastewater treatment scenarios. This is due to its potent photocatalytic performance under visible light and its impressive durability. Investigating the visible-light-induced photocatalytic behavior of SnO_2-ZnO quantum dots attached to g-C_3N_4 nanosheets was the goal of Vattikuti et al. The two main areas of concern were the degradation of contaminants and the creation of H_2. They successfully anchored SnO_2-ZnO quantum dots onto g-C_3N_4 nanosheets with their efforts. The resultant composite manifested heightened photocatalytic prowess when exposed to visible light, especially evident in its commendable degradation rates for contaminants like RhB and phenol. When the data were analyzed, it was discovered that the composite's RhB degradation rate was 3.5 times greater than that of pure g-C_3N_4. Similar to this, phenol's rate of degradation was 2.8 times more rapid than that of g-C_3N_4. Additionally, it was found that the composite's capacity to produce H_2 was astonishingly 4.6 times greater than that of pure g-C_3N_4. From these results, it is clear that the SnO_2-ZnO quantum dots, when attached to g-C_3N_4 nanosheets, significantly improve photocatalytic activity when exposed to visible light. This composite not only excels at degrading pollutants but also evinces significant potential in H_2 production. Such attributes earmark it as a viable solution for tasks ranging from environmental purification to fostering sustainable energy methodologies [58].

3.2. Bi-Based QD-g-C_3N_4 Nanocomposites

Bismuth-based quantum dots (QDs) have emerged as a captivating class of nanostructured semiconductors, drawing substantial interest because of their unique electronic, optical, and photocatalytic characteristics. Their size-dependent bandgaps offer specific tunability for photocatalytic reactions. $BiVO_4$ QDs [59], for instance, display improved light absorption due to quantum confinement effects, and they are recognized for their proficiency in visible-light-driven water splitting and pollutant degradation (Figure 4). Bismuth oxide (Bi_2O_3) QDs exhibit enhanced optical attributes and charge transfer characteristics, positioning them as formidable catalysts for UV and visible light pollutant degradation [31,60]. Bi_2WO_6 QDs, with their increased electron–hole separation at the quantum level, stand out in degrading diverse organic pollutants under visible light [61]. While Bi_2O_4 QDs are relatively less explored, they have demonstrated potential with amplified light interaction at the nanoscale and offer efficient photocatalytic reactions. Lastly, Bi_2MoO_6 quantum dots, known for their extended photogenerated charge carrier lifetimes, are potential frontrunners for organic compound degradation and hydrogen evolution tasks [62]. In a nutshell, the nano-dimensionality of bismuth-based QDs accentuates their photocatalytic performance by amplifying light absorption, optimizing charge transfer, and minimizing recombination, making them versatile contenders for an array of photocatalytic applications. The scientists wanted to create a ternary heterostructure comprising C60, g-C_3N_4, and $BiVO_4$ quantum dots as a photocatalyst. This was carried out to increase the photocatalytic activity when exposed to visible light. Under visible light irradiation, the synthesized ternary heterostructure shown improved photocatalytic activity. Compared to binary heterostructures and pure g-C_3N_4, the ternary heterostructure showed a greater photocatalytic degradation rate of Rhodamine B (RhB) [59]. The higher charge separation efficiency and expanded light absorption range were credited with the better photocatalytic performance. The potential mechanisms underlying the ternary heterostructure's improved photocatalytic activity were discussed by the researchers. They emphasized how C60 and BiVO$_4$ quantum dots worked together to promote charge separation and lessen recombination. The enhanced photocatalytic efficiency of the ternary heterostructure was also attributed to a wider light absorption range. In comparison to other structures, the ternary heterostructure of BiVO$_4$ quantum dots/C60/g-C$_3$N$_4$ was effectively constructed and showed higher photocatalytic activity. The study offers suggestions for creating effective photocatalysts for a variety of uses, including environmental cleanup.

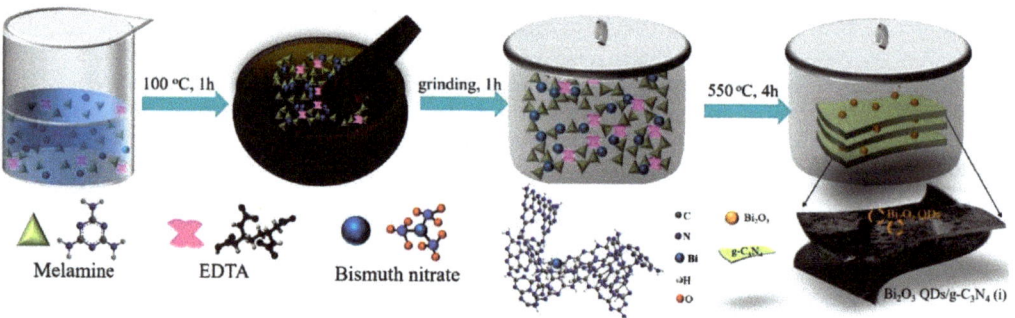

Figure 4. Schematic diagram of the synthesis process for Bi$_2$O$_3$ QD/g-C$_3$N$_4$ nanocomposite [60]. Copyright 2021, Elsevier.

Liang et al. looked at how well Bi$_2$O$_3$ QDs/g-C$_3$N$_4$ performed as a photocatalyst for both organic and inorganic contaminants. Tetracycline (TC) and Cr (VI) were chosen as representative environmental pollutants to assess the effectiveness of the samples' photocatalytic reduction and oxidation [60]. Under light illumination, the photocurrent density of the Bi$_2$O$_3$ QDs/g-C$_3$N$_4$ (ii) was noticeably higher than that of the g-C$_3$N$_4$, E-g-C$_3$N$_4$, and Bi$_2$O$_3$ QDs/g-C$_3$N$_4$, showing better charge transfer efficiency. Bi$_2$O$_3$ QDs/g-C$_3$N$_4$ had the shortest semicircular arc diameter, according to EIS Nyquist plots, indicating the lowest charge transfer resistance and quickest interfacial charge transport. Scavengers had an impact on the effectiveness of TC's degradation, proving that certain radicals were involved in the process. Byproducts of TC were produced using photocatalytic mineralization processes, with some intermediates demonstrating decreased toxicity after photocatalytic degradation. Bi$_2$O$_3$ QDs/g-C$_3$N$_4$ underwent photoinduction to improve the separation and transfer of photogenerated charges. As model environmental pollutants, TC and Cr (VI) were used to assess the photocatalytic performance. For Bi$_2$O$_3$ QDs/g-C$_3$N$_4$, the results showed increased charge transfer effectiveness and quicker interfacial charge transport. The research showed that Bi$_2$O$_3$ QDs/g-C$_3$N$_4$ had the potential to be an efficient photocatalyst for the oxidation of both organic and inorganic contaminants. Its better photocatalytic performance was aided by its improved charge transfer efficiency and decreased charge transfer resistance. The research offers a potential method for creating highly scattered metal oxides on 2D lamella semiconductors, expanding the photocatalyst's usefulness for removing a variety of environmental pollutants. Zeng et al. concentrated on the logical application of quantum dots (QDs) and graphitic carbon nitride (g-C$_3$N$_4$) semiconductors to increase their effectiveness as photocatalysts. The integration of g-C$_3$N$_4$ with QDs was intended to increase photogenerated electron transfer efficiency and produce significant photocatalytic activity. The researchers created brand-new Bi$_2$WO$_6$ QD/g-C$_3$N$_4$ nanocomposites with attapulgite (ATP) penetration [61]. The outcomes demonstrated that the ATP with a nanorod shape served as bridges to intercalate into the interlayers, thus enlarging the g-C$_3$N$_4$ inner space. With the use of an HPLC-MS system, the samples' photocatalytic degradation activities were examined. Using a specified formula, the degrading effectiveness of MBT in the solution was determined. The MBT solution was broken down while being exposed to radiation to gauge the photocatalytic performance. We examined the rate of MBT degradation in several samples and investigated the photocatalytic degradation processes. The BCA5 sample exhibited favorable photoelectric characteristics, which indicated quick interfacial charge movement and low resistance for the production of charge carriers. As a potential strategy for future photocatalytic applications, the integration of g-C_3N_4 and QDs with the interpenetration of ATP led to improved photocatalytic performance. In order to create a 2D-0D g-C_3N_4/Bi_2WO_6-OV composite catalyst, Cheng et al. combined two-dimensional (2D) graphite carbon nitride (g-C_3N_4) nanosheets with oxygen-containing vacancy zero-dimensional (0D) Bi_2WO_6 (BWO-OV) quantum dots [31]. The goal was to improve the catalyst's catalytic activity, increase the formation of photogenerated carriers, and improve light absorption. Utilizing Bi_2WO_6 with oxygen vacancies, which improved light absorption while simultaneously increasing the production of photogenerated carriers, was the novel method. The vacancy structure of Bi_2WO_6 and the heterojunction's creation both contributed to the photogenerated carriers' longer longevity. The composite of CN/BWO-OV-10 displayed the maximum intensity, indicating a greater capacity for NO degradation. The outcomes of the trapping tests revealed that superoxide radicals, holes, and electrons all contribute significantly to the photocatalytic reaction. Furthermore, hydroxyl is recognized as a less potent active free radical. By eliminating NO, the photocatalytic effectiveness was evaluated. The efficiency peaked at 61.2% when BWO-OV was 10% by mass of the total amount of CN. The CN/BWO composite demonstrated the superiority of CN/BWO-OV-10 with a rise in efficiency of 3.2%, achieving a degradation efficiency of 58%. At room temperature, the composite g-C_3N_4/Bi_2WO_6-OV structure removed nitric oxide (NO) at a rapid rate despite its low concentration. The composite catalyst's efficiency was higher than that of g-C_3N_4 or BWO-OV and superior to that of g-C_3N_4/Bi_2WO_6 without oxygen vacancies. The best catalytic activity was demonstrated with the composite g-C_3N_4/Bi_2WO_6-OV-10, reaching up to 61.2%. The substance also demonstrated outstanding stability throughout several iterations of experimentation.

Ding and co. investigated Bi_2MoO_6 QDs/g-C_3N_4 with heterojunctions for their potential in the selective oxidation of aromatic alkanes into aldehydes under visible-light-driven catalysis [62]. The study offers fresh insights into the manufacturing of 0D/2D photocatalysts with heterojunctions for effective selective oxidation of C(sp3)-H bonds (Figure 5). It also presents a novel structure that improves the separation of charge carriers. Outstanding visible-light-driven catalytic performance was shown with the Bi_2MoO_6 QD/g-C_3N_4 heterojunction in the selective oxidation of aromatic alkanes into aldehydes. The heterojunction's special structure, which facilitates the effective separation of charge carriers, is said to be responsible for the increased photocatalytic activity. Superior photocatalytic activity was demonstrated with the heterojunction of Bi_2MoO_6 QDs/g-C_3N_4, particularly in the selective oxidation of aromatic alkanes. The distinctive structure that enables greater charge carrier separation is responsible for this performance. Under visible light, the Bi_2MoO_6 QD/g-C_3N_4 heterojunction exhibits outstanding photocatalytic activity in the selective oxidation of aromatic alkanes to aldehydes. The distinctive structure, which improves charge carrier separation, is credited with the efficiency. The potential of 0D/2D photocatalysts with heterojunctions for the selective oxidation of C(sp^3)-H bonds is crucial information provided in this study.

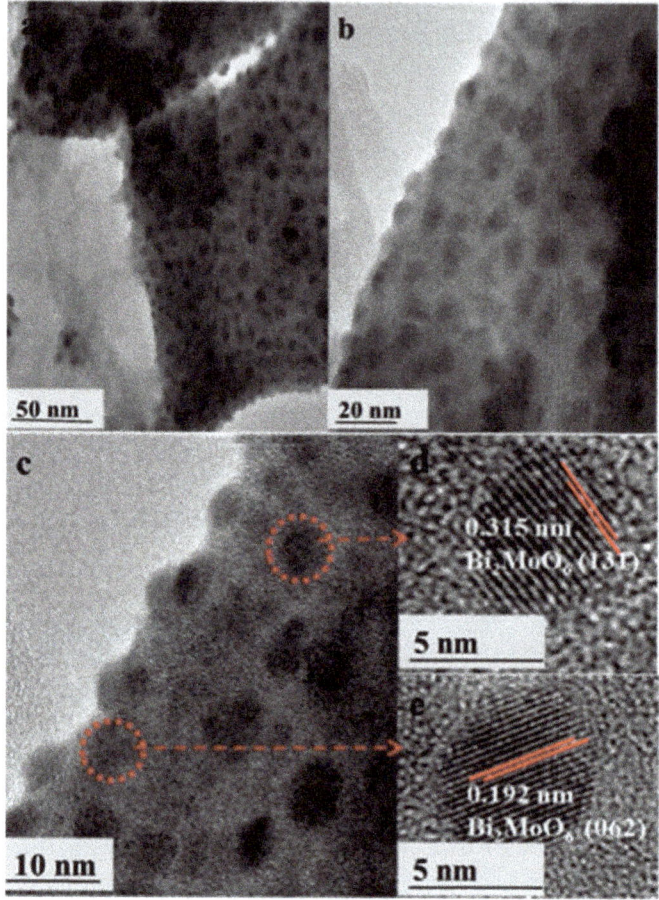

Figure 5. (**a,b**) TEM and (**c–e**) HRTEM images of Bi_2MoO_6 QD/g-C_3N_4 nanocomposite [62]. Copyright 2019, Elsevier.

3.3. Other Metal Oxide QD-g-C_3N_4 Nanocomposites

Sun et al. sought to create CeO_2 quantum dots anchored on g-C_3N_4 (CeO_2/g-C_3N_4) and examine the photocatalytic performance of the material [63]. The creation of CeO_2 quantum dots anchored on g-C_3N_4—which are anticipated to have improved photocatalytic properties—represents the work's originality (Figure 6). It is anticipated that the combination of these materials will enhance charge separation and increase the light absorption range. XRD, FTIR, SEM, TEM, XPS, and PL were used to characterize the synthesized CeO_2/g-C3N4. The outcomes demonstrated a homogeneous distribution of CeO_2 quantum dots on the g-C_3N_4 nanosheets. Better light absorption was discovered to be indicated with the bandgap of CeO_2/g-C_3N_4 being narrower than that of pure g-C_3N_4. By observing the degradation of Rhodamine B (RhB) under visible light irradiation, the photocatalytic performance was assessed. Compared to pure g-C_3N_4, the CeO_2/g-C_3N_4 demonstrated improved photocatalytic activity. The better charge separation and expanded light absorption range brought on with the presence of CeO_2 quantum dots are credited with the improved performance. Rhodamine B degradation under visible light showed higher photocatalytic performance from the CeO_2/g-C_3N_4 combination. This improved performance is a result of the cooperative action of g-C_3N_4 and CeO_2 quantum dots, which improves charge separation and light absorption. Although the exact findings section was not directly extracted, it may be deduced from the information given that the researchers were successful in creating CeO_2 quantum dots anchored on g-C_3N_4 with enhanced photocatalytic characteristics. Its improved Rhodamine B degradation under visible light made the composite material an attractive option for photocatalytic applications.

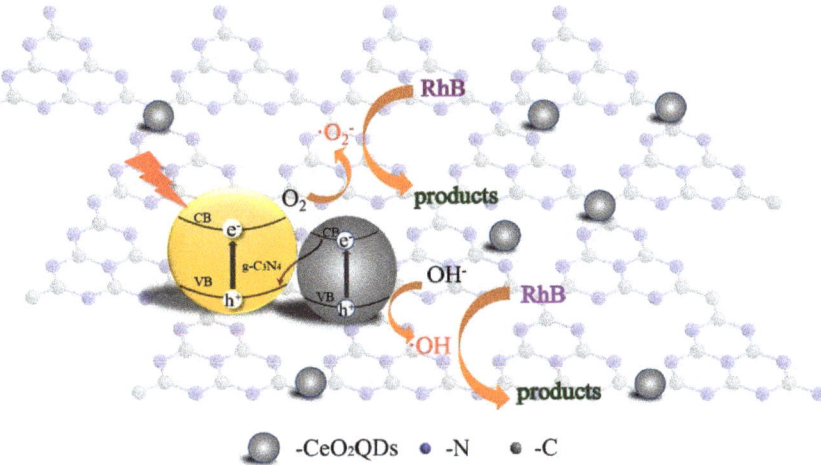

Figure 6. The photocatalytic RhB degradation mechanism over CeO_2 QD-modified g-C_3N_4 nanocomposite [63]. Copyright 2022, Elsevier.

In order to create a novel photocatalyst, Zhu et al. combined g-C_3N_4 nanosheets and MoO_3 quantum dots (QDs). This mixture was created with the goal of improving photocatalytic activity for the reduction in U(VI) under visible light. The addition of MoO_3 QDs to g-C_3N_4 nanosheets is what makes this work novel [19]. Following their easy hydrothermal synthesis, the MoO_3 QDs were loaded onto g-C_3N_4 nanosheets using a straightforward ultrasonic dispersion technique. This mixture was predicted to facilitate photogenerated electron–hole pair separation and improve photocatalytic activity. The MoO_3 QD/g-C_3N_4 nanosheets showed remarkable photocatalytic performance for U(VI) reduction when exposed to visible light in their as-prepared state. It was discovered that MoO_3 QDs should be loaded at a rate of 2%. The efficient separation of photogenerated electron–hole pairs and the expanded light absorption range were credited with the improved photocatalytic activity. Various characterization approaches were used to support the postulated photocatalytic mechanism. By measuring the concentration of U(VI) in aqueous solutions while they were exposed to visible light, the photocatalytic performance was assessed. The outcomes demonstrated that the photocatalytic reduction rate for U(VI) was greater in the MoO_3 QD/g-C_3N_4 nanosheets than in pure g-C_3N_4. It can be concluded that g-C_3N_4 nanosheets in combination with MoO_3 QDs present a viable method for improving the photocatalytic reduction in U(VI) under visible light. The proposed mechanism and the innovative photocatalyst design offer important new perspectives for this field's future study. The detailed description of recent research on the photocatalytic abilities of nanocomposites based on metal oxide QD's-g-C_3N_4 were provided in Table 1.

Table 1. Detailed description of recent research on the photocatalytic abilities of nanocomposites based on metal oxide QDs-g-C_3N_4.

Photocatalyst	Pollutant	Dosage	Light Source	Efficiency	Ref.
SnO_2 QDs/g-C_3N_4	RhB	50 mg/L	Visible light	95% in 60 min	[51]
TiO_2 QDs/g-C_3N_4	RhB	10 mg/L	Visible light	99% in 75 min	[44]
TiO_2 QDs/g-C_3N_4	Phenol	10 mg/L	Visible light	98% in 100 min	[44]
TiO_2 QDs/g-C_3N_4	Cr (VI)	20 mg/L	Visible light	99% in 60 min	[44]
TiO_2 QDs/g-C_3N_4	MO	10 mg/L	Solar light	98% in 120 min	[20]
CeO_2 QDs/g-C_3N_4	RhB	10 mg/L	Visible light	80% in 180 min	[63]
CeO_2 QDs/g-C_3N_4	MO	10 mg/L	Visible light	82% in 180 min	[63]
CeO_2 QDs/g-C_3N_4	MB	10 mg/L	Visible light	74% in 180 min	[63]
$BiVO_4$ QDs-g-C_3N_4	RhB	10 mg/L	Visible light	81% in 90 min	[59]
Bi_2O_3 QDs/g-C_3N_4	TC	10 mg/L	Visible light	83% in 120 min	[60]
Bi_2O_3 QDs/g-C_3N_4	Cr (VI)	10 mg/L	Visible light	88% in 60 min	[60]
Bi_2O_4 QDs/g-C_3N_4	RhB	10 mg/L	Visible light	78% in 60 min	[64]
Bi_2WO_6 QDs/g-C_3N_4	MBT	20 mg/L	Visible light	99% in 80 min	[61]
MoO_3 QDs/g-C_3N_4	U (VI)	50 mg/L	Visible light	96% in 150 min	[19]
Co_3O_4 QDs-g-C_3N_4	MTZ	10 mg/L	Visible light	77% in 180 min	[18]
Bi_2WO_6 QDs/g-C_3N_4	NO	0.5 ppm	Visible light	61%	[31]
SnO_2 QDs/g-C_3N_4	NO	600 ppb	Visible light	32%	[29]
SnO_2 NPs/g-C_3N_4	RhB	10 ppm	Visible light	97% in 50 min	[65]
SnO_2 NPs/g-C_3N_4	MB	10 ppm	Visible light	99% in 90 min	[66]
SnO_2 NPs/g-C_3N_4	CR	10 ppm	Visible light	96% in 90 min	[66]
SnO_2 NPs/g-C_3N_4	NO	500 ppb	Visible light	40%	[67]
SnO_2 NPs/g-C_3N_4	RhB	10 ppm	Solar light	86% in 240 min	[68]
TiO_2 NPs/g-C_3N_4	TC	20 ppm	Visible light	99% in 120 min	[69]
TiO_2 NPs/g-C_3N_4	TC	20 ppm	UV light	96% in 90 min	[70]
TiO_2 NPs/g-C_3N_4	TC	20 ppm	Visible light	90% in 120 min	[71]
TiO_2 NPs/g-C_3N_4	MB	20 ppm	Solar light	80% in 180 min	[72]
TiO_2 NPs/g-C_3N_4	TC	100 ppm	Visible light	80% in 100 min	[73]
ZnO NPs/g-C_3N_4	MB	10 ppm	Visible light	98% in 150 min	[74]
ZnO NPs/g-C_3N_4	MB	50 ppm	Visible light	98% in 180 min	[75]
ZnO NPs/g-C_3N_4	MB	10 ppm	Visible light	60% in 120 min	[76]
ZnO NPs/g-C_3N_4	MB	10 ppm	Visible light	92% in 120 min	[77]
ZnO NPs/g-C_3N_4	CR	10 ppm	Visible light	70% in 45 min	[78]
CeO_2 NPs/g-C_3N_4	MO	10 ppm	Visible light	96% in 100 min	[79]
CeO_2 NPs/g-C_3N_4	RhB	10 ppm	Visible light	96% in 60 min	[80]
CeO_2 NPs/g-C_3N_4	Cr	20 ppm	Visible light	96% in 100 min	[81]
CeO_2 QDs/g-C_3N_4	TC	10 ppm	Visible light	78% in 160 min	[82]
CeO_2 NPs/g-C_3N_4	MB	10 ppm	Visible light	70% in 180 min	[83]
$BiVO_4$ NPs/g-C_3N_4	4-CP	20 ppm	Visible light	95% in 100 min	[84]
$BiVO_4$ NPs/g-C_3N_4	MO	20 ppm	Visible light	82% in 60 min	[85]
$BiVO_4$ NPs/g-C_3N_4	MB	10 ppm	Visible light	88% in 120 min	[86]
$BiVO_4$ NPs/g-C_3N_4	TC	10 ppm	Visible light	89% in 120 min	[86]
Bi_2WO_6 NPs/g-C_3N_4	CIP	15 ppm	Visible light	98% in 120 min	[87]
Bi_2WO_6 NPs/g-C_3N_4	Diuron	20 ppm	Visible light	75% in 120 min	[88]
Bi_2WO_6 NPs/g-C_3N_4	ADN	10 ppm	Visible light	98% in 80 min	[89]
Co_3O_4 NPs/g-C_3N_4	Atrazine	-	Visible light	78% in 35 min	[90]
MoO_3 NPs/g-C_3N_4	TC	10 ppm	Visible light	86% in 100 min	[91]
MoO_3 NPs/g-C_3N_4	RhB	10 ppm	Visible light	99% in 25 min	[92]

Zhao et al. set out to create phosphorus-doped g-C_3N_4/Co_3O_4 quantum dots in a single step using vitamin B12. They sought to increase the synthesized substance's visible-light photocatalytic activity for the destruction of metronidazole (MTZ) [18]. The innovative aspect of the process is the one-step production of phosphorus-doped g-C_3N_4/Co_3O_4 quantum dots using vitamin B12. This technique is distinctive because it enhances the photocatalytic performance of g-C_3N_4 by combining the traits of both phosphorus doping and Co_3O_4 quantum dots. Several approaches were used to characterize the synthesized composites. The photodegradation of MTZ under visible light irradiation was used to assess the photocatalytic activities of the composites. The explanation for the increased photocatalytic activity was determined to be the synergistic interaction between the produced Co_3O_4 quantum dots and P-doped g-C_3N_4. This interaction improved photo-induced electron and hole separation efficiency, prevented their recombination, and reduced band gap energy. The generated Co_3O_4 quantum dots and P-doped g-C_3N_4 worked together to enhance the photocatalytic performance of the created material. As a result of this synergy, photoinduced electrons and holes were separated and transferred more effectively, which increased the photocatalytic degradation of MTZ. The study successfully demonstrated the synthesis of g-C_3N_4/Co_3O_4 quantum dots doped with phosphorus utilizing a one-step procedure and vitamin B12. The created substance demonstrated improved visible-light photocatalytic activity for the breakdown of MTZ. The fundamental causes of the enhanced photocatalytic performance were determined to be the synergistic interactions between the produced Co_3O_4 quantum dots and P-doped g-C_3N_4.

The heterojunctions formed between traditional QDs like CdS/CdSe and g-C_3N_4 have demonstrated effective charge separation due to their staggered band alignment, which minimizes recombination and enhances photocatalytic performance. We discussed how these strategies of interface engineering can be applied to metal oxide QDs to optimize their interactions with g-C_3N_4, focusing on creating synergistic band alignments that facilitate charge transfer and extend light absorption. In-depth analyses of the photocatalytic mechanisms in CdS/CdSe and Ag-In-Zn-S QDs combined with g-C3N4 have revealed critical factors such as quantum confinement effects, surface states, and the role of co-catalysts in improving photocatalytic activity. By integrating these insights, we will elaborate on how similar principles might govern the activity of metal oxide QDs, and how understanding these mechanisms can guide the optimization of their photocatalytic performance. Drawing parallels between the successes of these traditional QDs and our subject metal oxide QDs, we will discuss how strategies like precise size control, doping, or the introduction of defects, successful in traditional QDs, can be mirrored in metal oxides to modulate band structure, enhance light absorption, and improve charge carrier dynamics. While acknowledging the efficiency of traditional QDs, we also recognize concerns regarding their toxicity and environmental impact, particularly for Cd-based QDs. This contrast presents an opportunity to highlight the relative environmental friendliness of metal oxide QDs and the importance of pursuing these materials for sustainable photocatalysis.

Carbon quantum dots (CQDs) have emerged as a captivating class of carbon nanomaterials, characterized by their unique optical, electrical, and physicochemical properties. These properties, combined with their aqueous stability, low toxicity, high surface area, economic feasibility, and tunable photoluminescence behavior, make them promising candidates for photocatalytic applications. On the other hand, graphitic carbon nitride (g-C_3N_4) has gained attention as a stable carbon-based polymer with potential applications in various fields. The combination of CQDs and g-C_3N_4 offers a synergistic effect, enhancing the adsorptive and photocatalytic activity of the resulting nanocomposite. This is attributed to the broader visible-light absorption, increased specific surface area, and enhanced electron–hole pair migration and separation efficiency of the composite. Comparatively, while metal oxide quantum dots also present potential in photocatalytic applications, the CQDs and g-C_3N_4 combination stands out due to its non-toxic nature, economic feasibility, and enhanced photoluminescence properties. The interaction within this multicomponent photocatalyst promotes photocatalytic performance, making it a superior choice for wastewater treatment and other environmental applications. In essence, the amalgamation of CQDs and g-C_3N_4 presents a novel and efficient approach to address the challenges of wastewater treatment, emphasizing the importance of continued research in this domain [93].

4. Conclusions and Perspectives

Quantum dots (QDs) and graphitic carbon nitride (g-C_3N_4) together highlight a developing area of study with the potential to advance the photocatalytic frontier. Harnessing the unique electronic, optical, and photocatalytic properties intrinsic to QDs such as ZnO,

SnO_2, TiO_2, CeO_2, CO_3O_4, MoO_3, $BiVO_4$, Bi_2O_3, Bi_2WO_6, Bi_2O_4, and Bi_2MoO_6, when juxtaposed with the properties of g-C_3N_4, promises a synergy that could redefine the boundaries of photocatalysis. As we reflect on our discussions, let us draw some conclusions and speculate on future avenues.

Synergy of QDs with g-C_3N_4: quantum dots and g-C_3N_4 can be used to enhance light absorption, improve charge transfer, and reduce electron–hole recombination. This enhances their overall photocatalytic efficacy and paves the way for efficient light-driven reactions, particularly those aimed at energy conversion and environmental remediation.

Diverse Quantum Dot Landscape: each QD brings its unique characteristics. For instance, ZnO and TiO_2 QDs have been heralded for their UV light-driven photocatalytic activities, while the bismuth-based QDs such as $BiVO_4$, Bi_2O_3, and Bi_2WO_6 exhibit visible light-driven capabilities. The inherent bandgap variations, coupled with different photocorrosion resistances and surface chemistries, provide a diverse landscape for tailoring the desired photocatalytic response.

Enhanced Stability and Sustainability: one of the perennial challenges with photocatalysts is their stability and recyclability. g-C_3N_4's structural robustness, when combined with the protective attributes of QDs, leads to prolonged catalyst life, thereby elevating the potential for sustainable and scalable applications.

Heterostructuring and Multifunctionality: in addition to enhancing the activity of the quantum dots, the addition of g-C_3N_4 also brings multifunctionality. The heterostructures formed can serve as platforms for multiple simultaneous reactions, like water splitting alongside organic pollutant degradation, creating avenues for multifaceted photocatalytic systems.

5. Future Perspectives

Tailored Photocatalytic Systems: given the plethora of QDs available, future research should focus on systematically tailoring QD-g-C_3N_4 combinations to target specific photocatalytic reactions. This tailoring could lead to breakthroughs in reaction efficiency and selectivity.

Deep Dive into Charge Dynamics: to optimize the QD-g-C_3N_4 interfaces further, a comprehensive understanding of charge dynamics, including the rates of charge transfer, recombination, and trapping, is crucial. Advanced spectroscopic and microscopic techniques could elucidate these intricacies.

Scale-up and Real-world Applications: the transition from lab-scale research to real-world applications requires addressing challenges related to catalyst scale-up, stability under fluctuating environmental conditions, and integration with existing industrial setups.

Holistic Environmental Impact: as we push the boundaries of photocatalysis, it is imperative to ensure the green synthesis of these QD-g-C_3N_4 systems, minimizing any adverse environmental footprints and maximizing their eco-friendly applications.

Z-Schemes and Ternary Nanocomposites: researchers have the opportunity to explore and construct Z-scheme-based systems using QDs and g-C_3N_4. Additionally, there is potential for designing ternary nanocomposites, such as metal oxide QD-g-C_3N_4-noble metals, metal oxide QD-g-C_3N_4-carbon-related materials, and more.

In closing, the fusion of quantum dots with g-C_3N_4 heralds a new era in photocatalysis. The potential breakthroughs discussed here are just the tip of the iceberg. As research intensifies, the full spectrum of possibilities will undoubtedly be unveiled, driving a cleaner, greener, and energy-efficient future.

Author Contributions: Conceptualization, methodology, validation, writing—original draft preparation, T.T.S.; formal analysis, resources, data curation, writing—original draft preparation, T.R.G.; writing—review and editing, supervision, B.B.; supervision, project administration, funding acquisition, K.Y. All authors have read and agreed to the published version of the manuscript.

Funding: This work was supported with a National Research Foundation of Korea (NRF) grant funded by the Korean government (MSIT) (No. 2021R1C1C1011089).

Data Availability Statement: Not applicable.

Conflicts of Interest: The authors declare that they have no conflict of interest regarding the publication of this article, financial and/or otherwise.

References

1. Opoku, E.E.O.; Boachie, M.K. The environmental impact of industrialization and foreign direct investment. *Energy Policy* **2020**, *137*, 111178. [CrossRef]
2. Tran, V.V.; Park, D.; Lee, Y.-C. Indoor Air Pollution, Related Human Diseases, and Recent Trends in the Control and Improvement of Indoor Air Quality. *Int. J. Environ. Res. Public Health* **2020**, *17*, 2927. [CrossRef] [PubMed]
3. Antoci, A.; Galeotti, M.; Sordi, S. Environmental pollution as engine of industrialization. *Commun. Nonlinear Sci. Numer. Simul.* **2018**, *58*, 262–273. [CrossRef]
4. Siddiqua, A.; Hahladakis, J.N.; Al-Attiya, W.A.K.A. An overview of the environmental pollution and health effects associated with waste landfilling and open dumping. *Environ. Sci. Pollut. Res.* **2022**, *29*, 58514–58536. [CrossRef]
5. Manisalidis, I.; Stavropoulou, E.; Stavropoulos, A.; Bezirtzoglou, E. Environmental and Health Impacts of Air Pollution: A Review. *Front. Public Health* **2020**, *8*, 14. [CrossRef]
6. He, J.; Kumar, A.; Khan, M.; Lo, I.M.C. Critical review of photocatalytic disinfection of bacteria: From noble metals- and carbon nanomaterials-TiO_2 composites to challenges of water characteristics and strategic solutions. *Sci. Total Environ.* **2021**, *758*, 143953. [CrossRef]
7. Wei, Z.; Liu, J.; Shangguan, W. A review on photocatalysis in antibiotic wastewater: Pollutant degradation and hydrogen production. *Chin. J. Catal.* **2020**, *41*, 1440–1450. [CrossRef]
8. Zhang, F.; Wang, X.; Liu, H.; Liu, C.; Wan, Y.; Long, Y.; Cai, Z. Recent Advances and Applications of Semiconductor Photocatalytic Technology. *Appl. Sci.* **2019**, *9*, 2489. [CrossRef]
9. Guo, Q.; Zhou, C.; Ma, Z.; Yang, X. Fundamentals of TiO_2 Photocatalysis: Concepts, Mechanisms, and Challenges. *Adv. Mater.* **2019**, *31*, 1901997. [CrossRef]
10. Wang, J.; Guo, R.-T.; Bi, Z.-X.; Chen, X.; Hu, X.; Pan, W.-G. A review on TiO_2−x-based materials for photocatalytic CO_2 reduction. *Nanoscale* **2022**, *14*, 11512–11528. [CrossRef]
11. Zhao, Y.; Zhang, S.; Shi, R.; Waterhouse, G.I.N.; Tang, J.; Zhang, T. Two-dimensional photocatalyst design: A critical review of recent experimental and computational advances. *Mater. Today* **2020**, *34*, 78–91. [CrossRef]
12. Li, X.; Xiong, J.; Gao, X.; Huang, J.; Feng, Z.; Chen, Z.; Zhu, Y. Recent advances in 3D g-C_3N_4 composite photocatalysts for photocatalytic water splitting, degradation of pollutants and CO_2 reduction. *J. Alloys Compd.* **2019**, *802*, 196–209. [CrossRef]
13. Luo, Y.; Zhu, Y.; Han, Y.; Ye, H.; Liu, R.; Lan, Y.; Xue, M.; Xie, X.; Yu, S.; Zhang, L.; et al. g-C_3N_4-based photocatalysts for organic pollutant removal: A critical review. *Carbon Res.* **2023**, *2*, 14. [CrossRef]
14. Chen, L.; Maigbay, M.A.; Li, M.; Qiu, X. Synthesis and modification strategies of g-C_3N_4 nanosheets for photocatalytic applications. *Adv. Powder Mater.* **2023**, 100150, in press. [CrossRef]
15. Zhu, W.; Yue, Y.; Wang, H.; Zhang, B.; Hou, R.; Xiao, J.; Huang, X.; Ishag, A.; Sun, Y. Recent advances on energy and environmental application of graphitic carbon nitride (g-C_3N_4)-based photocatalysts: A review. *J. Environ. Chem. Eng.* **2023**, *11*, 110164. [CrossRef]
16. Babu, B.; Koutavarapu, R.; Shim, J.; Yoo, K. Enhanced visible-light-driven photoelectrochemical and photocatalytic performance of Au-SnO_2 quantum dot-anchored g-C_3N_4 nanosheets. *Sep. Purif. Technol.* **2020**, *240*, 116652. [CrossRef]
17. Jin, T.; Liu, C.B.; Chen, F.; Qian, J.C.; Qiu, Y.B.; Meng, X.R.; Chen, Z.G. Synthesis of g-C_3N_4/CQDs composite and its photocatalytic degradation property for Rhodamine B. *Carbon Lett.* **2022**, *32*, 1451–1462. [CrossRef]
18. Zhao, Z.W.; Fan, J.Y.; Deng, X.Y.; Liu, J. One-step synthesis of phosphorus-doped g-C_3N_4/Co_3O_4 quantum dots from vitamin B12 with enhanced visible-light photocatalytic activity for metronidazole degradation. *Chem. Eng. J.* **2019**, *360*, 1517–1529. [CrossRef]
19. Zhu, X.; Dong, Z.M.; Xu, J.D.; Lin, S.Y.; Liu, J.Y.; Cheng, Z.P.; Cao, X.H.; Wang, Y.Q.; Liu, Y.H.; Zhang, Z.B. Visible-light induced electron-transfer in MoO_3 QDs/g-C_3N_4 nanosheets for efficient photocatalytic reduction of U(VI). *J. Alloys Compd.* **2022**, *926*, 10. [CrossRef]
20. Guo, R.T.; Liu, X.Y.; Qin, H.; Ang, Z.Y.; Shi, X.; Pan, W.G.; Fu, Z.G.; Tang, J.Y.; Jia, P.Y.; Miao, Y.F.; et al. Photocatalytic reduction of CO_2 into CO over nanostructure Bi_2S_3 quantum dots/g-C_3N_4 composites with Z-scheme mechanism. *Appl. Surf. Sci* **2020**, *500*, 144059. [CrossRef]
21. Wang, J.M.; Yu, L.M.; Wang, Z.J.; Wei, W.; Wang, K.F.; Wei, X.H. Constructing 0D/2D Z-Scheme Heterojunction of CdS/g-C_3N_4 with Enhanced Photocatalytic Activity for H-2 Evolution. *Catal. Lett.* **2021**, *151*, 3550–3561. [CrossRef]
22. Li, G.; Huang, J.M.; Wang, N.N.; Huang, J.L.; Zheng, Y.M.; Zhan, G.W.; Li, Q.B. Carbon quantum dots functionalized g-C_3N_4 nanosheets as enhanced visible-light photocatalysts for water splitting. *Diam. Relat. Mat.* **2021**, *116*, 9. [CrossRef]
23. Huo, Y.; Ding, X.H.; Zhang, X.Y.; Ren, M.; Sang, L.B.; Wen, S.Y.; Song, D.Y.; Yang, Y.X. Graphene quantum dot implanted supramolecular carbon nitrides with robust photocatalytic activity against recalcitrant contaminants. *Catal. Sci. Technol.* **2022**, *12*, 3937–3946. [CrossRef]

24. Hu, X.Y.; Yu, Y.T.; Chen, D.D.; Xu, W.C.; Fang, J.Z.; Liu, Z.; Li, R.Q.; Yao, L.; Qin, J.J.; Fang, Z.Q. Anatase/Rutile homojunction quantum dots anchored on g-C_3N_4 nanosheets for antibiotics degradation in seawater matrice via coupled adsorption-photocatalysis: Mechanism insight and toxicity evaluation. *Chem. Eng. J.* **2022**, *432*, 15. [CrossRef]
25. Zhao, T.Y.; Xing, Z.P.; Xiu, Z.Y.; Li, Z.Z.; Yang, S.L.; Zhou, W. Oxygen-Doped MoS_2 Nanospheres/CdS Quantum Dots/g-C_3N_4 Nanosheets Super-Architectures for Prolonged Charge Lifetime and Enhanced Visible-Light-Driven Photocatalytic Performance. *ACS Appl. Mater. Interfaces* **2019**, *11*, 7104–7111. [CrossRef]
26. Wu, H.H.; Yu, S.Y.; Wang, Y.; Han, J.; Wang, L.; Song, N.; Dong, H.J.; Li, C.M. A facile one-step strategy to construct 0D/2D SnO_2/g-C_3N_4 heterojunction photocatalyst for high-efficiency hydrogen production performance from water splitting. *Int. J. Hydrogen Energy* **2020**, *45*, 30142–30152. [CrossRef]
27. Babu, B.; Cho, M.; Byon, C.; Shim, J. Sunlight-driven photocatalytic activity of SnO_2 QDs-g-C_3N_4 nanolayers. *Mater. Lett.* **2018**, *212*, 327–331. [CrossRef]
28. Li, N.X.; Liu, X.C.; Zhou, J.C.; Chen, W.S.; Liu, M.C. Encapsulating CuO quantum dots in MIL-125(Ti) coupled with g-C_3N_4 for efficient photocatalytic CO_2 reduction. *Chem. Eng. J.* **2020**, *399*, 11. [CrossRef]
29. Zou, Y.Z.; Xie, Y.; Yu, S.; Chen, L.; Cui, W.; Dong, F.; Zhou, Y. SnO_2 quantum dots anchored on g-C_3N_4 for enhanced visible-light photocatalytic removal of NO and toxic NO_2 inhibition. *Appl. Surf. Sci.* **2019**, *496*, 8. [CrossRef]
30. Zhu, Z.; Tang, X.; Fan, W.Q.; Liu, Z.; Huo, P.W.; Wang, T.S.; Yan, Y.S.; Li, C.X. Studying of Co-doped g-C_3N_4 and modified with Fe_3O_4 quantum dots on removing tetracycline. *J. Alloys Compd.* **2019**, *775*, 248–258. [CrossRef]
31. Cheng, C.; Chen, D.Y.; Li, N.J.; Li, H.; Xu, Q.F.; He, J.H.; Lu, J.M. Bi_2WO_6 quantum dots with oxygen vacancies combined with g-C_3N_4 for NO removal. *J. Colloid Interface Sci.* **2022**, *609*, 447–455. [CrossRef] [PubMed]
32. Paul, D.R.; Sharma, R.; Nehra, S.P.; Sharma, A. Effect of calcination temperature, pH and catalyst loading on photodegradation efficiency of urea derived graphitic carbon nitride towards methylene blue dye solution. *RSC Adv.* **2019**, *9*, 15381–15391. [CrossRef] [PubMed]
33. Surender, S.; Balakumar, S. Insight into the melamine-derived freeze-dried nanostructured g-C_3N_4 for expeditious photocatalytic degradation of dye pollutants. *Diam. Relat. Mat.* **2022**, *128*, 109269. [CrossRef]
34. Raaja Rajeshwari, M.; Kokilavani, S.; Sudheer Khan, S. Recent developments in architecturing the g-C_3N_4 based nanostructured photocatalysts: Synthesis, modifications and applications in water treatment. *Chemosphere* **2022**, *291*, 132735. [CrossRef]
35. Fernández-Catalá, J.; Greco, R.; Navlani-García, M.; Cao, W.; Berenguer-Murcia, Á.; Cazorla-Amorós, D. g-C_3N_4-Based Direct Z-Scheme Photocatalysts for Environmental Applications. *Catalysts* **2022**, *12*, 1137. [CrossRef]
36. Hayat, A.; Sohail, M.; El Jery, A.; Al-Zaydi, K.M.; Alshammari, K.F.; Khan, J.; Ali, H.; Ajmal, Z.; Taha, T.A.; Ud Din, I.; et al. Different Dimensionalities, Morphological Advancements and Engineering of g-C_3N_4-Based Nanomaterials for Energy Conversion and Storage. *Chem. Rec.* **2023**, *23*, e202200171. [CrossRef]
37. Linh, P.H.; Do Chung, P.; Van Khien, N.; Oanh, L.T.M.; Thu, V.T.; Bach, T.N.; Hang, L.T.; Hung, N.M.; Lam, V.D. A simple approach for controlling the morphology of g-C_3N_4 nanosheets with enhanced photocatalytic properties. *Diam. Relat. Mat.* **2021**, *111*, 108214. [CrossRef]
38. Zhang, M.; Yang, Y.; An, X.; Zhao, J.; Bao, Y.; Hou, L.-A. Exfoliation method matters: The microstructure-dependent photoactivity of g-C_3N_4 nanosheets for water purification. *J. Hazard. Mater.* **2022**, *424*, 127424. [CrossRef]
39. Chen, Y.; Yang, B.; Xie, W.; Zhao, X.; Wang, Z.; Su, X.; Yang, C. Combined soft templating with thermal exfoliation toward synthesis of porous g-C_3N_4 nanosheets for improved photocatalytic hydrogen evolution. *J. Mater. Res. Technol.* **2021**, *13*, 301–310. [CrossRef]
40. Wu, X.; Gao, D.; Wang, P.; Yu, H.; Yu, J. NH_4Cl-induced low-temperature formation of nitrogen-rich g-C_3N_4 nanosheets with improved photocatalytic hydrogen evolution. *Carbon* **2019**, *153*, 757–766. [CrossRef]
41. Zhou, L.; Shen, Z.; Wang, S.; Gao, J.; Tang, L.; Li, J.; Dong, Y.; Wang, Z.; Lyu, J. Construction of quantum-scale catalytic regions on anatase TiO_2 nanoparticles by loading TiO_2 quantum dots for the photocatalytic degradation of VOCs. *Ceram. Int.* **2021**, *47*, 21090–21098. [CrossRef]
42. Javed, S.; Islam, M.; Mujahid, M. Synthesis and characterization of TiO_2 quantum dots by sol gel reflux condensation method. *Ceram. Int.* **2019**, *45*, 2676–2679. [CrossRef]
43. Wang, S.; Wang, F.; Su, Z.; Wang, X.; Han, Y.; Zhang, L.; Xiang, J.; Du, W.; Tang, N. Controllable Fabrication of Heterogeneous p-TiO_2 QDs@g-C_3N_4 p-n Junction for Efficient Photocatalysis. *Catalysts* **2019**, *9*, 439. [CrossRef]
44. Wang, J.; Lin, W.; Hu, H.; Liu, C.; Cai, Q.; Zhou, S.; Kong, Y. Engineering Z-system hybrids of 0D/2D F-TiO_2 quantum dots/g-C_3N_4 heterostructures through chemical bonds with enhanced visible-light photocatalytic performance. *New J. Chem.* **2021**, *45*, 3067–3078. [CrossRef]
45. Lee, J.-H.; Jeong, S.-Y.; Son, Y.-D.; Lee, S.-W. Facile Fabrication of TiO_2 Quantum Dots-Anchored g-C_3N_4 Nanosheets as 0D/2D Heterojunction Nanocomposite for Accelerating Solar-Driven Photocatalysis. *Nanomaterials* **2023**, *13*, 1565. [CrossRef]
46. Kumari, K.; Ahmaruzzaman, M. SnO_2 quantum dots (QDs): Synthesis and potential applications in energy storage and environmental remediation. *Mater. Res. Bull.* **2023**, *168*, 112446. [CrossRef]
47. Yu, B.; Li, Y.; Wang, Y.; Li, H.; Zhang, R. Facile hydrothermal synthesis of SnO_2 quantum dots with enhanced photocatalytic degradation activity: Role of surface modification with chloroacetic acid. *J. Environ. Chem. Eng.* **2021**, *9*, 105618. [CrossRef]
48. Xu, Z.; Jiang, Y.; Li, Z.; Chen, C.; Kong, X.; Chen, Y.; Zhou, G.; Liu, J.-M.; Kempa, K.; Gao, J. Rapid Microwave-Assisted Synthesis of SnO_2 Quantum Dots for Efficient Planar Perovskite Solar Cells. *ACS Appl. Energy Mater.* **2021**, *4*, 1887–1893. [CrossRef]

49. Bathula, B.; Gurugubelli, T.R.; Yoo, J.; Yoo, K. Recent Progress in the Use of SnO_2 Quantum Dots: From Synthesis to Photocatalytic Applications. *Catalysts* **2023**, *13*, 765. [CrossRef]
50. Wu, Z.; Jing, H.; Zhao, Y.; Lu, K.; Liu, B.; Yu, J.; Xia, X.; Lei, W.; Hao, Q. Grain boundary and interface interaction Co-regulation promotes SnO_2 quantum dots for efficient CO_2 reduction. *Chem. Eng. J.* **2023**, *451*, 138477. [CrossRef]
51. Yousaf, M.U.; Pervaiz, E.; Minallah, S.; Afzal, M.J.; Honghong, L.; Yang, M. Tin oxide quantum dots decorated graphitic carbon nitride for enhanced removal of organic components from water: Green process. *Results Phys.* **2019**, *14*, 102455. [CrossRef]
52. Ji, H.; Fan, Y.; Yan, J.; Xu, Y.; She, X.; Gu, J.; Fei, T.; Xu, H.; Li, H. Construction of SnO_2/graphene-like $g-C_3N_4$ with enhanced visible light photocatalytic activity. *RSC Adv.* **2017**, *7*, 36101–36111. [CrossRef]
53. Mohamed, K.M.; Benitto, J.J.; Vijaya, J.J.; Bououdina, M. Recent Advances in ZnO-Based Nanostructures for the Photocatalytic Degradation of Hazardous, Non-Biodegradable Medicines. *Crystals* **2023**, *13*, 329. [CrossRef]
54. Mohamed, W.A.A.; Handal, H.T.; Ibrahem, I.A.; Galal, H.R.; Mousa, H.A.; Labib, A.A. Recycling for solar photocatalytic activity of Dianix blue dye and real industrial wastewater treatment process by zinc oxide quantum dots synthesized by solvothermal method. *J. Hazard. Mater.* **2021**, *404*, 123962. [CrossRef]
55. Mohamed, W.A.A.; Ibrahem, I.A.; El-Sayed, A.M.; Galal, H.R.; Handal, H.; Mousa, H.A.; Labib, A.A. Zinc oxide quantum dots for textile dyes and real industrial wastewater treatment: Solar photocatalytic activity, photoluminescence properties and recycling process. *Adv. Powder Technol.* **2020**, *31*, 2555–2565. [CrossRef]
56. Ye, Q.; Xu, L.; Xia, Y.; Gang, R.; Xie, C. Zinc oxide quantum dots/graphitic carbon nitride nanosheets based visible-light photocatalyst for efficient tetracycline hydrochloride degradation. *J. Porous Mat.* **2022**, *29*, 571–581. [CrossRef]
57. Ren, Z.; Ma, H.; Geng, J.; Liu, C.; Song, C.; Lv, Y. ZnO QDs/GO/$g-C_3N_4$ Preparation and Photocatalytic Properties of Composites. *Micromachines* **2023**, *14*, 1501. [CrossRef]
58. Vattikuti, S.V.P.; Reddy, P.A.K.; Shim, J.; Byon, C. Visible-Light-Driven Photocatalytic Activity of SnO_2–ZnO Quantum Dots Anchored on $g-C_3N_4$ Nanosheets for Photocatalytic Pollutant Degradation and H_2 Production. *ACS Omega* **2018**, *3*, 7587–7602. [CrossRef]
59. Wang, J.B.; Liu, C.; Yang, S.; Lin, X.; Shi, W.L. Fabrication of a ternary heterostructure $BiVO_4$ quantum dots/C-60/$g-C_3N_4$ photocatalyst with enhanced photocatalytic activity. *J. Phys. Chem. Solids* **2020**, *136*, 7. [CrossRef]
60. Liang, Y.; Xu, W.; Fang, J.; Liu, Z.; Chen, D.; Pan, T.; Yu, Y.; Fang, Z. Highly dispersed bismuth oxide quantum dots/graphite carbon nitride nanosheets heterojunctions for visible light photocatalytic redox degradation of environmental pollutants. *Appl. Catal. B* **2021**, *295*, 120279. [CrossRef]
61. Zeng, Y.; Yin, Q.; Liu, Z.; Dong, H. Attapulgite-interpenetrated $g-C_3N_4$/Bi_2WO_6 quantum-dots Z-scheme heterojunction for 2-mercaptobenzothiazole degradation with mechanism insight. *Chem. Eng. J.* **2022**, *435*, 134918. [CrossRef]
62. Ding, F.; Chen, P.; Liu, F.; Chen, L.; Guo, J.-K.; Shen, S.; Zhang, Q.; Meng, L.-H.; Au, C.-T.; Yin, S.-F. Bi_2MoO_6/$g-C_3N_4$ of 0D/2D heterostructure as efficient photocatalyst for selective oxidation of aromatic alkanes. *Appl. Surf. Sci.* **2019**, *490*, 102–108. [CrossRef]
63. Sun, Y.; Yuan, X.; Wang, Y.; Zhang, W.; Li, Y.; Zhang, Z.; Su, J.; Zhang, J.; Hu, S. CeO_2 quantum dots anchored $g-C_3N_4$: Synthesis, characterization and photocatalytic performance. *Appl. Surf. Sci.* **2022**, *576*, 151901. [CrossRef]
64. Qin, Y.Y.; Li, H.; Lu, J.; Ma, C.C.; Liu, X.L.; Meng, M.J.; Yan, Y.S. Fabrication of magnetic quantum dots modified Z-scheme Bi_2O_4/$g-C_3N_4$ photocatalysts with superior hydroxyl radical productivity for the degradation of rhodamine B. *Appl. Surf. Sci.* **2019**, *493*, 458–469. [CrossRef]
65. Sun, C.; Yang, J.; Zhu, Y.; Xu, M.; Cui, Y.; Liu, L.; Ren, W.; Zhao, H.; Liang, B. Synthesis of 0D SnO_2 nanoparticles/2D $g-C_3N_4$ nanosheets heterojunction: Improved charge transfer and separation for visible-light photocatalytic performance. *J. Alloys Compd.* **2021**, *871*, 159561. [CrossRef]
66. Mohammad, A.; Khan, M.E.; Karim, M.R.; Cho, M.H. Synergistically effective and highly visible light responsive SnO_2-$g-C_3N_4$ nanostructures for improved photocatalytic and photoelectrochemical performance. *Appl. Surf. Sci.* **2019**, *495*, 143432. [CrossRef]
67. Van Pham, V.; Mai, D.-Q.; Bui, D.-P.; Van Man, T.; Zhu, B.; Zhang, L.; Sangkaworn, J.; Tantirungrotechai, J.; Reutrakul, V.; Cao, T.M. Emerging 2D/0D $g-C_3N_4$/SnO_2 S-scheme photocatalyst: New generation architectural structure of heterojunctions toward visible-light-driven NO degradation. *Environ. Pollut.* **2021**, *286*, 117510. [CrossRef]
68. Wang, X.; He, Y.; Xu, L.; Xia, Y.; Gang, R. SnO_2 particles as efficient photocatalysts for organic dye degradation grown in-situ on $g-C_3N_4$ nanosheets by microwave-assisted hydrothermal method. *Mater. Sci. Semicond. Process.* **2021**, *121*, 105298. [CrossRef]
69. Zhang, B.; He, X.; Yu, C.; Liu, G.; Ma, D.; Cui, C.; Yan, Q.; Zhang, Y.; Zhang, G.; Ma, J.; et al. Degradation of tetracycline hydrochloride by ultrafine TiO_2 nanoparticles modified $g-C_3N_4$ heterojunction photocatalyst: Influencing factors, products and mechanism insight. *Chin. Chem. Lett.* **2022**, *33*, 1337–1342. [CrossRef]
70. Ni, S.; Fu, Z.; Li, L.; Ma, M.; Liu, Y. Step-scheme heterojunction $g-C_3N_4$/TiO_2 for efficient photocatalytic degradation of tetracycline hydrochloride under UV light. *Colloids Surf. A Physicochem. Eng. Asp.* **2022**, *649*, 129475. [CrossRef]
71. Guo, F.; Sun, H.; Huang, X.; Shi, W.; Yan, C. Fabrication of TiO_2/high-crystalline $g-C_3N_4$ composite with enhanced visible-light photocatalytic performance for tetracycline degradation. *J. Chem. Technol. Biotechnol.* **2020**, *95*, 2684–2693. [CrossRef]
72. Gündoğmuş, P.; Park, J.; Öztürk, A. Preparation and photocatalytic activity of $g-C_3N_4$/TiO_2 heterojunctions under solar light illumination. *Ceram. Int.* **2020**, *46*, 21431–21438. [CrossRef]
73. Li, Y.; Zhang, Q.; Lu, Y.; Song, Z.; Wang, C.; Li, D.; Tang, X.; Zhou, X. Surface hydroxylation of TiO_2/$g-C_3N_4$ photocatalyst for photo-Fenton degradation of tetracycline. *Ceram. Int.* **2022**, *48*, 1306–1313. [CrossRef]

74. Jung, H.; Pham, T.-T.; Shin, E.W. Effect of g-C3N4 precursors on the morphological structures of g-C_3N_4/ZnO composite photocatalysts. *J. Alloys Compd.* **2019**, *788*, 1084–1092. [CrossRef]
75. Ngullie, R.C.; Alaswad, S.O.; Bhuvaneswari, K.; Shanmugam, P.; Pazhanivel, T.; Arunachalam, P. Synthesis and Characterization of Efficient ZnO/g-C_3N_4 Nanocomposites Photocatalyst for Photocatalytic Degradation of Methylene Blue. *Coatings* **2020**, *10*, 500. [CrossRef]
76. Tan, X.; Wang, X.; Hang, H.; Zhang, D.; Zhang, N.; Xiao, Z.; Tao, H. Self-assembly method assisted synthesis of g-C_3N_4/ZnO heterostructure nanocomposites with enhanced photocatalytic performance. *Opt. Mater.* **2019**, *96*, 109266. [CrossRef]
77. Naseri, A.; Samadi, M.; Pourjavadi, A.; Ramakrishna, S.; Moshfegh, A.Z. Enhanced photocatalytic activity of ZnO/g-C_3N_4 nanofibers constituting carbonaceous species under simulated sunlight for organic dye removal. *Ceram. Int.* **2021**, *47*, 26185–26196. [CrossRef]
78. Mathialagan, A.; Manavalan, M.; Venkatachalam, K.; Mohammad, F.; Oh, W.C.; Sagadevan, S. Fabrication and physicochemical characterization of g-C_3N_4/ZnO composite with enhanced photocatalytic activity under visible light. *Opt. Mater.* **2020**, *100*, 109643. [CrossRef]
79. Xu, X.; Huang, T.; Xu, Y.; Hu, H.; Liao, S.; Hu, X.; Chen, D.; Zhang, M. Highly dispersed CeO_{2-x} nanoparticles with rich oxygen vacancies enhance photocatalytic performance of g-C_3N_4 toward methyl orange degradation under visible light irradiation. *J. Rare Earths* **2022**, *40*, 1255–1263. [CrossRef]
80. Subashini, A.; Varun Prasath, P.; Sagadevan, S.; Anita Lett, J.; Fatimah, I.; Mohammad, F.; Al-Lohedan, H.A.; Alshahateet, S.F.; Chun Oh, W. Enhanced photocatalytic degradation efficiency of graphitic carbon nitride-loaded CeO_2 nanoparticles. *Chem. Phys. Lett.* **2021**, *769*, 138441. [CrossRef]
81. Barathi, D.; Rajalakshmi, N.; Ranjith, R.; Sangeetha, R.; Meyvel, S. Controllable synthesis of CeO_2/g-C_3N_4 hybrid catalysts and its structural, optical and visible light photocatalytic activity. *Diam. Relat. Mat.* **2021**, *111*, 108161. [CrossRef]
82. Kaur, M.; Singh, S.; Mehta, S.K.; Kansal, S.K.; Umar, A.; Ibrahim, A.A.; Baskoutas, S. CeO_2 quantum dots decorated g-C_3N_4 nanosheets: A potential scaffold for fluorescence sensing of heavy metals and visible-light driven photocatalyst. *J. Alloys Compd.* **2023**, *960*, 170637. [CrossRef]
83. Wei, X.; Wang, X.; Pu, Y.; Liu, A.; Chen, C.; Zou, W.; Zheng, Y.; Huang, J.; Zhang, Y.; Yang, Y.; et al. Facile ball-milling synthesis of CeO_2/g-C_3N_4 Z-scheme heterojunction for synergistic adsorption and photodegradation of methylene blue: Characteristics, kinetics, models, and mechanisms. *Chem. Eng. J.* **2021**, *420*, 127719. [CrossRef]
84. Rathi, V.; Panneerselvam, A.; Sathiyapriya, R. A novel hydrothermal induced $BiVO_4$/g-C_3N_4 heterojunctions visible-light photocatalyst for effective elimination of aqueous organic pollutants. *Vacuum* **2020**, *180*, 109458. [CrossRef]
85. Xu, X.; Zhang, J.; Tao, F.; Dong, Y.; Wang, L.; Hong, T. Facile construction of Z-scheme g-C_3N_4/$BiVO_4$ heterojunctions for boosting visible-light photocatalytic activity. *Mater. Sci. Eng. B* **2022**, *279*, 115676. [CrossRef]
86. Reddy, C.V.; Nagar, A.; Shetti, N.P.; Reddy, I.N.; Basu, S.; Shim, J.; Kakarla, R.R. Novel g-C_3N_4/$BiVO_4$ heterostructured nanohybrids for high efficiency photocatalytic degradation of toxic chemical pollutants. *Chemosphere* **2023**, *322*, 138146. [CrossRef] [PubMed]
87. Mao, J.; Hong, B.; Wei, J.; Xu, J.; Han, Y.; Jin, H.; Jin, D.; Peng, X.; Li, J.; Yang, Y.; et al. Enhanced Ciprofloxacin Photodegradation of Visible-Light-Driven Z-Scheme g-C_3N_4/Bi_2WO_6 Nanocomposites and Interface Effect. *ChemistrySelect* **2019**, *4*, 13716–13723. [CrossRef]
88. Yasmeen, H.; Zada, A.; Li, W.; Xu, M.; Liu, S. Suitable energy platform of Bi_2WO_6 significantly improves visible-light degradation activity of g-C_3N_4 for highly toxic diuron pollutant. *Mater. Sci. Semicond. Process.* **2019**, *102*, 104598. [CrossRef]
89. Lian, X.; Xue, W.; Dong, S.; Liu, E.; Li, H.; Xu, K. Construction of S-scheme Bi_2WO_6/g-C_3N_4 heterostructure nanosheets with enhanced visible-light photocatalytic degradation for ammonium dinitramide. *J. Hazard. Mater.* **2021**, *412*, 125217. [CrossRef]
90. Yang, Q.; An, J.; Xu, Z.; Liang, S.; Wang, H. Performance and mechanism of atrazine degradation using Co_3O_4/g-C_3N_4 hybrid photocatalyst with peroxymonosulfate under visible light irradiation. *Colloids Surf. A Physicochem. Eng. Asp.* **2021**, *614*, 126161. [CrossRef]
91. Liu, L.; Huang, J.; Yu, H.; Wan, J.; Liu, L.; Yi, K.; Zhang, W.; Zhang, C. Construction of MoO_3 nanopaticles/g-C_3N_4 nanosheets 0D/2D heterojunction photocatalysts for enhanced photocatalytic degradation of antibiotic pollutant. *Chemosphere* **2021**, *282*, 131049. [CrossRef] [PubMed]
92. Xue, S.; Wu, C.; Pu, S.; Hou, Y.; Tong, T.; Yang, G.; Qin, Z.; Wang, Z.; Bao, J. Direct Z-Scheme charge transfer in heterostructured MoO_3/g-C_3N_4 photocatalysts and the generation of active radicals in photocatalytic dye degradations. *Environ. Pollut.* **2019**, *250*, 338–345. [CrossRef] [PubMed]
93. Liu, Y.; Huang, H.; Cao, W.; Mao, B.; Liu, Y.; Kang, Z. Advances in carbon dots: From the perspective of traditional quantum dots. *Mat. Chem. Front.* **2020**, *4*, 1586–1613. [CrossRef]

Disclaimer/Publisher's Note: The statements, opinions and data contained in all publications are solely those of the individual author(s) and contributor(s) and not of MDPI and/or the editor(s). MDPI and/or the editor(s) disclaim responsibility for any injury to people or property resulting from any ideas, methods, instructions or products referred to in the content.

Review

Recent Advances in g-C₃N₄ Photocatalysts: A Review of Reaction Parameters, Structure Design and Exfoliation Methods

Junxiang Pei, Haofeng Li, Songlin Zhuang, Dawei Zhang and Dechao Yu *

Engineering Research Center of Optical Instrument and System, Ministry of Education and Shanghai Key Laboratory of Modern Optical System, University of Shanghai for Science and Technology, Shanghai 200093, China; jxpei@usst.edu.cn (J.P.); lhf09042023@163.com (H.L.); slzhuangx@aliyun.com (S.Z.); dwzhang@usst.edu.cn (D.Z.)
* Correspondence: d.yu@usst.edu.cn

Abstract: Graphitized carbon nitride (g-C₃N₄), as a metal-free, visible-light-responsive photocatalyst, has a very broad application prospect in the fields of solar energy conversion and environmental remediation. The g-C₃N₄ photocatalyst owns a series of conspicuous characteristics, such as very suitable band structure, strong physicochemical stability, abundant reserves, low cost, etc. Research on the g-C₃N₄ or g-C₃N₄-based photocatalysts for real applications has become a competitive hot topic and a frontier area with thousands of publications over the past 17 years. In this paper, we carefully reviewed the recent advances in the synthesis and structural design of g-C₃N₄ materials for efficient photocatalysts. First, the crucial synthesis parameters of g-C₃N₄ were fully discussed, including the categories of g-C₃N₄ precursors, reaction temperature, reaction atmosphere and reaction duration. Second, the construction approaches of various nanostructures were surveyed in detail, such as hard and soft template, supramolecular preorganization and template-free approaches. Third, the characteristics of different exfoliation methods were compared and summarized. At the end, the problems of g-C₃N₄ materials in photocatalysis and the prospect of further development were disclosed and proposed to provide some key guidance for designing more efficient and applicable g-C₃N₄ or g-C₃N₄-based photocatalysts.

Keywords: photocatalyst; g-C₃N₄; reaction parameters; structure design; exfoliation

1. Introduction

In order to alleviate the problem of global warming and energy shortage, the development of renewable energy has become a major practical problem to be urgently solved by researchers all over the world [1]. As the most important renewable energy on the Earth, solar energy can be said to be inexhaustible. But so far, limited by various energy conversion technologies, the development and utilization of solar energy in the field of photocatalysis is far from enough. Recently, graphitic carbon nitride (g-C₃N₄) has attracted extremely wide attentions in photocatalysis due to its special band structure, stable properties, low price, and easy preparation [2–6]. The g-C₃N₄ is comprised of only carbon and nitrogen elements, which are very abundant on the Earth. Importantly, the g-C₃N₄ materials can be easily fabricated by thermal polymerization of abundant nitrogen-rich precursors such as melamine [7–16], dicyandiamide [17–22], cyanamide [23–25], urea [18,26,27], thiourea [28–30], ammonium thiocyanate [31–33], etc. Because the band gap of g-C₃N₄ is 2.7 eV, it can absorb visible light shorter than 450 nm effectively, implying broad prospects in solar energy conversion applications. Due to the aromatic C-N heterocycles, g-C₃N₄ is thermally stable up to 600 °C in air. Moreover, g-C₃N₄ is insoluble in acids, bases or organic solvents, exhibiting good chemical stability.

However, some bottlenecks in the photocatalytic activity of g-C₃N₄ still exist, such as fast photogenerated carrier recombination, limited active site, small specific surface area, low light absorption capacity, unsatisfactory crystallinity and unignorable surface defects. How to promote the efficient migration and separation of photogenerated carriers, expand the spectral response range and increase the specific surface area of g-C_3N_4 is the core problem to achieve high energy conversion efficiency. In practice, the introduction of impurities into the g-C_3N_4 matrix through copolymerization and doping has become an effective strategy to change the electronic structure and band structure of g-C_3N_4. On the other hand, numerous research works have demonstrated that the physicochemical properties and photocatalytic efficiency of the polymer g-C_3N_4 can be significantly improved by optimizing synthesis techniques such as supramolecular and copolymerization techniques with identical structural and nano-structural designs, or by template-assisted methods to improve porosity and surface area [34–39]. Amongst various modification approaches, designing and constructing a more suitable band structure is the most important prerequisite to improve the charge separation efficiency, thereby enhancing the photocatalytic performance.

In this review, the synthesis parameters of g-C_3N_4 are discussed first, mainly including the types of g-C_3N_4 precursors, reaction temperature, reaction atmosphere and reaction duration. The influence of different synthesis parameters of g-C_3N_4 are compared and summarized in detail here. Then, the construction approaches of various nanostructures are reviewed, such as hard and soft template, supramolecular preorganization, template-free approaches, etc. It again manifests that the specific surface area and photocatalytic efficiency of g-C_3N_4 can be directly manipulated by means of different nanostructure design approaches. Furthermore, the characteristics of different exfoliation methods are summarized for stark comparisons. Liquid exfoliation of bulk g-C_3N_4 has gradually become the most popular exfoliation method. The overall framework of synthesis and properties of g-C_3N_4 for enhanced photocatalytic performance are illustrated in Figure 1. Finally, the problems of g-C_3N_4 materials in photocatalysis and the prospect of further development are proposed, which may be favorable to the design of more efficient and practical g-C_3N_4 or g-C_3N_4-based photocatalyst.

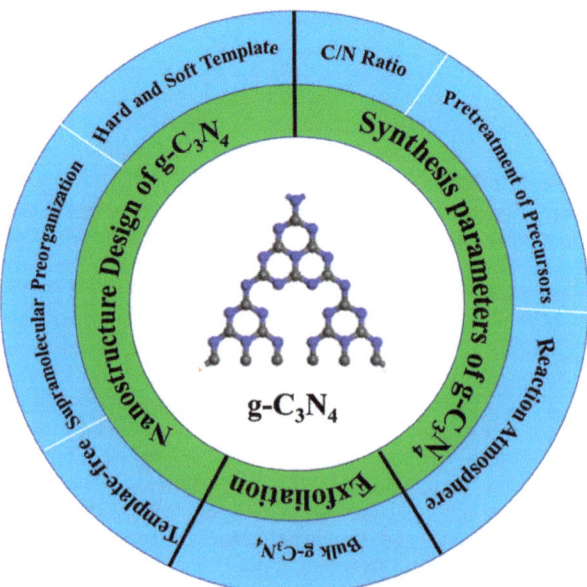

Figure 1. Overall framework of synthesis and properties of g-C_3N_4 for enhanced photocatalytic performance.

2. Influence of Synthesis Parameters

2.1. Precursors and Reaction Temperature

The first reported g-C_3N_4 as a heterogeneous catalysis was present in 2006 [40]. Subsequently, various precursors such as urea [41], thiourea [42], cyanamide [43,44] and dicyandiamide [45,46] have been employed to synthesize g-C_3N_4 by thermal treatment methods. In 2009, Wang et al. firstly used cyanamide as the precursor of g-C_3N_4 for producing hydrogen from water under visible-light irradiation in the presence of a sacrificial donor [47]. This pioneering work represents an important first step towards photosynthesis in general, where artificial conjugated polymer semiconductors can be used as energy transducers. In order to demonstrate the reaction intermediate compounds, characterization techniques such as thermogravimetric analysis (TGA) and X-ray diffraction (XRD) are used to characterize the reaction. Figure 2a displayed that the graphitic planes are constructed from tri-s-triazine units connected by planar amino groups. Figure 2b is the XRD pattern of the obtained g-C_3N_4 powder. From the ultraviolet-visible spectrum (Figure 2c), it can be seen that the band gap of g-C_3N_4 is 2.7 eV. The synthesis of g-C_3N_4 was a combination of polyaddition and polycondensation. At a reaction temperature of 203 and 234 °C, the cyanamide molecules can be condensed to dicyandiamide and melamine, respectively. The ammonia is then removed by condensation. When the temperature reaches 335 °C, large amounts of melamine products are detected. When further heating to 390 °C, the rearrangements of melamine will result in the formation of tri-s-triazine units. Finally, when heating to 520 °C, the polymeric g-C_3N_4 are synthesized via the further condensation of the unit. However, g-C_3N_4 will be unstable at above 600 °C. Furthermore, when the reaction degree is higher than 700 °C, g-C_3N_4 will decompose. Figure 2d shows the structural phase transition process from cyanamide to g-C_3N_4 at different temperatures. In addition to in-situ characterization experiments to verify the reaction process, it can also be demonstrated by relevant simulation calculations. The first-principles DFT calculations were performed using a plane wave basis set with a 550 eV energy cutoff [40]. The calculation results showed that the cohesion energy increased under the addition of multiple reaction pathways, which confirmed that melamine was produced upon heating the cyanamide, as shown in Figure 2e. In another work, Ang et al. demonstrated that when thiourea is used as the precursor and TiO_2 or SiO_2 are used as the inorganic substrate, the melon nanocomposites can be formed at a low temperature of 400 °C [48]. However, the degree of polymerization of g-C_3N_4 in the nanocomposites is low, so its photocatalytic performance is moderate. Following this, Zhang et al. reported a simple method of g-C_3N_4 synthesis from thiourea without the aid of any substrates [49]. It was found that the obtained g-C_3N_4 exhibited a good condensation likely due to the presence of sulfur species in thiourea, thus accelerating the degree of polymerization and condensation of thiourea at high temperatures. Having insights into the results of different reports, it can be found that different precursors have distinct characteristics and advantages. In order to obtain well-condensed g-C_3N_4 with high quality, the most crucial point is to select the best reaction temperature corresponding to the selected precursor.

In order to prove that g-C_3N_4 has been successfully prepared, a variety of analytical measurements can be used, such as X-ray photoelectron spectroscopy (XPS), XRD and Fourier transform infrared (FTIR) spectroscopy. The basic experimental procedures for the synthesis, characteristics of different precursors and characterization of g-C_3N_4 are described above. What effects other synthesis parameters have on g-C_3N_4 will be compared and discussed in detail in the following sections.

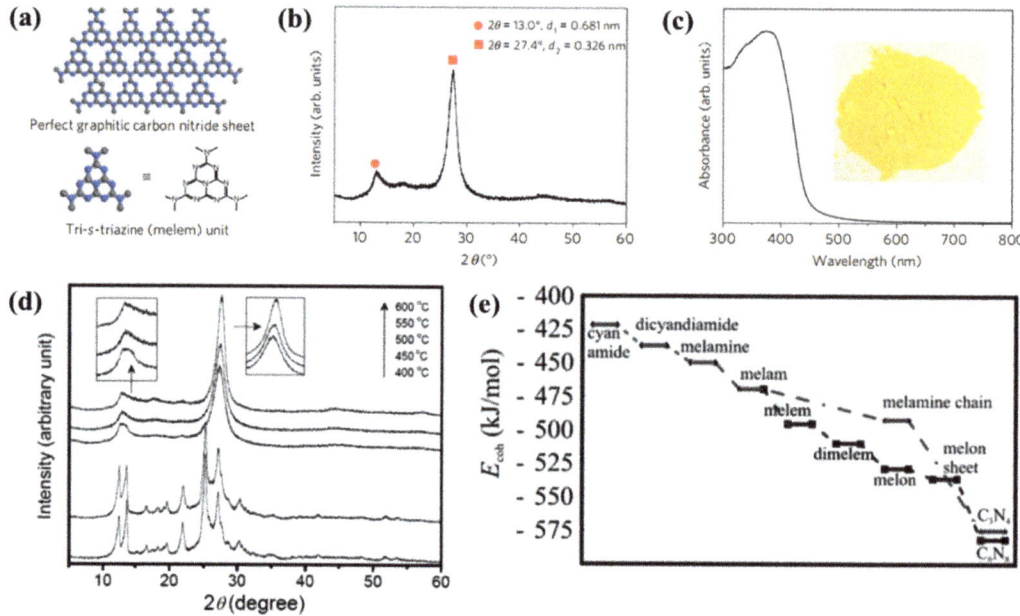

Figure 2. Crystal structure and optical properties of g-C_3N_4. (**a**) Schematic diagram of a perfect g-C_3N_4 sheet constructed from melem units. (**b**) Experimental XRD pattern of polymeric carbon nitride, revealing a graphitic structure with an interplanar stacking distance of the aromatic unit (0.326 nm). (**c**) Diffuse reflectance spectrum of the polymeric carbon nitride. Inset: Photograph of the photocatalyst. (**d**) XRD patterns of g-C_3N_4 treated at different temperatures. (**e**) Calculated energy diagram for the development of g-C_3N_4 using cyanamide precursor [47].

2.2. C/N Ratio

Generally, g-C_3N_4 exhibits a high physicochemical stability and ideal band structure, due to the high condensation degree and the presence of the heptazine ring structure. When the appropriate precursor and condensation method are selected, the C/N ratio in layered g-C_3N_4 is about 0.75. Many studies have confirmed that when the precursors and synthesis parameters of g-C_3N_4 are changed, the physicochemical properties of g-C_3N_4 will be significantly affected, such as band gap width, specific surface area, C/N ratio, etc., which will directly affect the photocatalytic efficiency and other applications' performance [50–54]. Yan et al. synthesized g-C_3N_4 by directly heating the low-cost melamine, and they change the C/N ratio by controlling different heating temperatures [55]. The research showed that when the heating temperatures increased from 500 to 580 °C, the ratio of C/N increased from 0.721 to 0.742. Meanwhile, the band gaps of g-C_3N_4 decreased from 2.8 to 2.75 eV. Apparently, increasing the heat temperature will decrease the photooxidation ability of g-C_3N_4. The C/N ratio, i.e., the degree of condensation, is inconsistent with the structural integrality. Therefore, in order to obtain a better photocatalytic efficiency, reasonable optimization measures must be taken to make the C/N ratio closer to the ideal ratio (0.75) [56]. Therefore, the presence of trace amino groups is actually conducive to improving the surface activity of g-C_3N_4, thus exhibiting better interaction with the reactant molecules [8,57,58]. However, due to incomplete concentration, the C/N ratio is lower than the ideal value (0.75), resulting in a large number of defects, inhibiting the rate of carrier transport and separation, and seriously reducing the photocatalytic efficiency. Therefore, in order to obtain a relatively high photocatalytic efficiency, the optimal C/N ratio is 0.75, which can usually reflect that g-C_3N_4 has a good concentration and stability.

2.3. Pretreatment of Precursors

It has been shown that modification and pretreatment of nitrogen-rich precursors before thermal annealing can effectively improve the physicochemical properties of g-C_3N_4. One of the effective pretreatment methods is by acid treatment. Yan et al. reported the synthesis of g-C_3N_4 by directly heating the sulfuric-acid-treated melamine precursor [59]. It is worth noting that the carbon nitride synthesized from sulfuric acid treated melamine (15.6 m^2/g) shows relatively higher BET surface area than that of samples synthesized from untreated melamine (8.6 m^2/g). The reason can be attributed to the effect of pretreatment of melamine with H_2SO_4 on its condensation process, during which sublimation of melamine is inhibited significantly. In addition to the pretreatment of melamine with H_2SO_4, HCl and HNO_3 also exhibited good pretreatment effects on melamine [60–63].

In addition to acid precursors, pretreatment methods of sulfur-mediated synthesis can also be used to regulate the structure and physicochemical properties of g-C_3N_4 [64]. The fundamental reason is that the presence of the sulfur group in the sulfur-containing thiourea provides an additional chemical pathway to regulate the degree of condensation and polymerization of g-C_3N_4 because it is easy to leave the -SH groups. Zhang et al. advanced this strategy by employing cheap and easily available elemental sulfur as the external sulfur species instead of sulfur-containing precursors for the sulfur-mediated synthesis of g-C_3N_4 photocatalysts [65]. In comparison with unmodified g-C_3N_4, the vibrations of g-C_3N_4-S_x are less intensive when increasing the amount of elemental sulfur (S_8), especially for the broad band at 2900–3300 cm^{-1}. Thus, S_8-mediated synthesis helps to advance the polymerization of melamine precursors, leaving fewer amino-containing groups as surface defects. In this report, the structure, electronic and optical properties of g-C_3N_4 have been effectively modified, and its physicochemical properties have also been significantly improved. Under visible light irradiation at 420 nm, the photocatalytic activity of water reduction and oxidation was enhanced.

2.4. Reaction Atmosphere

In addition to the types of precursors, reaction temperature and duration, the physicochemical properties and structure of g-C_3N_4 are also strongly influenced by the reaction atmosphere, because the reaction atmosphere can induce a variety of defects and carbon and nitrogen vacancies. In fact, defects are essential for catalytic reactions because they can act as active sites for reactant molecules and change the band structure by introducing additional energy levels in the forbidden band, thus extending the spectral absorption range [66–69]. By controlling the polycondensation temperature of a dicyandiamide precursor in the preparation of g-C_3N_4, Niu et al. introduced nitrogen vacancies in the framework of g-C_3N_4 [70]. The excess electrons caused by nitrogen loss in g-C_3N_4 lead to a large number of C^{3+} states associated with nitrogen vacancies in the band gap, thus reducing the intrinsic band gap from 2.74 eV to 2.66 eV. Steady and time-resolved fluorescence emission spectra show that, due to the existence of abundant nitrogen vacancies, the intrinsic radiative recombination of electrons and holes in g-C_3N_4 is greatly restrained, and the population of short-lived and long-lived charge carriers is decreased and increased, respectively. In another study, Niu et al. produced a novel visible light photocatalyst R-melon by heating the melon in a hydrogen atmosphere [71]. Compared to the pristine melon with a bandgap of 2.78 eV, the resultant R-melon with a bandgap of 2.03 eV has a widened visible light absorption range and suppressed radiative recombination of photo-excited charge carriers due to homogeneous self-modification with nitrogen vacancies. Table 1 shows the summary of the specific surface areas and band gaps of g-C_3N_4 photocatalysts developed from various precursors and different reaction parameters. The results demonstrated that the band structure, electronic properties and specific surface area of g-C_3N_4 can be changed by adjusting the reaction parameters and precursors, so as to improve its photocatalytic performance.

Table 1. Precursors and Reaction Parameters Employed in the g-C_3N_4 Synthesis.

Precursors	Synthesis Parameters	Surface Area ($m^2\,g^{-1}$)	Band Gap Energy (eV)	Ref.
Cyanamide	550 °C, 4 h, air	10	2.7	[47]
Dicyandiamide	550 °C, 3 h, air	12.3	2.66	[56]
Dicyandiamide	600 °C, 4 h, air	12.8	2.75	[72]
Dicyandiamide	550 °C, 4 h, H_2	20.91	2.0	[73]
Melamine	500 °C, 2 h, air	7.1	2.83	[74]
Melamine	520 °C, 2 h, N_2	17.4	2.74	[75]
H_2SO_4-treated melamine	600 °C, 4 h, Ar	15.6	2.69	[59]
HCl-treated melamine	550 °C, 4 h, air	26.2	2.73	[61]
Thiourea	550 °C, 3 h, air	11.3	2.6	[56]
Urea	450 °C, 2 h, air	135.6	2.76	[76]

3. Morphology and Structure Design of g-C_3N_4

3.1. Hard and Soft Template Approach

Apart from regulating the synthesis parameters, introducing nano-templates and nano-casting with different morphology and ordered porosity on the basis of bulk g-C_3N_4 is another promising method to change the morphology and structural characteristics of g-C_3N_4 structure and the interlayer interaction. As a matter of fact, researchers have effectively designed controllable nanostructures for g-C_3N_4 through hard template or soft template methods, such as porous g-C_3N_4, one-dimensional nanostructures, hollow g-C_3N_4 nanospheres, etc [11,77–85]. It has been proven that the porosity, structure, morphology, surface area and size can be easily controlled by adjusting the appropriate template. Moreover, the larger surface area and more active sites are generally more favorable for photocatalytic applications of g-C_3N_4.

The hard template method is almost identical to the traditional casting process and is one of the most common techniques for developing nanostructured g-C_3N_4 materials. In this way, the various structures and geometries of g-C_3N_4 can be designed using hard templates as needed, and their length scales are usually around nanometers and microns. The most typical structure-oriented agent is a silica template with a controllable nanostructure. The early study on the mesoporous g-C_3N_4 synthesized using cyanamide as a precursor and silica nanoparticles with a size of 12 nm as a template was reported by Goettmann et al. [40]. The results show that the silica nanoparticles can be uniformly dispersed in the cyanamide monomer, which is due to the appropriate surface interaction between the silica surface and the amine and aromatic nitrogen groups. After heating treatment and cyanamide condensation, the g-C_3N_4/silica hybrid is formed, and well-dispersed silica nanoparticles are preserved in the g-C_3N_4 matrix. Ammonium hydrogen fluoride (NH_4HF_2) solution can be used to remove the silicon template. The average diameter is 12 nm, and the surface area in the range of 86 to 439 $m^2\,g^{-1}$ can be regulated by adjusting the mass ratios of silica/cyanamide from 0 to 1.6. In another work carried out by Yuki Fukasawa et al., uniform-sized silica nanospheres (SNSs) assembled into close-packed structures were used as a primary template for ordered porous g-C_3N_4, which was subsequently used as a hard template to generate regularly arranged Ta_3N_5 nanoparticles of well-controlled size [86]. The cyanamide is infiltrated and polymerized in the narrow void of SNSs to form porous g-C_3N_4, and then the SNSs is removed by HF treatment, as shown in Figure 3a. Therefore, the resulting g-C_3N_4 has an anti-opal structure, and the size of the spherical hole indicates the size of the SNSs used, as shown in the SEM images of Figure 3b–e. In this study, the pore size of g-C_3N_4 was between 50 and 80 nm. In spite of the silica hard template, Chen et al. reported the synthesis of porous g-C_3N_4 by using multi-walled carbon nanotube (CNT) as a novel hard template [87]. Unlike other hard templates, CNT can be easily removed and recovered by ultrasonic methods, resulting in a relatively simple preparation of porous g-C_3N_4.

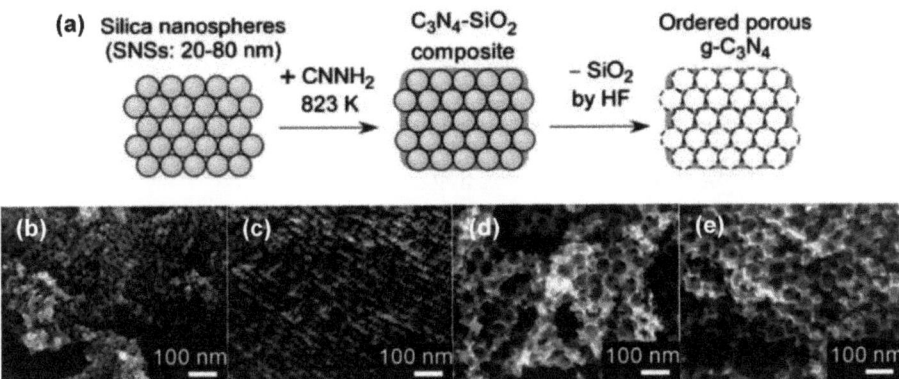

Figure 3. (**a**) Synthesis procedure of ordered porous g-C_3N_4. Field emission SEM (FESEM) images of porous g-C_3N_4 prepared using silica spheres with various diameters: (**b**) 20, (**c**) 30, (**d**) 50, and (**e**) 80 nm [86].

It can be seen that during the synthesis of g-C_3N_4 through hard templates, extremely dangerous, toxic and expensive fluorine-containing etchers (such as HF and NH_4HF_2) are used to remove the sacrificial templates. This greatly limits the practical application of the method in large-scale industrial processes. Therefore, apart from the hard template synthesis method of g-C_3N_4, the relatively "environmentally friendly" soft template process can not only change the morphology and structure of g-C_3N_4 through the selection of multiple soft templates, but also simplify the synthesis route of g-C_3N_4 [88,89]. Different from the hard template method, in the soft template method, the nano-structure g-C_3N_4 is synthesized by soft structure guiding agents such as ionic liquid, amphiphilic block polymer and surfactant, so as to rationally design g-C_3N_4 with a highly porous nanostructure [90]. For instance, bimodal mesoporous g-C_3N_4 is synthesized using Triton X-100 as a soft-template and melamine and glutaraldehyde as precursors through polymerization and carbonization [91]. The results show that the mesopore sizes in the g-C_3N_4 are centered at 3.8 nm and 10–40 nm. The former was attributed to the removal of the Triton X-100, while the latter was ascribed to the aggregates of plate-like g-C_3N_4. In another study reported by Wang et al., a variety of soft-templates (e.g., Triton X-100, P123, F127, Brij30, Brij58, and Brij76) as well as some ionic surfactants are tested as structure-directing agents for the synthesis of mesoporous g-C_3N_4 [90]. Most as-prepared g-C_3N_4 materials possess a high surface area. Moreover, this work points out two problems. The first was that only a few selected soft templates (Triton X-100 and ionic liquids) led to the presence of mesoporous g-C_3N_4 structures with high surface areas. Due to the premature decomposition of the template material, the pores of g-C_3N_4 are easily resealed. The second problem is that the obtained g-C_3N_4 contains a large amount of carbon elements from the template polymer, which significantly changes the morphology and structure of g-C_3N_4, thereby reducing its photocatalytic activity.

Overall, it is not difficult to find that the synthesis of g-C_3N_4 with various nanostructures assisted by hard and soft templates is a facile and efficient approach. Nowadays, researchers are still developing a variety of new templates to achieve more interesting g-C_3N_4 nanostructures with high photocatalytic efficiency.

3.2. Supramolecular Preorganization Approach

In contrast to the previously discussed hard and soft template synthesis approaches, molecular self-assembly is a self-templating approach (namely supramolecular preorganization approach) in which molecules spontaneously form a stable g-C_3N_4 structure from non-covalent bonds under equilibrium conditions in the absence of an external template [80,92–94]. Recently, supramolecular preassembly of triazine molecules has become an interesting method to regulate the structural, textural, optical, and electronic features of g-C_3N_4, thus affecting its photocatalytic activity [95–98]. For example, nanostructured g-C_3N_4 materials can be developed by supramolecular preorganization of melamine precursors to triazine derivatives to form hydrogen bond molecular assemblies, i.e., melamine–cyanuric acid, melamine–trithiocyanuric acid mixtures or their derivatives [99–101]. Jun et al. first synthesized g-C_3N_4 by molecular cooperative assembly between triazine molecules [102]. Flower-like, layered spherical aggregates of melamine cyanuric acid complex (MCA) are formed by precipitation from equimolecular mixtures in dimethyl sulfoxide (DMSO). The oxygen-containing intermolecular structure connected by hydrogen bonding and stacked in graphitic fashion facilitates the condensation process and enables structural perfection. The obtained material synthesized by this supramolecular preorganization approach has stronger spectral absorption (an increased band gap of 0.16 eV), and the photogenerated carrier lifetime is extended nearly twice as long as that of bulk g-C_3N_4.

It can be seen that the combination of two or more monomers in different solvents can form supramolecular complexes. These supramolecular complexes are usually linked by hydrogen bonds. Therefore, it is expected that the addition of new monomers to hydrogen-bonded supramolecular complexes as "terminators" will be an attractive technique to further adjust the morphology, photophysical properties, and electronic band structure of g-C_3N_4.

3.3. Template-Free Approach

Compared with the hard and soft template synthesis approaches, the template-free approach has unique advantages, such as no need for various high-cost and dangerous templates containing fluorine, and no residue of any template components. Indeed, many studies have proven that g-C_3N_4 nanostructure designs with a variety of morphologies and desired sizes, such as nanorods, quantum dots, microspheres, nanofibers, etc., can also be achieved using a template-free approach. Bai et al. reported that the transformation of g-C_3N_4 from nanoplates to nanorods was realized by a simple reflux method [103]. Various aspect ratios were achieved by changing the reflux duration and solvent ratio without the assistance of any templates. As per the TEM images shown in Figure 4a–d, most irregular nanoplates are observed in the untreated g-C_3N_4 sample, while the refluxed g-C_3N_4 sample mainly contains nanorods. The length of g-C_3N_4 nanorods is 0.5–3 μm. The growth mechanism and process can be summarized as follows (Figure 4e): under ultrasonic action, CH_3OH can strip g-C_3N_4 bulks into nanosheets. The reflux process can remove surface defects and transform them into nanorods. After reflux treatment, the g-C_3N_4 nanosheets showed the metastable state. With the increase of time, the layered structure of g-C_3N_4 will curl and finally grow into nanorods. Finally, the shorter g-C_3N_4 nanorods are redissolved into the solution, while the longer nanorods will continue to grow during reflux treatment. The increase in visible light degradation activity of methylene blue can be induced for the increase of active lattice face and elimination of surface defects.

Additionally, Wang et al. described a facile and generally feasible method to synthesize nanotube-type g-C_3N_4 by directly heating melamine packed in an appropriate compact degree without templates [104]. A certain amount of melamine was placed into a semi-closed alumina crucible followed by consecutively shaking the crucible using a vibrator at a fast rate to achieve a moderately compact packing degree. This process is very crucial for the synthesis of nanotube-type g-C_3N_4. TEM images show that the wall thickness of the nanotubes in the bulk phase is about 15 ± 2 nm, while the inner diameter is about 18 ± 2 nm. In this report, the formation process and mechanism of nanotubes have been studied in detail. During the pyrolysis process, melamine releases NH_3 gas, which passes through the stacked melamine layers to form rolled g-C_3N_4 nanosheets. Due to the need to reduce the surface free energy, the nanosheets eventually bend into the form of nanotubes. It is worth pointing out that when the melamine layer is loosely stacked, g-C_3N_4 cannot form a nanotube structure, which indicates that obtaining the appropriate packing degree of g-C_3N_4 is crucial for the formation of the nanotube structure.

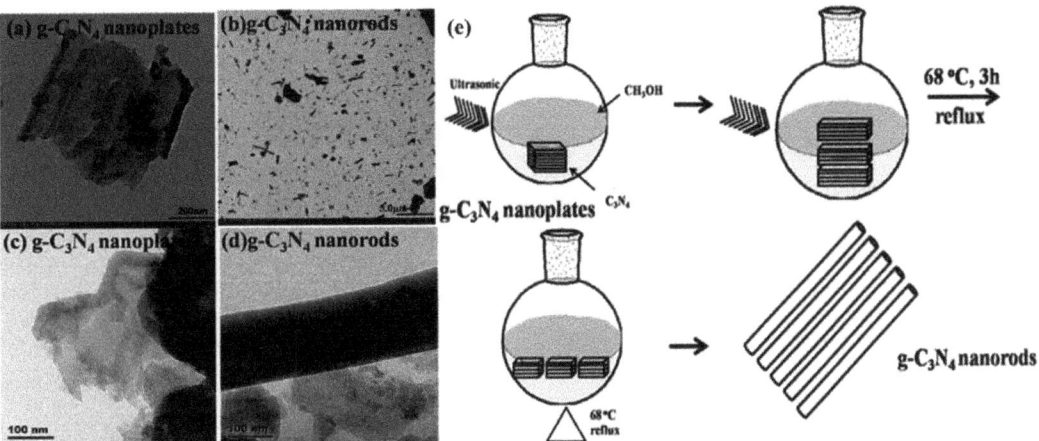

Figure 4. TEM images of g-C$_3$N$_4$ nanoplates (**a**,**c**) and nanorods (**b**,**d**). (**e**) Schematic illustration of the formation processes of g-C$_3$N$_4$ nanorods from g-C$_3$N$_4$ nanoplates [103].

4. Exfoliation of Bulk g-C$_3$N$_4$

Although the specific surface area of the monolayer g-C$_3$N$_4$ is theoretically large, the specific surface area of the block is very low indeed, usually less than 10 m^2 g^{-1}, due to the stacking of g-C$_3$N$_4$ layers [105]. Therefore, delaminating g-C$_3$N$_4$ into several layers is a promising way to improve photocatalytic performance and produce more interesting surface, optical and electronic properties [106–109]. There are many methods of exfoliating g-C$_3$N$_4$, such as ultrasonication-assisted liquid exfoliation, the liquid ammonia-assisted lithiation and the post-thermal oxidation etching route [17,108–115]. Similar to most two-dimensional materials, there is a weak van der Waals force between the layers of g-C$_3$N$_4$. Thus, the van der Waals force can be effectively overcome and the separation between layers can be achieved by means of energy assistance such as ultrasound in an appropriate solvent.

Liquid exfoliation is simple and convenient, and has gradually become the most commonly used exfoliation method by most researchers. Yang et al. demonstrated the synthesis of free-standing g-C$_3$N$_4$ nanosheets by liquid phase exfoliation [110]. The method uses g-C$_3$N$_4$ powder as a starting material and various organic solvents (such as isopropanol (IPA), N-methyl-pyrrolidone (NMP), acetone, and ethanol) as dispersing media. As shown in Figure 5a,b, many nanosheets with laminar morphology like silk veil can be observed, which is in stark contrast from that of bulk g-C$_3$N$_4$. The photographs in Figure 5e clearly demonstrated that NMP is a promising dispersing solvent among various dispersing solvents. It can stably disperse individual nanosheets; however, the disadvantage is that the boiling point of NMP is relatively high, which is difficult to remove. The agglomeration phenomenon of g-C$_3$N$_4$ layer can be observed in NMP. In comparison, low-boiling-point IPA is the best medium for preparing g-C$_3$N$_4$. The obtained g-C$_3$N$_4$ in IPA exhibited a surface area of up to 384 m^2 g^{-1} and a thickness of about 2 nm, as shown in Figure 5c,d. Moreover, the stability of IPA is superior, and there is no precipitation phenomenon after 4 months of storage (Figure 5f).

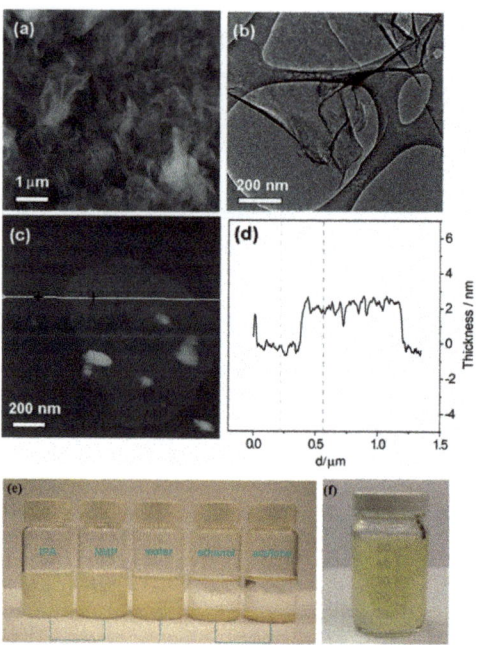

Figure 5. Typical FESEM (**a**) and TEM (**b**) images unveiling the flexible g-C$_3$N$_4$ nanosheets with the size from 500 nm to several micrometers. Representative AFM image (**c**) and corresponding thickness analysis (**d**) taken around the white line (**c**) revealing a uniform thickness of about 2 nm for g-C$_3$N$_4$ nanosheets. (**e**) Photographs of the g-C$_3$N$_4$ nanosheet dispersions in IPA, NMP, water, ethanol and acetone, respectively, after 2 days storage under ambient conditions, and (**f**) in IPA solvent after 4 months storage [110].

5. Doping of g-C$_3$N$_4$

It is well-known that g-C$_3$N$_4$ is a metal-free n-type semiconductor. Due to the high ionization energy and high electronegativity of metal-free semiconductors, it is easy for them to form covalent bonds with other compounds by obtaining electrons during the reaction. In order to maintain this unique advantage of metal-free semiconductors, researchers have implemented a series of non-metal doping g-C$_3$N$_4$, including oxygen, phosphorus, sulfur, carbon, halogen, nitrogen and boron [116–121]. For instance, O-doping is a facile method to improve the photocatalytic ability of g-C$_3$N$_4$. Zeng et al. synthesized one-dimensional porous architectural g-C$_3$N$_4$ nanorods by direct calcination of hydrous melamine nanofibers precipitated from an aqueous solution of melamine [122]. The porous structure increases the specific surface area, enhances the light absorption capacity and improves the catalytic reaction rate. At the same time, doping oxygen atoms into the g-C$_3$N$_4$ matrix breaks the symmetry of the pristine structure, making more efficient separation of electron/hole pairs. In general, non-metallic doping usually changes the surface morphology and structure of g-C$_3$N$_4$, thereby affecting the light absorption efficiency and regulating the catalytic efficiency.

Different from non-metal doping, metal-doped g-C$_3$N$_4$ has the advantages of reducing band gap to enhance visible light absorption and to improve catalytic performance. Commonly used doped metals include alkali metals (Li, Na, K) and transition metals (Fe, Cu, and W) [123–127]. For example, Xiong et al. reported that the doping of alkali metals (K, Na) in the g-C$_3$N$_4$ framework significantly increases the transfer and separation rates of photogenerated carriers and induces more efficient redox catalytic reactions [128]. DFT calculation results showed that K or Na doping can reduce the band gap energy of g-C_3N_4. Generally, the introduction of metal ions can form new energy levels, increase the specific surface area, and sometimes resist the recombination of charge carriers produced by photons. The catalytic efficiency of noble metal doping is usually higher, but the photodegradation efficiency of some ordinary transition-metal-doped g-C_3N_4 for some pollutants can be comparable to that of noble-metal-doped g-C_3N_4 when the appropriate doping amount and mode are selected.

6. Applications of g-C_3N_4

Due to its moderate energy gap, excellent electronic properties, rich functional groups and surface defects, g-C_3N_4 can be widely used in environmental treatment and pollutant degradation, including water splitting, hydrogen generation, CO_2 conversion and organic pollutants degradation [129–132].

As reported, the pristine g-C_3N_4 has limitations such as small specific surface area and fast charge recombination rate, which leads to a low water splitting ability of g-C_3N_4. To solve this problem, Chen et al. improved the water splitting capacity via adjusting the dimension of g-C_3N_4 [133]. As shown in Figure 6a,b, they demonstrated that the evolution rates of H_2 and O_2 of three-dimensional porous g-C_3N_4 in visible light are significantly higher than those of the pristine g-C_3N_4, reaching 101.4 and 49.1 μmol g^{-1} h^{-1}, respectively. Fu et al. reported oxygen-doped g-C_3N_4 [134]. As exhibited in Figure 6c-d, the O-doped g-C_3N_4 has a narrower band gap and greater CO_2 affinity, which significantly improves the photogenerated carrier separation efficiency and CO_2 conversion ability. In addition, a lot of efforts have been made to enhance its photocatalytic activity to improve its pollutant degradation ability. As shown in Figure 6e-f, Dou et al. reported that mesoporous g-C_3N_4 has a strong ability to remove antibiotics under visible light [135], which is mainly due to the porous structure that improved the utilization of light.

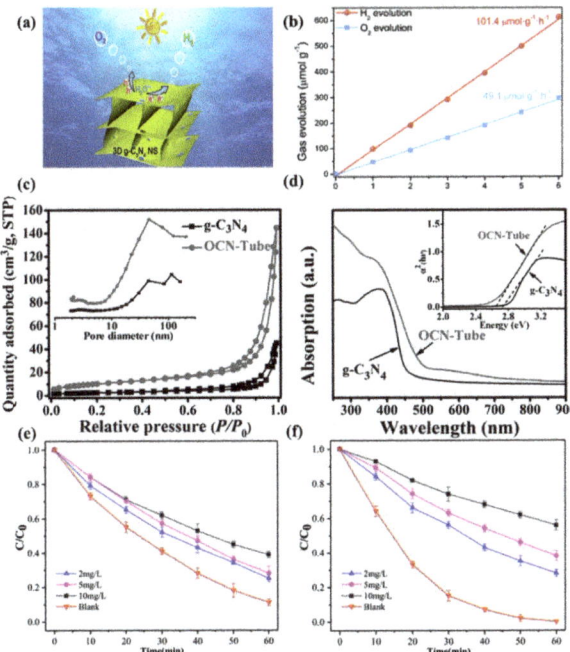

Figure 6. (a) Water splitting for H_2 and O_2 Evolution. (b) Time−dependent overall water splitting over 3D g-C_3N_4 [133]. (c) N_2 adsorption–desorption isotherms and corresponding pore size distribution curves. (d) UV–vis diffuse reflectance spectra [134]. The effects of humic acid on (e) amoxicillin and (f) cefotaxime photodegradation by mesoporous carbon nitride (initial pH = 7) [135].

7. Conclusions and Prospects

In summary, due to the merits of low cost, high stability and visible light response, g-C_3N_4 is one of the most promising photocatalytic materials to replace TiO_2. This review mainly introduces the synthesis of g-C_3N_4, the improvement of the g-C_3N_4 crystal structure, the light absorption enhancement, structure design optimization, as well as the improvement of electronic properties and optimization of energy band, so as to promote the photocatalytic application of g-C_3N_4. As summarized above, the appropriate reaction temperature and duration of the condensation process are beneficial to improve the crystallinity of g-C_3N_4. Various desirable nanostructures of g-C_3N_4 can be constructed via hard and soft template approaches, supramolecular preorganization approach, and template-free approach. Liquid exfoliation of bulk g-C_3N_4 has becoming the most facile and promising method to improve the surface area of g-C_3N_4.

Therefore, in order to synthesize the ideal g-C_3N_4 with high photocatalytic efficiency, it is necessary to pay attention to the following crucial elements: (i) Controlling the corresponding reaction temperature and reaction time according to the selected precursor material; (ii) Controlling the C/N ratio close to 0.75 and the band gap to 2.7 eV; (iii) Extending the specific surface area by selecting suitable nanostructure design approaches.

It can be assumed that in the future, g-C_3N_4, a booming photocatalytic hot spot material, will face unlimited opportunities and challenges. Although g-C_3N_4 can be easily synthesized by thermal polymerization of nitrogen-rich precursors, its photocatalytic efficiency is not high due to its small specific surface area, limited surface reaction sites, and insufficient utilization of broad-spectrum sunlight. Here, some pivotal issues are elaborated in the following:

(i) So far, the use of visible and near-infrared light is far from sufficient. On the one hand, it depends on the synthesis and structural design of the g-C_3N_4 material; on the other hand, a third or more component with a suitable band structure can also be cleverly added to design the interface electronic structure and expand the optical absorption region.

(ii) Due to the limited surface area of g-C_3N_4 and the inevitable defects on the surface, the smooth transfer of photogenerated carriers is hindered. Therefore, the synthesis approached needs to be further developed. In order to optimize photocatalytic performance, some environmentally friendly, simple and efficient synthesis routes are urgently required.

(iii) The integration of photocatalytic applications in many fields into one photocatalytic system is a widely used catalytic process. This requires the hybridization of multifunctional materials with reasonable energy structure. How to optimize the energy band structure and surface topography of g-C_3N_4 reasonably to improve the compatibility of energy conversion and environmental protection is very promising.

(iv) More theoretical studies about g-C_3N_4 need to be combined with practical catalytic applications. It is certain that in-depth fundamental theory based on physical chemistry research, in collaboration with laboratory findings, will positively promote the advances in materials science and technology.

Author Contributions: Conceptualization, S.Z. and D.Z.; investigation, H.L.; writing—original draft preparation, J.P. and D.Y. All authors have read and agreed to the published version of the manuscript.

Funding: This research received no external funding.

Conflicts of Interest: The authors declare no conflict of interest.

References

1. Wang, S.; Sun, L.; Iqbal, S. Green financing role on renewable energy dependence and energy transition in E7 economies. *Renew. Energy* **2022**, *200*, 1561–1572. [CrossRef]

2. Xu, M.-L.; Lu, M.; Qin, G.-Y.; Wu, X.-M.; Yu, T.; Zhang, L.-N.; Li, K.; Cheng, X.; Lan, Y.-Q. Piezo-Photocatalytic Synergy in BiFeO$_3$@COF Z-Scheme Heterostructures for High-Efficiency Overall Water Splitting. *Angew. Chem.-Int. Ed.* **2022**, *61*, e202210700. [CrossRef] [PubMed]
3. Miao, Z.; Wang, Q.; Zhang, Y.; Meng, L.; Wang, X. In situ construction of S-scheme AgBr/BiOBr heterojunction with surface oxygen vacancy for boosting photocatalytic CO$_2$ reduction with H$_2$O. *Appl. Catal. B-Environ.* **2022**, *301*, 120802. [CrossRef]
4. Peng, X.; Wu, J.; Zhao, Z.; Wang, X.; Dai, H.; Wei, Y.; Xu, G.; Hu, F. Activation of peroxymonosulfate by single atom Co-N-C catalysts for high-efficient removal of chloroquine phosphate via non-radical pathways: Electron-transfer mechanism. *Chem. Eng. J.* **2022**, *429*, 132245. [CrossRef]
5. Bai, J.; Shen, R.; Jiang, Z.; Zhang, P.; Li, Y.; Li, X. Integration of 2D layered CdS/WO$_3$ S-scheme heterojunctions and metallic Ti$_3$C$_2$ MXene-based Ohmic junctions for effective photocatalytic H$_2$ generation. *Chin. J. Catal.* **2022**, *43*, 359–369. [CrossRef]
6. Chen, T.; Yu, K.; Dong, C.; Yuan, X.; Gong, X.; Lian, J.; Cao, X.; Li, M.; Zhou, L.; Hu, B.; et al. Advanced photocatalysts for uranium extraction: Elaborate design and future perspectives. *Coord. Chem. Rev.* **2022**, *467*, 214615. [CrossRef]
7. Guo, S.; Deng, Z.; Li, M.; Jiang, B.; Tian, C.; Pan, Q.; Fu, H. Phosphorus-Doped Carbon Nitride Tubes with a Layered Micro-nanostructure for Enhanced Visible-Light Photocatalytic Hydrogen Evolution. *Angew. Chem.-Int. Ed.* **2016**, *55*, 1830–1834. [CrossRef]
8. Guo, S.; Tang, Y.; Xie, Y.; Tian, C.; Feng, Q.; Zhou, W.; Jiang, B. P-doped tubular g-C$_3$N$_4$ with surface carbon defects: Universal synthesis and enhanced visible-light photocatalytic hydrogen production. *Appl. Catal. B-Environ.* **2017**, *218*, 664–671. [CrossRef]
9. Liu, H.; Chen, D.; Wang, Z.; Jing, H.; Zhang, R. Microwave-assisted molten-salt rapid synthesis of isotype triazine-/heptazine based g-C$_3$N$_4$ heterojunctions with highly enhanced photocatalytic hydrogen evolution performance. *Appl. Catal. B-Environ.* **2017**, *203*, 300–313. [CrossRef]
10. Mo, Z.; Xu, H.; Chen, Z.; She, X.; Song, Y.; Wu, J.; Yan, P.; Xu, L.; Leia, Y.; Yuan, S.; et al. Self-assembled synthesis of defect-engineered graphitic carbon nitride nanotubes for efficient conversion of solar energy. *Appl. Catal. B-Environ.* **2018**, *225*, 154–161. [CrossRef]
11. Zhang, J.-W.; Gong, S.; Mahmood, N.; Pan, L.; Zhang, X.; Zou, J.-J. Oxygen-doped nanoporous carbon nitride via water-based homogeneous supramolecular assembly for photocatalytic hydrogen evolution. *Appl. Catal. B-Environ.* **2018**, *221*, 9–16. [CrossRef]
12. Zhu, B.; Xia, P.; Ho, W.; Yu, J. Isoelectric point and adsorption activity of porous g-C$_3$N$_4$. *Appl. Surf. Sci.* **2015**, *344*, 188–195. [CrossRef]
13. Huang, H.; Xiao, K.; Tian, N.; Dong, F.; Zhang, T.; Du, X.; Zhang, Y. Template-free precursor-surface-etching route to porous, thin g-C$_3$N$_4$ nanosheets for enhancing photocatalytic reduction and oxidation activity. *J. Mater. Chem. A* **2017**, *5*, 17452–17463. [CrossRef]
14. Tian, N.; Zhang, Y.; Li, X.; Xiao, K.; Du, X.; Dong, F.; Waterhouse, G.I.N.; Zhang, T.; Huang, H. Precursor-reforming protocol to 3D mesoporous g-C$_3$N$_4$ established by ultrathin self-doped nanosheets for superior hydrogen evolution. *Nano Energy* **2017**, *38*, 72–81. [CrossRef]
15. Wang, X.; Zhou, C.; Shi, R.; Liu, Q.; Waterhouse, G.I.N.; Wu, L.; Tung, C.-H.; Zhang, T. Supramolecular precursor strategy for the synthesis of holey graphitic carbon nitride nanotubes with enhanced photocatalytic hydrogen evolution performance. *Nano Res.* **2019**, *12*, 2385–2389. [CrossRef]
16. Zhou, C.; Shi, R.; Shang, L.; Wu, L.-Z.; Tung, C.-H.; Zhang, T. Template-free large-scale synthesis of g-C$_3$N$_4$ microtubes for enhanced visible light-driven photocatalytic H$_2$ production. *Nano Res.* **2018**, *11*, 3462–3468. [CrossRef]
17. Han, Q.; Wang, B.; Gao, J.; Cheng, Z.; Zhao, Y.; Zhang, Z.; Qu, L. Atomically Thin Mesoporous Nanomesh of Graphitic C$_3$N$_4$ for High-Efficiency Photocatalytic Hydrogen Evolution. *Acs Nano* **2016**, *10*, 2745–2751. [CrossRef]
18. Yang, P.; Ou, H.; Fang, Y.; Wang, X. A Facile Steam Reforming Strategy to Delaminate Layered Carbon Nitride Semiconductors for Photoredox Catalysis. *Angew. Chem.-Int. Ed.* **2017**, *56*, 3992–3996. [CrossRef]
19. Lan, Z.-A.; Zhang, G.; Wang, X. A facile synthesis of Br-modified g-C$_3$N$_4$ semiconductors for photoredox water splitting. *Appl. Catal. B-Environ.* **2016**, *192*, 116–125. [CrossRef]
20. Zhang, M.; Bai, X.; Liu, D.; Wang, J.; Zhu, Y. Enhanced catalytic activity of potassium-doped graphitic carbon nitride induced by lower valence position. *Appl. Catal. B-Environ.* **2015**, *164*, 77–81. [CrossRef]
21. Hu, S.; Ma, L.; You, J.; Li, F.; Fan, Z.; Lu, G.; Liu, D.; Gui, J. Enhanced visible light photocatalytic performance of g-C$_3$N$_4$ photocatalysts co-doped with iron and phosphorus. *Appl. Surf. Sci.* **2014**, *311*, 164–171. [CrossRef]
22. Hu, S.; Li, F.; Fan, Z.; Wang, F.; Zhao, Y.; Lv, Z. Band gap-tunable potassium doped graphitic carbon nitride with enhanced mineralization ability. *Dalton Trans.* **2015**, *44*, 1084–1092. [CrossRef] [PubMed]
23. Lan, H.; Li, L.; An, X.; Liu, F.; Chen, C.; Liu, H.; Qu, J. Microstructure of carbon nitride affecting synergetic photocatalytic activity: Hydrogen bonds vs. structural defects. *Appl. Catal. B-Environ.* **2017**, *204*, 49–57. [CrossRef]
24. Shen, Y.; Guo, X.; Bo, X.; Wang, Y.; Guo, X.; Xie, M.; Guo, X. Effect of template-induced surface species on electronic structure and photocatalytic activity of g-C$_3$N$_4$. *Appl. Surf. Sci.* **2017**, *396*, 933–938. [CrossRef]
25. Perez-Molina, A.; Pastrana-Martinez, L.M.; Morales-Torres, S.; Maldonado-Hodar, F.J. Photodegradation of cytostatic drugs by g-C$_3$N$_4$: Synthesis, properties and performance fitted by selecting the appropriate precursor. *Catal. Today* **2023**, *418*, 114068. [CrossRef]

26. Gao, Y.; Zhu, Y.; Lyu, L.; Zeng, Q.; Xing, X.; Hu, C. Electronic Structure Modulation of Graphitic Carbon Nitride by Oxygen Doping for Enhanced Catalytic Degradation of Organic Pollutants through Peroxymonosulfate Activation. *Environ. Sci. Technol.* **2018**, *52*, 14371–14380. [CrossRef] [PubMed]
27. Fang, J.; Fan, H.; Li, M.; Long, C. Nitrogen self-doped graphitic carbon nitride as efficient visible light photocatalyst for hydrogen evolution. *J. Mater. Chem. A* **2015**, *3*, 13819–13826. [CrossRef]
28. Yu, H.; Shi, R.; Zhao, Y.; Bian, T.; Zhao, Y.; Zhou, C.; Waterhouse, G.I.N.; Wu, L.-Z.; Tung, C.-H.; Zhang, T. Alkali-Assisted Synthesis of Nitrogen Deficient Graphitic Carbon Nitride with Tunable Band Structures for Efficient Visible-Light-Driven Hydrogen Evolution. *Adv. Mater.* **2017**, *29*, 1605148. [CrossRef]
29. Li, X.; Zhang, J.; Huo, Y.; Dai, K.; Li, S.; Chen, S. Two-dimensional sulfur- and chlorine-codoped g-C_3N_4/CdSe-amine heterostructures nanocomposite with effective interfacial charge transfer and mechanism insight. *Appl. Catal. B-Environ.* **2021**, *280*, 119452. [CrossRef]
30. Li, G.; Tang, Y.; Fu, T.; Xiang, Y.; Xiong, Z.; Si, Y.; Guo, C.; Jiang, Z. S, N co-doped carbon nanotubes coupled with CoFe nanoparticles as an efficient bifunctional ORR/OER electrocatalyst for rechargeable Zn-air batteries. *Chem. Eng. J.* **2022**, *429*, 132174. [CrossRef]
31. Yu, B.; Shi, J.; Tan, S.; Cui, Y.; Zhao, W.; Wu, H.; Luo, Y.; Li, D.; Meng, Q. Efficient (>20%) and Stable All-Inorganic Cesium Lead Triiodide Solar Cell Enabled by Thiocyanate Molten Salts. *Angew. Chem.-Int. Ed.* **2021**, *60*, 13436–13443. [CrossRef] [PubMed]
32. Su, Y.; Zhang, Y.; Zhuang, X.; Li, S.; Wu, D.; Zhang, F.; Feng, X. Low-temperature synthesis of nitrogen/sulfur co-doped three-dimensional graphene frameworks as efficient metal-free electrocatalyst for oxygen reduction reaction. *Carbon* **2013**, *62*, 296–301. [CrossRef]
33. Zhang, L.; Xu, Z. A review of current progress of recycling technologies for metals from waste electrical and electronic equipment. *J. Clean. Prod.* **2016**, *127*, 19–36. [CrossRef]
34. Ye, C.; Li, J.-X.; Li, Z.-J.; Li, X.-B.; Fan, X.-B.; Zhang, L.-P.; Chen, B.; Tung, C.-H.; Wu, L.-Z. Enhanced Driving Force and Charge Separation Efficiency of Protonated g-C_3N_4 for Photocatalytic O_2 Evolution. *ACS Catal.* **2015**, *5*, 6973–6979. [CrossRef]
35. Zhang, G.; Zang, S.; Wang, X. Layered Co(OH)(2) Deposited Polymeric Carbon Nitrides for Photocatalytic Water Oxidation. *ACS Catal.* **2015**, *5*, 941–947. [CrossRef]
36. Hao, X.; Zhou, J.; Cui, Z.; Wang, Y.; Wang, Y.; Zou, Z. Zn-vacancy mediated electron-hole separation in ZnS/g-C_3N_4 heterojunction for efficient visible-light photocatalytic hydrogen production. *Appl. Catal. B-Environ.* **2018**, *229*, 41–51. [CrossRef]
37. Tang, J.-Y.; Guo, R.-T.; Zhou, W.-G.; Huang, C.-Y.; Pan, W.-G. Ball-flower like NiO/g-C_3N_4 heterojunction for efficient visible light photocatalytic CO_2 reduction. *Appl. Catal. B-Environ.* **2018**, *237*, 802–810. [CrossRef]
38. Zhang, H.; Zhao, L.; Geng, F.; Guo, L.-H.; Wan, B.; Yang, Y. Carbon dots decorated graphitic carbon nitride as an efficient metal-free photocatalyst for phenol degradation. *Appl. Catal. B-Environ.* **2016**, *180*, 656–662. [CrossRef]
39. Wang, Z.; Guan, W.; Sun, Y.; Dong, F.; Zhou, Y.; Ho, W.-K. Water-assisted production of honeycomb-like g-C_3N_4 with ultralong carrier lifetime and outstanding photocatalytic activity. *Nanoscale* **2015**, *7*, 2471–2479. [CrossRef]
40. Goettmann, F.; Fischer, A.; Antonietti, M.; Thomas, A. Metal-free catalysis of sustainable Friedel-Crafts reactions: Direct activation of benzene by carbon nitrides to avoid the use of metal chlorides and halogenated compounds. *Chem. Commun.* **2006**, 4530–4532. [CrossRef]
41. Niu, Y.; Hu, F.; Xu, H.; Zhang, S.; Song, B.; Wang, H.; Li, M.; Shao, G.; Wang, H.; Lu, H. Exploration for high performance g-C_3N_4 photocatalyst from different precursors. *Mater. Today Commun.* **2023**, *34*, 105040. [CrossRef]
42. Thi Kim Anh, N.; Thanh-Truc, P.; Huy, N.-P.; Shin, E.W. The effect of graphitic carbon nitride precursors on the photocatalytic dye degradation of water-dispersible graphitic carbon nitride photocatalysts. *Appl. Surf. Sci.* **2021**, *537*, 148027.
43. Cardenas, A.; Vazquez, A.; Obregon, S.; Ruiz-Gomez, M.A.; Rodriguez-Gonzalez, V. New insights into the fluorescent sensing of Fe^{3+} ions by g-C_3N_4 prepared from different precursors. *Mater. Res. Bull.* **2021**, *142*, 111385. [CrossRef]
44. Liu, X.; Xu, X.; Gan, H.; Yu, M.; Huang, Y. The Effect of Different g-C_3N_4 Precursor Nature on Its Structural Control and Photocatalytic Degradation Activity. *Catalysts* **2023**, *13*, 848. [CrossRef]
45. Catherine, H.N.; Chiu, W.-L.; Chang, L.-L.; Tung, K.-L.; Hu, C. Gel-like Ag-Dicyandiamide Metal-Organic Supramolecular Network-Derived g-C_3N_4 for Photocatalytic Hydrogen Generation. *Acs Sustain. Chem. Eng.* **2022**, *10*, 8360–8369. [CrossRef]
46. Ren, X.; Yu, Q.; Pan, J.; Wang, Q.; Li, Y.; Shi, N. Photocatalytic degradation of methylene blue by C/g-C_3N_4 composites formed by different carbon sources. *React. Kinet. Mech. Catal.* **2022**, *135*, 2279–2289. [CrossRef]
47. Wang, X.; Maeda, K.; Thomas, A.; Takanabe, K.; Xin, G.; Carlsson, J.M.; Domen, K.; Antonietti, M. A metal-free polymeric photocatalyst for hydrogen production from water under visible light. *Nat. Mater.* **2009**, *8*, 76–80. [CrossRef] [PubMed]
48. Ang, T.P.; Chan, Y.M. Comparison of the Melon Nanocomposites in Structural Properties and Photocatalytic Activities. *J. Phys. Chem. C* **2011**, *115*, 15965–15972. [CrossRef]
49. Zhang, G.; Zhang, J.; Zhang, M.; Wang, X. Polycondensation of thiourea into carbon nitride semiconductors as visible light photocatalysts. *J. Mater. Chem.* **2012**, *22*, 8083–8091. [CrossRef]
50. Xiao, H.; Wang, W.; Liu, G.; Chen, Z.; Lv, K.; Zhu, J. Photocatalytic performances of g-C_3N_4 based catalysts for RhB degradation: Effect of preparation conditions. *Appl. Surf. Sci.* **2015**, *358*, 313–318. [CrossRef]
51. Sun, T.; Li, C.; Bao, Y.; Fan, J.; Liu, E. S-Scheme $MnCo_2S_4$/g-C_3N_4 Heterojunction Photocatalyst for H_2 Production. *Acta Phys.-Chim. Sin.* **2023**, *39*, 2212009. [CrossRef]

52. Wang, J.; Wang, S. A critical review on graphitic carbon nitride (g-C$_3$N$_4$)-based materials: Preparation, modification and environmental application. *Coord. Chem. Rev.* **2022**, *453*, 214338. [CrossRef]
53. Zhang, X.; Ma, P.; Wang, C.; Gan, L.; Chen, X.; Zhang, P.; Wang, Y.; Li, H.; Wang, L.; Zhou, X.; et al. Unraveling the dual defect sites in graphite carbon nitride for ultra-high photocatalytic H$_2$O$_2$ evolution. *Energy Environ. Sci.* **2022**, *15*, 830–842. [CrossRef]
54. Baladi, E.; Davar, F.; Hojjati-Najafabadi, A. Synthesis and characterization of g-C$_3$N$_4$-CoFe$_2$O$_4$-ZnO magnetic nanocomposites for enhancing photocatalytic activity with visible light for degradation of penicillin G antibiotic. *Environ. Res.* **2022**, *215*, 114270. [CrossRef] [PubMed]
55. Yan, S.C.; Li, Z.S.; Zou, Z.G. Photodegradation Performance of g-C$_3$N$_4$ Fabricated by Directly Heating Melamine. *Langmuir* **2009**, *25*, 10397–10401. [CrossRef] [PubMed]
56. Zhang, Y.; Liu, J.; Wu, G.; Chen, W. Porous graphitic carbon nitride synthesized via direct polymerization of urea for efficient sunlight-driven photocatalytic hydrogen production. *Nanoscale* **2012**, *4*, 5300–5303. [CrossRef] [PubMed]
57. Cui, Y.; Zhang, G.; Lin, Z.; Wang, X. Condensed and low-defected graphitic carbon nitride with enhanced photocatalytic hydrogen evolution under visible light irradiation. *Appl. Catal. B-Environ.* **2016**, *181*, 413–419. [CrossRef]
58. Xu, Q.; Zhu, B.; Cheng, B.; Yu, J.; Zhou, M.; Ho, W. Photocatalytic H$_2$ evolution on graphdiyne/g-C$_3$N$_4$ hybrid nanocomposites. *Appl. Catal. B-Environ.* **2019**, *255*, 117770. [CrossRef]
59. Yan, H.; Chen, Y.; Xu, S. Synthesis of graphitic carbon nitride by directly heating sulfuric acid treated melamine for enhanced photocatalytic H$_2$ production from water under visible light. *Int. J. Hydrogen Energy* **2012**, *37*, 125–133. [CrossRef]
60. Dong, G.; Zhang, L. Porous structure dependent photoreactivity of graphitic carbon nitride under visible light. *J. Mater. Chem.* **2012**, *22*, 1160–1166. [CrossRef]
61. Zhang, X.-S.; Hu, J.-Y.; Jiang, H. Facile modification of a graphitic carbon nitride catalyst to improve its photoreactivity under visible light irradiation. *Chem. Eng. J.* **2014**, *256*, 230–237. [CrossRef]
62. Gao, J.; Zhou, Y.; Li, Z.; Yan, S.; Wang, N.; Zou, Z. High-yield synthesis of millimetre-long, semiconducting carbon nitride nanotubes with intense photoluminescence emission and reproducible photoconductivity. *Nanoscale* **2012**, *4*, 3687–3692. [CrossRef] [PubMed]
63. Zhong, Y.; Wang, Z.; Feng, J.; Yan, S.; Zhang, H.; Li, Z.; Zou, Z. Improvement in photocatalytic H$_2$ evolution over g-C$_3$N$_4$ prepared from protonated melamine. *Appl. Surf. Sci.* **2014**, *295*, 253–259. [CrossRef]
64. Zhang, J.; Sun, J.; Maeda, K.; Domen, K.; Liu, P.; Antonietti, M.; Fu, X.; Wang, X. Sulfur-mediated synthesis of carbon nitride: Band-gap engineering and improved functions for photocatalysis. *Energy Environ. Sci.* **2011**, *4*, 675–678. [CrossRef]
65. Zhang, J.; Zhang, M.; Zhang, G.; Wang, X. Synthesis of Carbon Nitride Semiconductors in Sulfur Flux for Water Photoredox Catalysis. *Acs Catal.* **2012**, *2*, 940–948. [CrossRef]
66. Gong, X.-Q.; Selloni, A.; Batzill, M.; Diebold, U. Steps on anatase TiO$_2$(101). *Nat. Mater.* **2006**, *5*, 665–670. [CrossRef] [PubMed]
67. Nowotny, M.K.; Sheppard, L.R.; Bak, T.; Nowotny, J. Defect chemistry of titanium dioxide. application of defect engineering in processing of TiO$_2$-based photocatalysts. *J. Phys. Chem. C* **2008**, *112*, 5275–5300. [CrossRef]
68. Hong, Z.; Shen, B.; Chen, Y.; Lin, B.; Gao, B. Enhancement of photocatalytic H-2 evolution over nitrogen-deficient graphitic carbon nitride. *J. Mater. Chem. A* **2013**, *1*, 11754–11761. [CrossRef]
69. Lau, V.W.-H.; Mesch, M.B.; Duppel, V.; Blum, V.; Senker, J.; Lotsch, B.V. Low-Molecular-Weight Carbon Nitrides for Solar Hydrogen Evolution. *J. Am. Chem. Soc.* **2015**, *137*, 1064–1072. [CrossRef]
70. Niu, P.; Liu, G.; Cheng, H.-M. Nitrogen Vacancy-Promoted Photocatalytic Activity of Graphitic Carbon Nitride. *J. Phys. Chem. C* **2012**, *116*, 11013–11018. [CrossRef]
71. Niu, P.; Yin, L.-C.; Yang, Y.-Q.; Liu, G.; Cheng, H.-M. Increasing the Visible Light Absorption of Graphitic Carbon Nitride (Melon) Photocatalysts by Homogeneous Self-Modification with Nitrogen Vacancies. *Adv. Mater.* **2014**, *26*, 8046–8052. [CrossRef] [PubMed]
72. Martin, D.J.; Qiu, K.; Shevlin, S.A.; Handoko, A.D.; Chen, X.; Guo, Z.; Tang, J. Highly Efficient Photocatalytic H$_2$ Evolution from Water using Visible Light and Structure-Controlled Graphitic Carbon Nitride. *Angew. Chem.-Int. Ed.* **2014**, *53*, 9240–9245. [CrossRef] [PubMed]
73. Tay, Q.; Kanhere, P.; Ng, C.F.; Chen, S.; Chakraborty, S.; Huan, A.C.H.; Sum, T.C.; Ahuja, R.; Chen, Z. Defect Engineered g-C$_3$N$_4$ for Efficient Visible Light Photocatalytic Hydrogen Production. *Chem. Mater.* **2015**, *27*, 4930–4933. [CrossRef]
74. Mo, Z.; She, X.; Li, Y.; Liu, L.; Huang, L.; Chen, Z.; Zhang, Q.; Xu, H.; Li, H. Synthesis of g-C$_3$N$_4$ at different temperatures for superior visible/UV photocatalytic performance and photoelectrochemical sensing of MB solution. *Rsc Adv.* **2015**, *5*, 101552–101562. [CrossRef]
75. Dong, G.; Ho, W.; Wang, C. Selective photocatalytic N$_2$ fixation dependent on g-C$_3$N$_4$ induced by nitrogen vacancies. *J. Mater. Chem. A* **2015**, *3*, 23435–23441. [CrossRef]
76. Wu, G.; Thind, S.S.; Wen, J.; Yan, K.; Chen, A. A novel nanoporous α-C$_3$N$_4$ photocatalyst with superior high visible light activity. *Appl. Catal. B-Environ.* **2013**, *142*, 590–597. [CrossRef]
77. Liu, J.; Wang, H.; Chen, Z.P.; Moehwald, H.; Fiechter, S.; van de Krol, R.; Wen, L.; Jiang, L.; Antonietti, M. Microcontact-Printing-Assisted Access of Graphitic Carbon Nitride Films with Favorable Textures toward Photoelectrochemical Application. *Adv. Mater.* **2015**, *27*, 712–718. [CrossRef] [PubMed]
78. Yang, Z.; Zhang, Y.; Schnepp, Z. Soft and hard templating of graphitic carbon nitride. *J. Mater. Chem. A* **2015**, *3*, 14081–14092. [CrossRef]

79. Tong, Z.; Yang, D.; Li, Z.; Nan, Y.; Ding, F.; Shen, Y.; Jiang, Z. Thylakoid-Inspired Multishell g-C_3N_4 Nanocapsules with Enhanced Visible-Light Harvesting and Electron Transfer Properties for High-Efficiency Photocatalysis. *ACS Nano* **2017**, *11*, 1103–1112. [CrossRef]
80. Ong, W.-J.; Tan, L.-L.; Ng, Y.H.; Yong, S.-T.; Chai, S.-P. Graphitic Carbon Nitride (g-C_3N_4)-Based Photocatalysts for Artificial Photosynthesis and Environmental Remediation: Are We a Step Closer To Achieving Sustainability? *Chem. Rev.* **2016**, *116*, 7159–7329. [CrossRef]
81. Dong, G.; Zhang, Y.; Pan, Q.; Qiu, J. A fantastic graphitic carbon nitride (g-C_3N_4) material: Electronic structure, photocatalytic and photoelectronic properties. *J. Photochem. Photobiol. C-Photochem. Rev.* **2014**, *20*, 33–50. [CrossRef]
82. Li, M.; Zhang, L.; Wu, M.; Du, Y.; Fan, X.; Wang, M.; Zhang, L.; Kong, Q.; Shi, J. Mesostructured CeO_2/g-C_3N_4 nanocomposites: Remarkably enhanced photocatalytic activity for CO_2 reduction by mutual component activations. *Nano Energy* **2016**, *19*, 145–155. [CrossRef]
83. Dong, J.; Zhang, Y.; Hussain, M.I.; Zhou, W.; Chen, Y.; Wang, L.-N. g-C_3N_4: Properties, Pore Modifications, and Photocatalytic Applications. *Nanomaterials* **2022**, *12*, 121. [CrossRef] [PubMed]
84. Zhao, S.; Zhang, Y.; Zhou, Y.; Wang, Y.; Qiu, K.; Zhang, C.; Fang, J.; Sheng, X. Facile one-step synthesis of hollow mesoporous g-C_3N_4 spheres with ultrathin nanosheets for photoredox water splitting. *Carbon* **2018**, *126*, 247–256. [CrossRef]
85. Hu, R.; Wang, X.; Dai, S.; Shao, D.; Hayat, T.; Alsaedi, A. Application of graphitic carbon nitride for the removal of Pb(II) and aniline from aqueous solutions. *Chem. Eng. J.* **2015**, *260*, 469–477. [CrossRef]
86. Fukasawa, Y.; Takanabe, K.; Shimojima, A.; Antonietti, M.; Domen, K.; Okubo, T. Synthesis of Ordered Porous Graphitic-C_3N_4 and Regularly Arranged Ta3N5 Nanoparticles by Using Self-Assembled Silica Nanospheres as a Primary Template. *Chem. Asian J.* **2011**, *6*, 103–109. [CrossRef] [PubMed]
87. Chen, X.; Wang, H.; Meng, R.; Chen, M. Porous Graphitic Carbon Nitride Synthesized via Using Carbon Nanotube as a Novel Recyclable Hard Template for Efficient Visible Light Photocatalytic Organic Pollutant Degradation. *Chemistryselect* **2019**, *4*, 6123–6129. [CrossRef]
88. Zhang, Y.; Schnepp, Z.; Cao, J.; Ouyang, S.; Li, Y.; Ye, J.; Liu, S. Biopolymer-Activated Graphitic Carbon Nitride towards a Sustainable Photocathode Material. *Sci. Rep.* **2013**, *3*, 2163. [CrossRef]
89. Xu, J.; Wang, Y.; Zhu, Y. Nanoporous Graphitic Carbon Nitride with Enhanced Photocatalytic Performance. *Langmuir* **2013**, *29*, 10566–10572. [CrossRef]
90. Wang, Y.; Wang, X.; Antonietti, M.; Zhang, Y. Facile One-Pot Synthesis of Nanoporous Carbon Nitride Solids by Using Soft Templates. *Chemsuschem* **2010**, *3*, 435–439. [CrossRef]
91. Shen, W.; Ren, L.; Zhou, H.; Zhang, S.; Fan, W. Facile one-pot synthesis of bimodal mesoporous carbon nitride and its function as a lipase immobilization support. *J. Mater. Chem.* **2011**, *21*, 3890–3894. [CrossRef]
92. Shalom, M.; Inal, S.; Fettkenhauer, C.; Neher, D.; Antonietti, M. Improving Carbon Nitride Photocatalysis by Supramolecular Preorganization of Monomers. *J. Am. Chem. Soc.* **2013**, *135*, 7118–7121. [CrossRef] [PubMed]
93. Zhang, M.; Yan, X.; Huang, F.; Niu, Z.; Gibson, H.W. Stimuli-Responsive Host-Guest Systems Based on the Recognition of Cryptands by Organic Guests. *Acc. Chem. Res.* **2014**, *47*, 1995–2005. [CrossRef] [PubMed]
94. Cui, Q.; Xu, J.; Wang, X.; Li, L.; Antonietti, M.; Shalom, M. Phenyl-Modified Carbon Nitride Quantum Dots with Distinct Photoluminescence Behavior. *Angew. Chem.-Int. Ed.* **2016**, *55*, 3672–3676. [CrossRef] [PubMed]
95. Bhunia, M.K.; Yamauchi, K.; Takanabe, K. Harvesting Solar Light with Crystalline Carbon Nitrides for Efficient Photocatalytic Hydrogen Evolution. *Angew. Chem.-Int. Ed.* **2014**, *53*, 11001–11005. [CrossRef] [PubMed]
96. Jordan, T.; Fechler, N.; Xu, J.; Brenner, T.J.K.; Antonietti, M.; Shalom, M. "Caffeine Doping" of Carbon/Nitrogen-Based Organic Catalysts: Caffeine as a Supramolecular Edge Modifier for the Synthesis of Photoactive Carbon Nitride Tubes. *Chemcatchem* **2015**, *7*, 2826–2830. [CrossRef]
97. Gao, J.; Wang, J.; Qian, X.; Dong, Y.; Xu, H.; Song, R.; Yan, C.; Zhu, H.; Zhong, Q.; Qian, G.; et al. One-pot synthesis of copper-doped graphitic carbon nitride nanosheet by heating Cu-melamine supramolecular network and its enhanced visible-light-driven photocatalysis. *J. Solid State Chem.* **2015**, *228*, 60–64. [CrossRef]
98. Chen, Z.P.; Antonietti, M.; Dontsova, D. Enhancement of the Photocatalytic Activity of Carbon Nitrides by Complex Templating. *Chem. A Eur. J.* **2015**, *21*, 10805–10811. [CrossRef]
99. Shalom, M.; Guttentag, M.; Fettkenhauer, C.; Inal, S.; Neher, D.; Llobet, A.; Antonietti, M. In Situ Formation of Heterojunctions in Modified Graphitic Carbon Nitride: Synthesis and Noble Metal Free Photocatalysis. *Chem. Mater.* **2014**, *26*, 5812–5818. [CrossRef]
100. Fan, X.; Xing, Z.; Shu, Z.; Zhang, L.; Wang, L.; Shi, J. Improved photocatalytic activity of g-C_3N_4 derived from cyanamide-urea solution. *RSC Adv.* **2015**, *5*, 8323–8328. [CrossRef]
101. Shalom, M.; Gimenez, S.; Schipper, F.; Herraiz-Cardona, I.; Bisquert, J.; Antonietti, M. Controlled Carbon Nitride Growth on Surfaces for Hydrogen Evolution Electrodes. *Angew. Chem.-Int. Ed.* **2014**, *53*, 3654–3658. [CrossRef] [PubMed]
102. Jun, Y.-S.; Lee, E.Z.; Wang, X.; Hong, W.H.; Stucky, G.D.; Thomas, A. From Melamine-Cyanuric Acid Supramolecular Aggregates to Carbon Nitride Hollow Spheres. *Adv. Funct. Mater.* **2013**, *23*, 3661–3667. [CrossRef]
103. Bai, X.; Wang, L.; Zong, R.; Zhu, Y. Photocatalytic Activity Enhanced via g-C_3N_4 Nanoplates to Nanorods. *J. Phys. Chem. C* **2013**, *117*, 9952–9961. [CrossRef]
104. Wang, S.; Li, C.; Wang, T.; Zhang, P.; Li, A.; Gong, J. Controllable synthesis of nanotube-type graphitic C_3N_4 and their visible-light photocatalytic and fluorescent properties. *J. Mater. Chem. A* **2014**, *2*, 2885–2890. [CrossRef]

105. Chen, D.; Wang, K.; Xiang, D.; Zong, R.; Yao, W.; Zhu, Y. Significantly enhancement of photocatalytic performances via core-shell structure of ZnO@mpg-C_3N_4. *Appl. Catal. B-Environ.* **2014**, *147*, 554–561. [CrossRef]
106. Cheng, F.; Wang, H.; Dong, X. The amphoteric properties of g-C_3N_4 nanosheets and fabrication of their relevant heterostructure photocatalysts by an electrostatic re-assembly route. *Chem. Commun.* **2015**, *51*, 7176–7179. [CrossRef] [PubMed]
107. Schwinghammer, K.; Mesch, M.B.; Duppel, V.; Ziegler, C.; Senker, J.; Lotsch, B.V. Crystalline Carbon Nitride Nanosheets for Improved Visible-Light Hydrogen Evolution. *J. Am. Chem. Soc.* **2014**, *136*, 1730–1733. [CrossRef] [PubMed]
108. Liu, G.; Wang, T.; Zhang, H.; Meng, X.; Hao, D.; Chang, K.; Li, P.; Kako, T.; Ye, J. Nature-Inspired Environmental "Phosphorylation" Boosts Photocatalytic H_2 Production over Carbon Nitride Nanosheets under Visible-Light Irradiation. *Angew. Chem.-Int. Ed.* **2015**, *54*, 13561–13565. [CrossRef]
109. Niu, P.; Zhang, L.; Liu, G.; Cheng, H.-M. Graphene-Like Carbon Nitride Nanosheets for Improved Photocatalytic Activities. *Adv. Funct. Mater.* **2012**, *22*, 4763–4770. [CrossRef]
110. Yang, S.; Gong, Y.; Zhang, J.; Zhan, L.; Ma, L.; Fang, Z.; Vajtai, R.; Wang, X.; Ajayan, P.M. Exfoliated Graphitic Carbon Nitride Nanosheets as Efficient Catalysts for Hydrogen Evolution under Visible Light. *Adv. Mater.* **2013**, *25*, 2452–2456. [CrossRef]
111. Zhao, H.; Yu, H.; Quan, X.; Chen, S.; Zhang, Y.; Zhao, H.; Wang, H. Fabrication of atomic single layer graphitic-C_3N_4 and its high performance of photocatalytic disinfection under visible light irradiation. *Appl. Catal. B-Environ.* **2014**, *152*, 46–50. [CrossRef]
112. Yin, Y.; Han, J.; Zhang, X.; Zhang, Y.; Zhou, J.; Muir, D.; Sutarto, R.; Zhang, Z.; Liu, S.; Song, B. Facile synthesis of few-layer-thick carbon nitride nanosheets by liquid ammonia-assisted lithiation method and their photocatalytic redox properties. *RSC Adv.* **2014**, *4*, 32690–32697. [CrossRef]
113. Ma, L.; Fan, H.; Li, M.; Tian, H.; Fang, J.; Dong, G. A simple melamine-assisted exfoliation of polymeric graphitic carbon nitrides for highly efficient hydrogen production from water under visible light. *J. Mater. Chem. A* **2015**, *3*, 22404–22412. [CrossRef]
114. Xu, H.; Yan, J.; She, X.; Xu, L.; Xia, J.; Xu, Y.; Song, Y.; Huang, L.; Li, H. Graphene-analogue carbon nitride: Novel exfoliation synthesis and its application in photocatalysis and photoelectrochemical selective detection of trace amount of Cu^{2+}. *Nanoscale* **2014**, *6*, 1406–1415. [CrossRef] [PubMed]
115. Ma, L.; Fan, H.; Wang, J.; Zhao, Y.; Tian, H.; Dong, G. Water-assisted ions in situ intercalation for porous polymeric graphitic carbon nitride nanosheets with superior photocatalytic hydrogen evolution performance. *Appl. Catal. B Environ.* **2016**, *190*, 93–102. [CrossRef]
116. Zhou, L.; Zhang, H.; Sun, H.; Liu, S.; Tade, M.O.; Wang, S.; Jin, W. Recent advances in non-metal modification of graphitic carbon nitride for photocatalysis: A historic review. *Catal. Sci. Technol.* **2016**, *6*, 7002–7023. [CrossRef]
117. Zhou, L.; Feng, J.; Qiu, B.; Zhou, Y.; Lei, J.; Xing, M.; Wang, L.; Zhou, Y.; Liu, Y.; Zhang, J. Ultrathin g-C_3N_4 nanosheet with hierarchical pores and desirable energy band for highly efficient H_2O_2 production. *Appl. Catal. B-Environ.* **2020**, *267*, 118396. [CrossRef]
118. Tang, C.; Cheng, M.; Lai, C.; Li, L.; Yang, X.; Du, L.; Zhang, G.; Wang, G.; Yang, L. Recent progress in the applications of non-metal modified graphitic carbon nitride in photocatalysis. *Coord. Chem. Rev.* **2023**, *474*, 214846. [CrossRef]
119. Sudhaik, A.; Raizada, P.; Shandilya, P.; Jeong, D.-Y.; Lim, J.-H.; Singh, P. Review on fabrication of graphitic carbon nitride based efficient nanocomposites for photodegradation of aqueous phase organic pollutants. *J. Ind. Eng. Chem.* **2018**, *67*, 28–51. [CrossRef]
120. Nasir, M.S.; Yang, G.; Ayub, I.; Wang, S.; Wang, L.; Yan, W.; Peng, S.; Ramakarishna, S. Recent development in graphitic carbon nitride based photocatalysis for hydrogen generation. *Appl. Catal. B-Environ.* **2019**, *257*, 117855. [CrossRef]
121. Hayat, A.; Syed, J.A.S.G.; Al-Sehemi, A.S.; El-Nasser, K.; Taha, T.A.A.; Al-Ghamdi, A.A.; Amin, M.; Ajmal, Z.; Iqbal, W.; Palamanit, A.; et al. State of the art advancement in rational design of g-C_3N_4 photocatalyst for efficient solar fuel transformation, environmental decontamination and future perspectives. *Int. J. Hydrogen Energy* **2022**, *47*, 10837–10867. [CrossRef]
122. Zeng, Y.; Liu, X.; Liu, C.; Wang, L.; Xia, Y.; Zhang, S.; Luo, S.; Pei, Y. Scalable one-step production of porous oxygen-doped g-C_3N_4 nanorods with effective electron separation for excellent visible-light photocatalytic activity. *Appl. Catal. B-Environ.* **2018**, *224*, 1–9. [CrossRef]
123. Liu, Q.; Zhang, J. Graphene Supported Co-g-C_3N_4 as a Novel Metal-Macrocyclic Electrocatalyst for the Oxygen Reduction Reaction in Fuel Cells. *Langmuir* **2013**, *29*, 3821–3828. [CrossRef] [PubMed]
124. Jiang, J.; Cao, S.; Hu, C.; Chen, C. A comparison study of alkali metal-doped g-C_3N_4 for visible-light photocatalytic hydrogen evolution. *Chin. J. Catal.* **2017**, *38*, 1981–1989. [CrossRef]
125. He, F.; Wang, Z.; Li, Y.; Peng, S.; Liu, B. The nonmetal modulation of composition and morphology of g-C_3N_4-based photocatalysts. *Appl. Catal. B-Environ.* **2020**, *269*, 118828. [CrossRef]
126. Gong, Y.; Zhao, X.; Zhang, H.; Yang, B.; Xiao, K.; Guo, T.; Zhang, J.; Shao, H.; Wang, Y.; Yu, G. MOF-derived nitrogen doped carbon modified g-C_3N_4 heterostructure composite with enhanced photocatalytic activity for bisphenol A degradation with peroxymonosulfate under visible light irradiation. *Appl. Catal. B-Environ.* **2018**, *233*, 35–45. [CrossRef]
127. Cao, S.; Low, J.; Yu, J.; Jaroniec, M. Polymeric Photocatalysts Based on Graphitic Carbon Nitride. *Adv. Mater.* **2015**, *27*, 2150–2176. [CrossRef] [PubMed]
128. Xiong, T.; Cen, W.; Zhang, Y.; Dong, F. Bridging the g-C_3N_4 Interlayers for Enhanced Photocatalysis. *Acs Catal.* **2016**, *6*, 2462–2472. [CrossRef]
129. Wen, J.; Zhou, L.; Tang, Q.; Xiao, X.; Sun, S. Photocatalytic degradation of organic pollutants by carbon quantum dots functionalized g-C_3N_4: A review. *Ecotoxicol. Environ. Saf.* **2023**, *262*, 115133. [CrossRef]

130. Wang, Y.; Zhong, S.; Niu, Z.; Dai, Y.; Li, J. Synthesis and up-to-date applications of 2D microporous g-C_3N_4 nanomaterials for sustainable development. *Chem. Commun.* **2023**, *59*, 10883–10911. [CrossRef]
131. Truong, H.B.; Hur, J.; Nguyen, X.C. Recent advances in g-C_3N_4-based photocatalysis for water treatment: Magnetic and floating photocatalysts, and applications of machine-learning techniques. *J. Environ. Manag.* **2023**, *345*, 118895. [CrossRef]
132. Ding, M.; Wei, Z.; Zhu, X.; Liu, D. Effect of Different g-C_3N_4 Content on Properties of $NiCo_2S_4$/g-C_3N_4 Composite as Electrode Material for Supercapacitor. *Int. J. Electrochem. Sci.* **2022**, *17*, 22121. [CrossRef]
133. Chen, X.; Shi, R.; Chen, Q.; Zhang, Z.; Jiang, W.; Zhu, Y.; Zhang, T. Three-dimensional porous g-C_3N_4 for highly efficient photocatalytic overall water splitting. *Nano Energy* **2019**, *59*, 644–650. [CrossRef]
134. Fu, J.; Zhu, B.; Jiang, C.; Cheng, B.; You, W.; Yu, J. Hierarchical Porous O-Doped g-C_3N_4 with Enhanced Photocatalytic CO_2 Reduction Activity. *Small* **2017**, *13*, 1603938. [CrossRef]
135. Dou, M.; Wang, J.; Gao, B.; Xu, C.; Yang, F. Photocatalytic difference of amoxicillin and cefotaxime under visible light by mesoporous g-C_3N_4: Mechanism, degradation pathway and DFT calculation. *Chem. Eng. J.* **2020**, *383*, 123134. [CrossRef]

Disclaimer/Publisher's Note: The statements, opinions and data contained in all publications are solely those of the individual author(s) and contributor(s) and not of MDPI and/or the editor(s). MDPI and/or the editor(s) disclaim responsibility for any injury to people or property resulting from any ideas, methods, instructions or products referred to in the content.

Article

A Novel Organic/Inorganic Dual Z-Scheme Photocatalyst with Visible-Light Response for Organic Pollutants Degradation

Taoming Yu [1,†], Doudou Wang [1,†], Lili Li [1], Wenjing Song [1], Xuan Pang [2,*] and Ce Liang [1,*]

[1] Key Laboratory of Automobile Materials, Ministry of Education, College of Materials Science and Engineering, Jilin University, Changchun 130022, China; ytm19990602@163.com (T.Y.); wangdd166@163.com (D.W.); lilyllee@jlu.edu.cn (L.L.); wjsong_22@163.com (W.S.)
[2] Key Laboratory of Polymer Ecomaterials, Changchun Institute of Applied Chemistry, Chinese Academy of Sciences, Changchun 130022, China
* Correspondence: xpang@ciac.ac.cn (X.P.); liangce@jlu.edu.cn (C.L.)
† These authors contributed equally to this work.

Abstract: The design of highly efficient organic/inorganic photocatalysts with visible-light response has attracted great attention for the removal of organic pollutants. In this work, the polyacrylonitrile (PAN) worked as the matrix polymer, while polyaniline (PANI) and Sb_2S_3–ZnO were used as organic/inorganic photocatalysts. The heterojunction PAN/PANI–Sb_2S_3–ZnO photocatalyst was prepared using electrospinning and surface ultrasound. PAN/PANI–Sb_2S_3–ZnO exhibited an excellent visible-light absorption intensity in the wavelength range of 400–700 nm. The maximum removal efficiencies of PAN/PANI–Sb_2S_3–ZnO for four organic dyes were all greater than 99%. The mechanism study showed that a dual Z-scheme could be constructed ingeniously because of the well-matched bandgaps between organic and inorganic components in the photocatalyst, which achieved efficient separation of photogenerated carriers and reserved photogenerated electrons (e^-) and holes (h^+) with strong redox ability. The active species •OH and •O_2^- played an important role in the photocatalytic process. The composite photocatalyst also had excellent stability and reusability. This work suggested a pathway for designing novel organic/inorganic composite photocatalysts with visible-light response.

Keywords: photocatalyst; Sb_2S_3; PANI; ZnO

1. Introduction

With the rapid development of industrialization, the discharge of dye wastewater has become one of the main sources of water pollution, which brings serious harm to the ecological environment and human health [1–3]. Photocatalysis has been widely studied for its low energy consumption, low cost, and absence of secondary pollution among many organic dye treatment methods [4]. However, the common inorganic photocatalysts have a narrow spectral absorption range and low utilization rate of visible light, which severely limits their wide application [5]. Hence, developing highly efficient visible-light-driven photocatalysts for environmental remediation has become an active research area.

In recent years, the organic conjugated polymers have attracted extensive attention from researchers because of their excellent visible-light absorption range, adjustable band gap, and great advantages compared with small molecules in forming and processing, which opens up broader application prospects for the development of photocatalysts [5–7]. The delocalized π–π conjugated structures have been proven to induce a wide spectral absorption range, which is beneficial to improve the fraction of the visible light used [8]. Among these delocalized conjugated materials, polyaniline (PANI) is a potential photocatalytic material due to its unique electrical conductivity, facile preparation, and chemical stability [9]. When PANI is exposed to light, the energy absorbed by electrons transitions from highest unoccupied molecular orbital (HOMO) to lowest unoccupied molecular orbital (LOMO) produced photogenic e^--h^+ pairs [10–12]. The presence of e^- or h^+ at the impurity level can improve the electrical conductivity of PANI, showing a photoelectric conversion effect.

However, the single component photocatalyst is always limited by the quick combination of photo-generated e^- and h^+, which results in a poor quantum efficiency and low photocatalytic activity [13–15]. To date, many efforts, such as non-metal doping, facet control, surface sensitization, and heterojunction construction, have been applied to limit the recombination of photogenerated carriers [16–19]. Among these approaches, heterojunction photocatalysts have been fabricated extensively to enhance the separation efficiency of photoexcited electron–hole pairs [20–22]. Ge et al. reported the preparation of g-C_3N_4/PANI composite photocatalysts, which displayed an enhanced photocatalytic activity for the photodegradation of methylene blue (MB) dye [23]. Wang et al. synthesized the PANI/BiOCl composite photocatalyst via BiOCl modified using PANI [24]. The photocatalytic activity of PANI/BiOCl composites was greater than BiOCl, which could be assigned to the synergistic effect between PANI and BiOCl. Sb_2S_3 and ZnO are also common inorganic photocatalysts. Sb_2S_3 has the advantages of a low cost, suitable band gap (1.5–2.2 eV), good visible-light absorption capacity and charge transport performance [25–27]. ZnO, with a band gap of 3.2 eV, has the advantage of a strong redox capacity of its photogenerated charge carriers [28–30]. It could be seen that the band structures of Sb_2S_3 and ZnO were both well-matched with PANI, so a heterojunction might be formed among PANI, Sb_2S_3, and ZnO. Membrane photocatalysts have emerged as a powerful and attractive way to remove dyes from water resources. Compared with a coating film, electrospinning has been increasingly employed in fabricating nanofiber membranes with high porosity and a large specific-surface area [31]. Meanwhile, the dispersity of functional nanoparticles on membrane photocatalysts could influence the photocatalytic performance of photocatalysts. Surface ultrasound is an effective method that can achieve uniform dispersion of functional particles on the membrane surface [32].

Herein, a dual Z-scheme photocatalyst of PAN/PANI–Sb_2S_3–ZnO was designed using electrospinning and surface ultrasound. The chemical structures, morphologies, and optical and electrochemical performance of photocatalysts were systemically analyzed. The photocatalytic performance of PAN/PANI–Sb_2S_3–ZnO was evaluated using the degradation of organic dyes under visible-light irradiation. The main active species were explored using free radical trapping experiments. Furthermore, a novel dual Z-scheme photocatalytic mechanism was proposed.

2. Results and Discussion
2.1. The Structure of the Composite Membranes

The FTIR spectra of PAN/PANI, PAN/PAN/PANI–Sb_2S_3, and PAN/PANI–Sb_2S_3–ZnO are shown in Figure 1. The characteristic peak at 1450 cm^{-1} was attributed to the tensile vibration of benzene ring C=C in PANI [33]. The peak at 1029 cm^{-1} was the vibration of C–H bonds on disubstituted benzene in the PANI [33]. The peak at 2943 cm^{-1} corresponded to the tensile vibration peak of –CH_2 in PAN. The peak at 1308 cm^{-1} was attributed to the in-plane flexural vibration of CH_2 in PAN. Compared with PAN/PANI, both PAN/PANI–Sb_2S_3 and PAN/PANI–Sb_2S_3–ZnO showed a new characteristic peak at 630 cm^{-1}, which was mainly attributed to the formation of Sb–S bonds, proving that Sb_2S_3 was successfully doped in PAN/PANI [34]. However, the tensile vibration of the Zn–O bond at 443 cm^{-1} in the fingerprint region [35] was not observed in PAN/PANI–Sb_2S_3–ZnO.

Figure 2 shows the XRD pattern of the samples. A diffraction peak appeared at the 2θ of 25.5°, corresponding to the semi-crystalline structure of the PANI [36]. The diffraction peaks appeared at 2θ of 40.6°, 43.8°, 46.6°, 49.1°, 65.6°, and 71.0° in PAN/PANI–Sb_2S_3, which was attributed to the (611), (323), (631), (513), (624), and (045) crystal planes of Sb_2S_3. Compared with PAN/PANI–Sb_2S_3, the PAN/PANI–Sb_2S_3–ZnO showed new diffraction peaks at 31.7°, 34.4°, 36.2°, 47.5°, 56.6°, 62.9°, 67.9°, 69.1°, and 76.9°, which corresponded to the (100), (002), (101), (102), (110), (103), (112), (201), and (202) crystal planes of ZnO. The results showed that PAN/PANI–Sb$_2$S$_3$–ZnO had been successfully prepared [37].

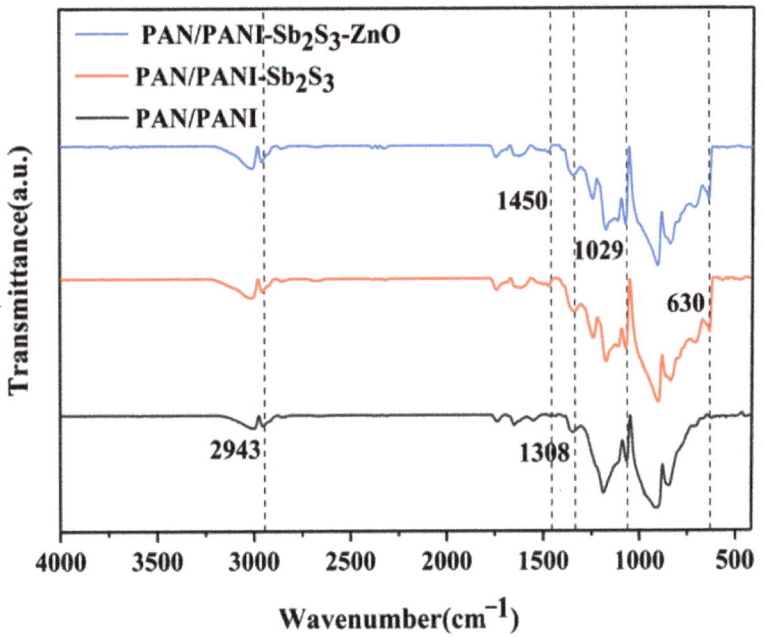

Figure 1. FTIR spectra of composite membranes.

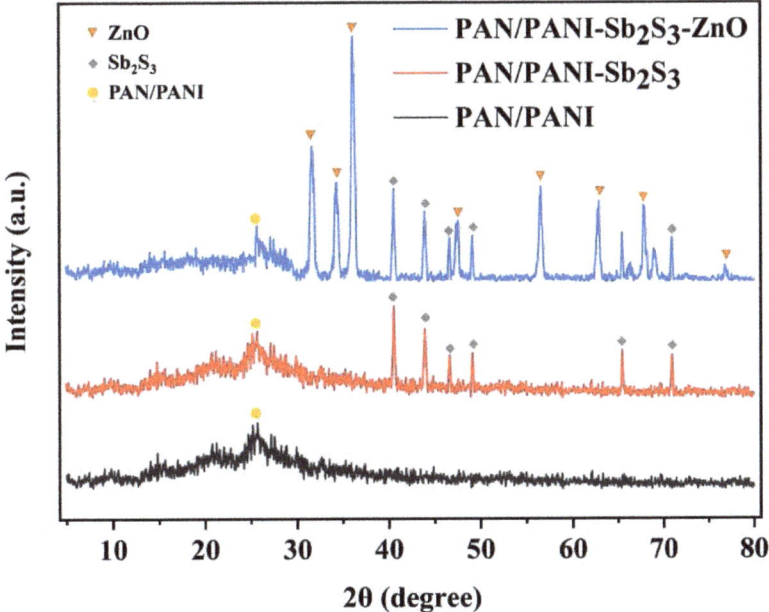

Figure 2. XRD spectra of composite membranes.

The surface chemical components of PAN/PANI, PAN/PANI–Sb$_2$S$_3$ and PAN/PANI–Sb$_2$S$_3$–ZnO composites were further analyzed using XPS. Figure 3 shows the full XPS spectra of the photocatalysts. It could be observed that the peaks appeared at 286 eV, 400 eV, and 531 eV, which corresponded to the peaks of C 1s, N 1s, and O 1s, respectively. The new peaks at 170 eV and 498 eV for PAN/PANI–Sb$_2$S$_3$ corresponded to the S 2p and Sb 3d peaks [38]. Compared with PAN/PANI–Sb$_2$S$_3$, PAN/PANI–Sb$_2$S$_3$–ZnO had new peaks at 11 eV, 89 eV, 139 eV, and 1022 eV, which were attributed to the Zn 3d, Zn 3p, Zn 3s, and Zn 2p, respectively [39]. The dispersion of ZnO on the surface of the fiber membrane was proved.

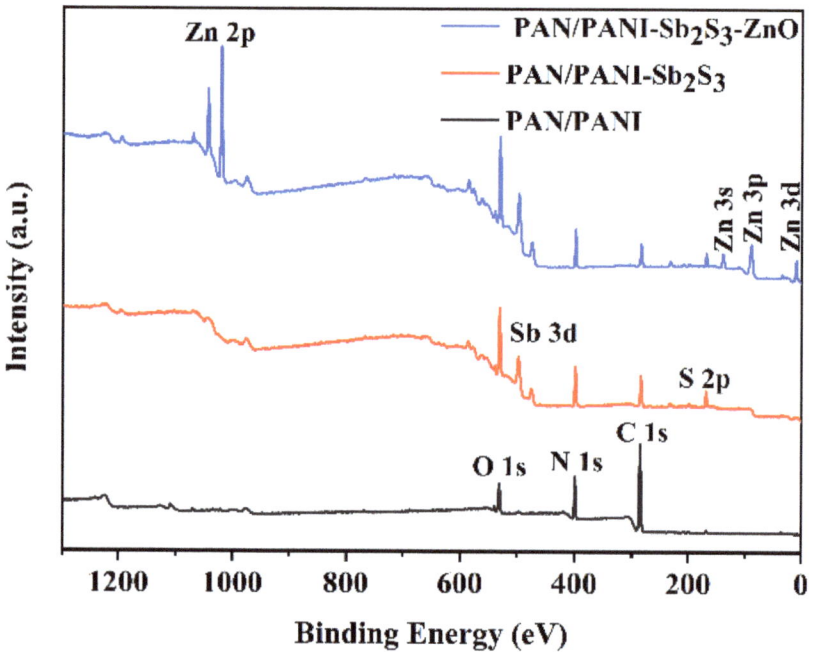

Figure 3. XPS full spectra of composite membranes.

Figure 4a shows the high-resolution XPS spectra of the C 1s of PAN/PANI–Sb$_2$S$_3$–ZnO. The peaks at 284.8 eV, 286.6 eV, and 288.6 eV corresponded to the C–H/C–C, C–N and C=N [40]. The N 1s spectra was shown in Figure 4b. The peaks at 399.5 eV and 401.7 eV represented the –NH and –N$^+$ = structure [36]. The high-resolution spectra of O 1 s is shown in Figure 4c, where the peak at 530.4 eV was attributed to oxygen ions in the Zn–O bond and the peak at 532.3 eV was attributed to the presence of a C=O bond. Figure 4d illustrates the spectra of Sb 2p. The peaks at 539.7 eV and 531.1 eV corresponded to Sb^{3+} 3d 3/2 and Sb^{3+} 3d 5/2, respectively [41]. Figure 4e shows the high-resolution spectra of S 2p. The peaks at 163.1 and 162.1 eV were attributed to S 2p 1/2 and S 2p 3/2, respectively [41]. The high-resolution spectra of Zn 2p are shown in Figure 4f, where the peaks at 1045.1 eV and 1021.9 eV corresponded to Zn 2p 1/2 and Zn 2p 3/2, respectively [42].

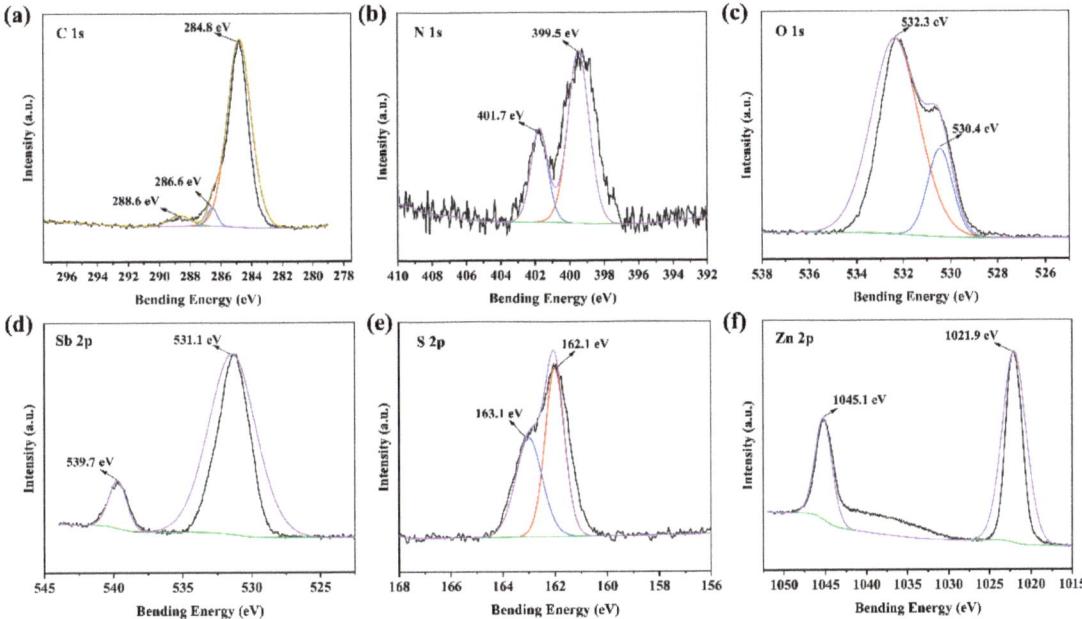

Figure 4. XPS high-resolution spectra of (**a**) C 1s, (**b**) N 1s, (**c**) O 1s, (**d**) Sb 2p, (**e**) S 2p, (**f**) Zn 2p of PAN/PANI–Sb_2S_3–ZnO.

2.2. The Morphologies of the Composite Membrane

Figure 5a shows an SEM image of the PAN/PANI. It could be seen that PAN/PANI fibers had a relatively uniform diameter distribution. In Figure 5b, PAN/PANI–Sb_2S_3 fibers were continuous and uniform, and the spheroidal Sb_2S_3 was dispersed on the fiber surface. Compared with PAN/PANI–Sb_2S_3, ZnO nanoparticles with a smaller diameter were evenly distributed along the fibers PAN/PANI–Sb_2S_3–ZnO (Figure 5c).

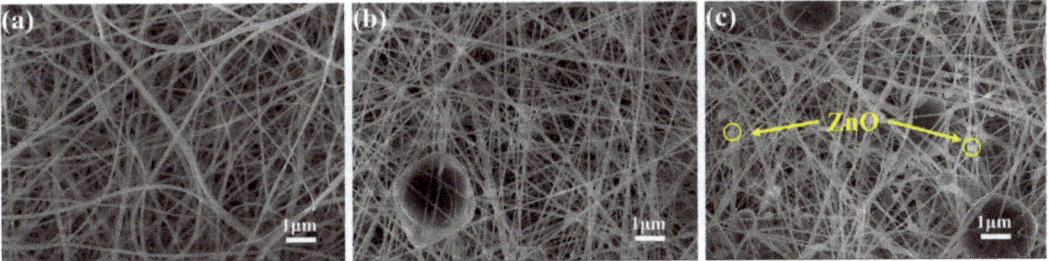

Figure 5. SEM images of (**a**) PAN/PANI, (**b**) PAN/PANI–Sb_2S_3, and (**c**) PAN/PANI–Sb_2S_3–ZnO.

A TEM image of the PAN/PANI–Sb_2S_3–ZnO composite membrane is depicted in Figure 6a. It could be seen that both Sb_2S_3 and ZnO had good interfacial contact with fibers. Figure 6b–g showed the EDS images of PAN/PANI–Sb_2S_3–ZnO. The distributions of the C, N, and O elements were uniformly dispersed. The distributions of the S, Sb, and Zn elements were consistent with the distributions of Sb_2S_3 and ZnO in the TEM image, which further proved the heterogeneous structures of composite photocatalysts.

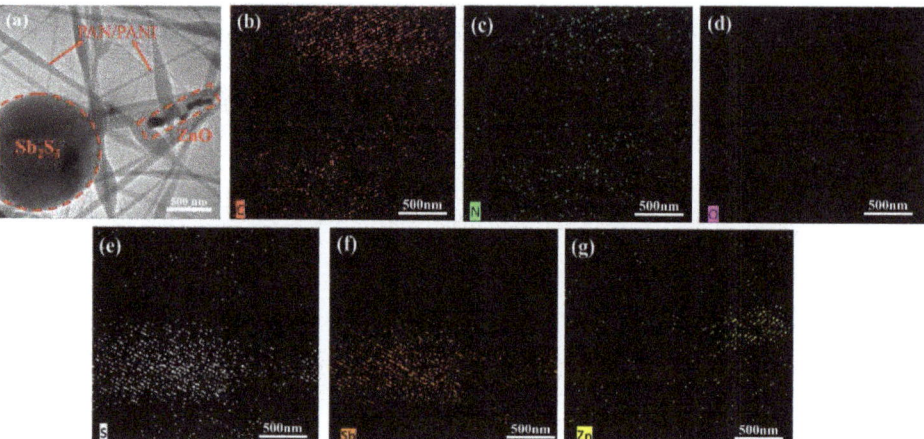

Figure 6. TEM (**a**) and EDS (**b**–**g**) images of the PAN/PANI–Sb_2S_3–ZnO membrane.

2.3. The Optical and Electrochemical Properties of Composite Membranes

The UV-vis DRS was used to analyze the optical properties of PAN/PANI, PAN/PANI–Sb_2S_3, and PAN/PANI–Sb_2S_3–ZnO (shown in Figure 7). It could be observed that all the samples showed a strong absorption capacity in the wavelength range of 400–700 nm. This was mainly due to the π-conjugated structures in the PANI molecular chain, which could provide a wide spectrum absorption range and high absorption coefficient [43–45]. It was worth noting that the light absorption intensity of the photocatalyst decreased with the addition of Sb_2S_3 and ZnO.

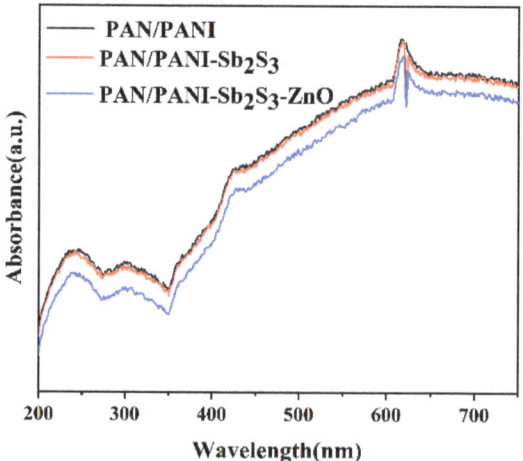

Figure 7. UV-vis DRS spectra of PAN/PANI, PAN/PANI–Sb_2S_3 and PA; N/PANI–Sb_2S_3–ZnO.

The carrier separation efficiencies of PAN/PANI, PAN/PANI–Sb_2S_3, and PAN/PANI–Sb_2S_3–ZnO were analyzed using PL spectroscopy, as shown in Figure 8a. It was observed that PAN/PANI exhibited a strong fluorescence peak, indicating that PAN/PANI had a high e^-h^+ recombination rate under light. The fluorescence peak of PAN/PANI–Sb_2S_3 was lower than that of PAN/PANI, indicating that the recombination of photogenerated carriers was inhibited after the addition of Sb_2S_3. The main reason for this was that PANI and Sb_2S_3 formed an effective heterojunction structure, which promoted carrier migration.

PAN/PANI–Sb$_2$S$_3$–ZnO exhibited the weakest fluorescence peak, which indicated that the addition of ZnO further inhibited the recombination of carriers and promoted the separation of e$^-$ and h$^+$. Time-resolved photoluminescence spectra (TRPS) were used to investigate carrier lifetime. τ_1 and τ_2 represented the lifetimes of radioactive and nonradioactive charges. As shown in Figure 8b, the average lifetimes (τ) of PAN/PANI, PAN/PANI–Sb$_2$S$_3$, and PAN/PANI–Sb$_2$S$_3$–ZnO were 4.91, 5.02, and 5.45 ns, respectively. The average carrier lifetime of PAN/PANI–Sb$_2$S$_3$–ZnO was longer than that of PAN/PANI and PAN/PANI–Sb$_2$S$_3$, which proved that PAN/PANI–Sb$_2$S$_3$–ZnO had the highest separation efficiency for photogenerated charge carriers.

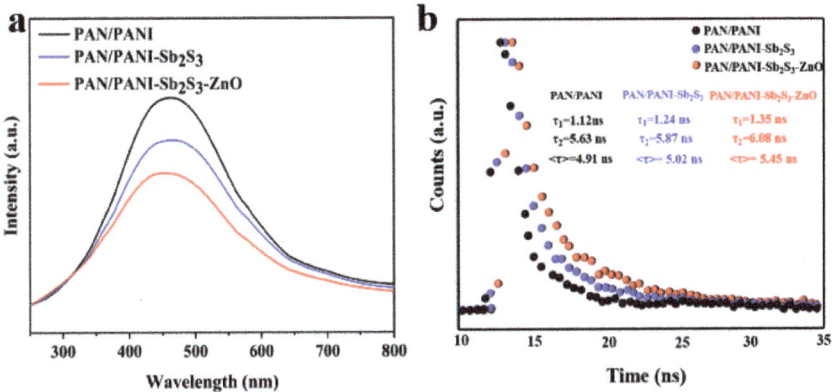

Figure 8. (a) PL spectra of PAN/PANI, PAN/PANI–Sb$_2$S$_3$, and PAN/PANI–Sb$_2$S$_3$–ZnO; and (b) TRPS of PAN/PANI, PAN/PANI–Sb$_2$S$_3$, and PAN/PANI–Sb$_2$S$_3$–ZnO.

2.4. Evaluation of Removal Performance of Organic Dyes

Figure 9 shows the kinetic curves for RhB with the PAN/PANI, PAN/PANI–Sb$_2$S$_3$, and PAN/PANI–Sb$_2$S$_3$–ZnO under simulated sunlight irradiation. Under 40 min simulated sunlight irradiation, the degradation efficiency of RhB dye with PAN/PANI was only 37.39%, indicating that PANI with a conjugated structure had the lowest photocatalytic degradation efficiency. After adding Sb$_2$S$_3$, the photocatalytic degradation efficiency of PAN/PANI–Sb$_2$S$_3$ increased to 85.41%. The remarkable improvement of photocatalytic performance might be because of the promoted carrier separation efficiency of the heterojunction structure in PAN/PANI–Sb$_2$S$_3$. The degradation efficiency of RhB dye with PAN/PANI–Sb$_2$S$_3$–ZnO further improved to 99.2%. The corresponding kinetic parameters are shown in Table S1. The determination coefficient values (R^2) proved that the photocatalytic reaction process of the composites was consistent with the pseudo-first order kinetic model [46,47]. The k_{app} values for PAN/PANI, PAN/PANI–Sb$_2$S$_3$, and PAN/PANI–Sb$_2$S$_3$–ZnO were 0.0323, 0.0429 and 0.0598 min^{-1}, respectively. The PAN/PANI–Sb$_2$S$_3$–ZnO had a maximum k_{app} value and the lowest $t_{1/2}$, which clarified the fastest photocatalytic degradation rate. Considering that the self-degradation of organic dyes, amoxicillin was selected as the target pollutant to further explore the photocatalytic degradation ability of PAN/PANI–Sb$_2$S$_3$–ZnO (see Figure S1). Under simulated sunlight irradiation for 40 min, the degradation efficiency of amoxicillin using PAN/PANI was only 38.12%. After adding Sb$_2$S$_3$, the photocatalytic degradation efficiency increased to 86.45%. The highest degradation efficiency of amoxicillin was obtained using PAN/PANI–Sb$_2$S$_3$–ZnO, with 94.13% degraded. The results showed that PAN/PANI–Sb$_2$S$_3$–ZnO had the best photodegradation ability for the pollutants.

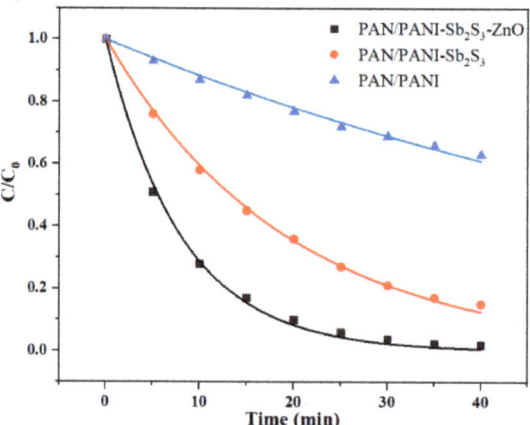

Figure 9. Kinetic curves from pseudo-first order kinetic model fitted to the experimental data of RhB photodegradation reaction using PAN/PANI, PAN/PANI–Sb_2S_3, and PAN/PANI–Sb_2S_3–ZnO.

Figure 10a shows the effect of pH on the degradation of RhB, MB, MO, and CR using PAN/PANI–Sb_2S_3–ZnO. With an increase in the pH value, the removal efficiency of RhB and MB improved and the removal efficiency of MO and CR decreased. The maximum removal efficiencies of 99.81% and 96.42% were achieved for RhB and MB at pH = 10.0, and the maximum removal efficiencies of 86.77% and 80.54% were achieved for MO and CR at pH = 2.0. As presented in Figure 10b, the isoelectric point of the PAN/PANI–Sb_2S_3–ZnO membrane was at pH = 5.0 (pH_{pzc} = 5.0). When pH < pH_{pzc}, PAN/PANI–Sb_2S_3–ZnO was positively charged. The removal efficiencies of anionic dyes were relatively high because of the electrostatic attraction between PAN/PANI–Sb_2S_3–ZnO and anionic dye. When pH > pH_{pzc}, PAN/PANI–Sb_2S_3–ZnO was negatively charged. The removal efficiencies of cationic dyes were relatively high because of the electrostatic attraction between PAN/PANI–Sb_2S_3–ZnO and cationic dye.

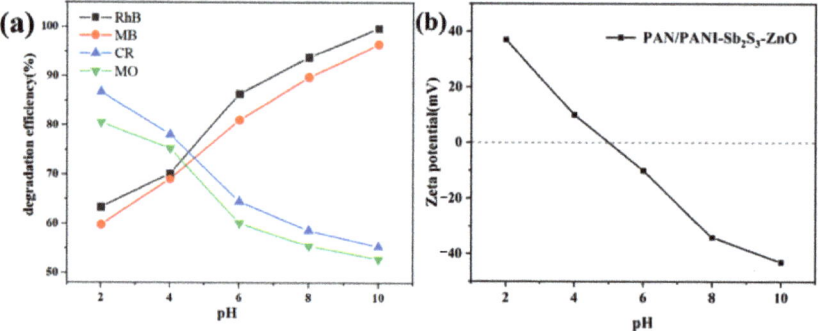

Figure 10. (a) Photocatalytic degradation efficiency of PAN/PANI–Sb_2S_3–ZnO photocatalyst for RhB, MB, CR and MO at different pH values and (b) dependence of the Zeta potential of the PAN/PANI–Sb_2S_3–ZnO membrane surface on pH.

Figure 11 shows the kinetic curves for RhB, MB, CR, and MO using PAN/PANI–Sb_2S_3–ZnO. Under 40 min simulated sunlight irradiation, the degradation efficiencies of PAN/PANI–Sb_2S_3–ZnO for RhB, MB, CR, and MO were 99.8%, 96.42%, 86.77%, and 78.54%. Obviously, the degradation efficiencies of PAN/PANI–Sb_2S_3–ZnO for cationic dyes (RhB and MB) were higher than those of anionic dyes (CR and MO). This was because the pH values of the four organic dye solutions were all greater than the pH_{pzc} of PAN/PANI–

Sb$_2$S$_3$–ZnO, so the membrane was electronegative in the dye solutions. Under the effect of electrostatic attraction, the negatively charged PAN/PANI–Sb$_2$S$_3$–ZnO was more favorable to fully contact with cationic dyes, which was conducive to the photocatalytic degradation process. The corresponding kinetic parameters are shown in Table S2. The k_{app} values of cationic RhB and MB were higher than those of anionic CR and MO, which further proved that PAN/PANI–Sb$_2$S$_3$–ZnO had a higher photocatalytic degradation rate for cationic dyes. PAN/PANI–Sb$_2$S$_3$–ZnO exhibited a higher photodegradation rate for RhB and MB than other reported heterostructure photocatalysts, as depicted in Table S3.

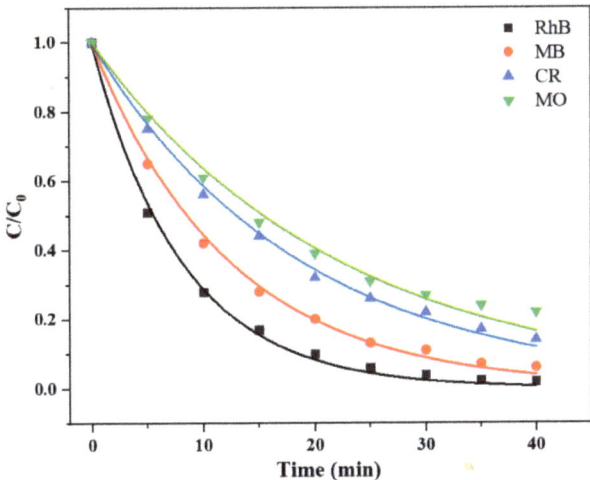

Figure 11. Kinetic curves from pseudo-first order kinetic model fitted to the experimental data of RhB, MB, CR, and MO photodegradation reaction under photocatalytic degradation using PAN/PANI–Sb$_2$S$_3$–ZnO. [Reaction conditions: initial concentration = 12 mg/L, photocatalyst = 0.2 g, pH value for MO and CR solutions = 2, pH value for RhB and MB solutions = 10, t = 40 min, simulated sunlight irradiation].

The Langmuir–Hinshelwood (L–H) model (presented in Supplementary Materials) was used to study the photodegradation kinetics of different initial concentrations of RhB using PAN/PANI–Sb$_2$S$_3$–ZnO. The validated linear relation for curves of $\ln(C_0/C_t)$ versus t are shown in Figure S1. The kinetic parameters of photocatalytic degradation are shown in Table S4, and the correlation coefficients (R^2) were all greater than 0.98. It could be seen that the k_{app} values decreased with the increasing initial concentration, which was possibly due to the difficulty of penetration of light irradiations into a large concentration gradient. When the initial concentration was 12 mg/L, the k_{app} values were the greatest.

The UV-vis absorbance changes of RhB, MB, CR, and MO dye solutions at different photocatalytic times are illustrated in Figure 12. The maximum UV-vis absorbances for RhB, MB, MO, and CR at 554 nm, 664 nm, 497 nm, and 460 nm were 2.2505, 1.9876, 2.9085, and 1.8492, respectively. With the increase of irradiation time, the UV-vis absorption peaks of the four organic dyes decreased. After 40 min simulated sunlight irradiation, the UV-vis absorbance of RhB, MB, CR, and MO decreased to 0.0426, 0.0592, 0.1975, and 0.3584, respectively, which proved that the organic dyes had been effectively degraded.

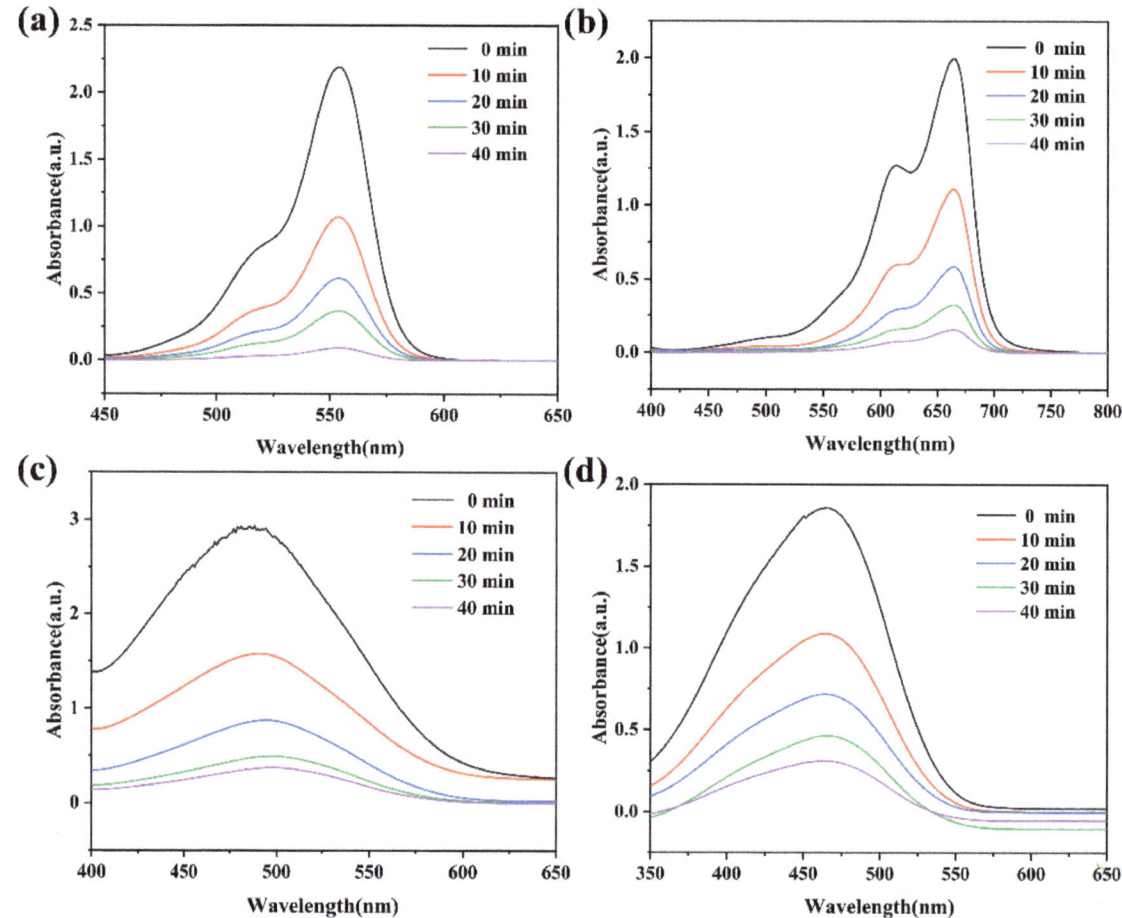

Figure 12. UV-vis absorbances of (**a**) RhB, (**b**) MB, (**c**) CR, and (**d**) MO with the variation in the irradiation time [Reaction conditions: initial concentration = 12 mg/L, photocatalyst = 0.2 g, solution pH of MO and CR: 2, solution pH of RhB and MB: 10, t = 40 min, simulated sunlight irradiation].

Figure 13 depicts the LC-MS analysis of RhB intermediates after photodegradation for 5 min, 10 min, 15 min, and 20 min. The *m/z* value determined the molecular weight of the intermediate product after photocatalytic degradation. It could be seen that after 5–20 min of photocatalytic degradation, the main peak of the *m/z* value of RhB dye decreased from 355.11 to 128.95. The RhB intermediates (*m/z* > 240) of larger molecular weight had been completely degraded into small molecules after 20 min of photocatalysis, which proved the occurrence of a photocatalytic redox reaction. It was concluded that the photocatalytic degradation process mainly involved the de-ethylation of RhB, the cracking and mineralization of chromophore [48].

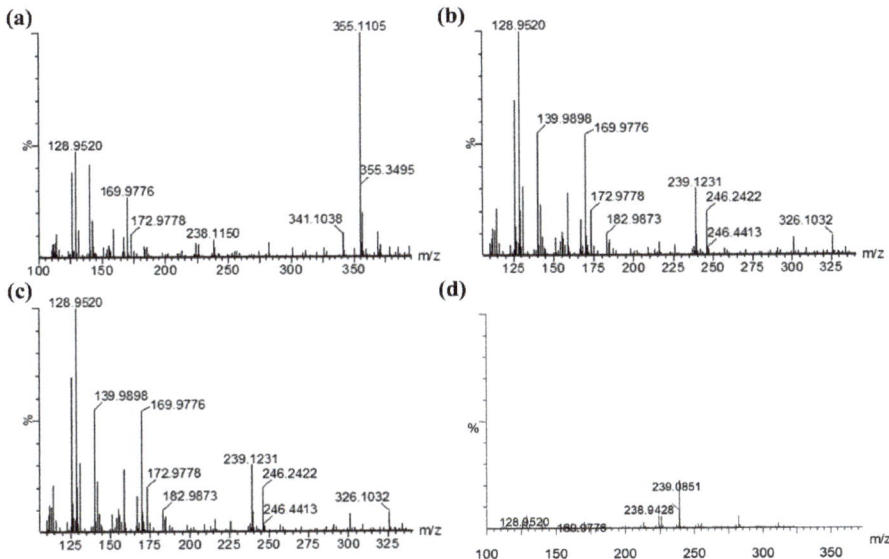

Figure 13. LC/MS spectra of RhB degradation with PAN/PANI–Sb$_2$S$_3$–ZnO after (**a**) 5 min, (**b**) 10 min, (**c**) 15 min, and (**d**) 20 min illumination.

2.5. The Photocatalytic Mechanism of PAN/PANI−Sb$_2$S$_3$−ZnO Membrane

The main active species produced by the PAN/PANI–Sb$_2$S$_3$–ZnO membrane during photocatalysis were investigated using free radical capture experiments, as shown in Figure 14a. •O$_2^-$, •OH, h$^+$, and e$^-$ were quenched by adding BQ, t-BuOH, OA, and AgNO$_3$, respectively [49–51]. It was observed that the photocatalytic degradation efficiency decreased after the addition of trapping agents. The dye removal efficiencies decreased to 45.64% and 52.9% after adding t-BuOH and BQ, which indicated that •OH and •O$_2^-$ were the main reactive species in the photocatalytic process. After adding OA and AgNO$_3$, the photocatalytic degradation effects were reduced to 76.03% and 87.64%, respectively, indicating that h$^+$ and e$^-$ played a supporting role in the photocatalytic reaction.

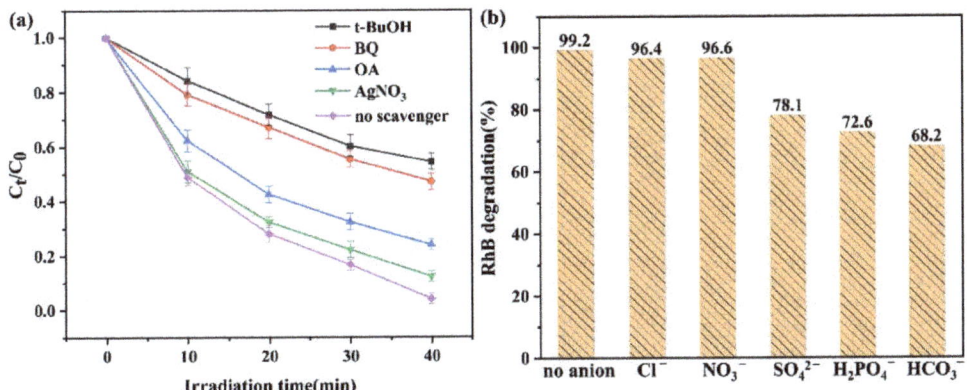

Figure 14. Degradation efficiency of PAN/PANI–Sb$_2$S$_3$–ZnO for RhB in the presence of (**a**) different scavengers and (**b**) different anions [Reaction conditions: initial concentration = 12 mg/L, photocatalyst = 0.2 g, pH value for RhB solutions = 10, t = 40 min, simulated sunlight irradiation].

The aqueous environment contained a variety of anions that had unpredictable effects on the photocatalytic activity. Thus, the influences of coexisting ions (NaCl (Cl$^-$), NaNO$_3$ (NO$_3^-$), Na$_3$PO$_4$ (H$_2$PO$_4^-$), Na$_2$SO$_4$ (SO$_4^{2-}$), and NaHCO$_3$ (HCO$_3^-$)), with a concentration of 30 mM in the PAN/PANI–Sb$_2$S$_3$–ZnO photocatalytic system, were evaluated. The dye degradation was slightly affected by Na$^+$ because of its inert nature [52]. As shown in Figure 14b, after the addition of Cl$^-$ and NO$_3^-$, the degradation efficiency of RhB did not decrease significantly. When SO$_4^{2-}$, H$_2$PO$_4^-$, and HCO$_3^-$ were added, the degradation efficiencies for RhB dye decreased to 78.1%, 72.6%, and 68.2%, respectively. According to the free radical trapping experiments, •OH and h$^+$ were active species in the photocatalytic reaction process. •OH or h$^+$ were consumed by SO$_4^{2-}$, H$_2$PO$_4^-$, and HCO$_3^-$, and produced •SO$_4^-$, •H$_2$PO$_4$, and •CO$_3^-$, as shown in Equations (1)–(5). Although the generated •SO$_4^-$, •H$_2$PO$_4$, and •CO$_3^-$ as active species could degrade organic dyes, their redox performance was lower than that of •OH and h$^+$ [53].

$$SO_4^{2-} + h^+ \rightarrow •SO_4^- \tag{1}$$

$$SO_4^{2-} + •OH \rightarrow •SO_4^- + OH^- \tag{2}$$

$$HCO_3^- + •OH \rightarrow •CO_3^- + H_2O \tag{3}$$

$$H_2PO_4^- + h^+ \rightarrow •H_2PO_4 \tag{4}$$

$$H_2PO_4^- + •OH \rightarrow •H_2PO_4 + OH^- \tag{5}$$

To better understand the heterojunction structure, Tauc plots and Mott–Schottky tests were conducted, as shown in Figure 15. The band gap energy was calculated using Equation (6):

$$(\alpha h\nu) = C(h\nu - E_g)^n \tag{6}$$

where α is the light absorption coefficient, $h\nu$ represents the incident photon energy, E_g is the band gap energy, and C is the light speed. The calculated E_g for PANI, Sb$_2$S$_3$, and ZnO was 2.80, 1.75, and 3.15 eV, respectively. The Mott–Schottky plots of PANI, Sb$_2$S$_3$, and ZnO are shown in Figure 16d–f. The flat band potential values (E_{fb}) of PANI, Sb$_2$S$_3$, and ZnO acquired from the x-intercept in Mott–Schottky plots were approximately 0.43, −0.72, and −0.47 (vs. Ag/AgCl reference). The normal hydrogen electrode potential (E_{NHE}) could be calculated with Equation (7):

$$E_{NHE} = E_{Ag/AgCl} + 0.197 \tag{7}$$

The highest occupied molecular orbital (HOMO) potential of PANI was equal to its standard hydrogen electrode potential, while the conduction band (CB) potential of Sb$_2$S$_3$ and ZnO was equal to the standard hydrogen electrode potential. Thus, the HOMO potential of PANI was 0.63 eV, and the CB potential of Sb$_2$S$_3$ and ZnO was −0.52 eV and −0.27 eV. According to Equation (8), the lowest unoccupied molecular orbital (LUMO) potential of PANI was −2.17 eV, and the valence band (VB) potential of Sb$_2$S$_3$ and ZnO was 1.23 eV and 2.88 eV, respectively.

$$E_{VB/HOMO} = E_{CB/LUMO} + E_g \tag{8}$$

Based on the above discussion, two possible photocatalytic mechanisms were proposed. It was assumed that the composite photocatalyst conformed to the e$^-$–h$^+$ migration mechanism of the traditional type II heterojunction, as shown in Figure 16a. Under simulated sunlight irradiation, the electrons in PANI, Sb$_2$S$_3$, and ZnO absorbed light energy and produced photogenerated e$^-$–h$^+$. The e$^-$ in the CB of PANI and Sb$_2$S$_3$ would migrate to the CB of ZnO, while h^+ in the VB of Sb_2S_3 and ZnO would migrate to the VB of PANI; thus, the separation efficiency of e^-–h^+ was promoted. However, the electrons and holes accumulated in the CB of ZnO and VB of PANI both reduced the generation of reactive oxygen species ($\bullet O_2^-$ and $\bullet OH$). At the same time, the CB position (−0.27 eV) was lower than the standard redox potential of $O_2/\bullet O_2^-$ (−0.33 eV), so the electrons accumulated in the CB of ZnO could not reduce O_2 to form $\bullet O_2^-$. Moreover, the addition of TEMPOL ($\bullet O_2^-$ scavenger) did not affect the degradation rate of RhB with pure ZnO (Figure S2), which further indicated that $\bullet O_2^-$ did not play a role in the degradation of RhB by ZnO. 5,5-dimethyl-1-pyrrolineN-oxide (DMPO) was used as the spin trap to detect the $\bullet O_2^-$ produced under light using electron spin resonance (ESR), which further verified that ZnO could not generate $\bullet O_2^-$. As shown in Figure S3, the signal of DMPO-$\bullet O_2^-$ was not observed in the systems with the photocatalysts of ZnO with light illumination, suggesting that no $\bullet O_2^-$ radicals were generated. In addition, the HOMO of PANI (0.63 eV) was lower than the standard redox potential of $H_2O/\bullet OH$ (2.7 eV) and $OH^-/\bullet OH$ (1.99 eV), so h^+ in PANI could not react with H_2O or OH^- to form $\bullet OH$. Therefore, according to the traditional type II heterojunction mechanism, e^- and h^+ were the main photocatalytic active species, which was inconsistent with the results of the free radical trapping experiments.

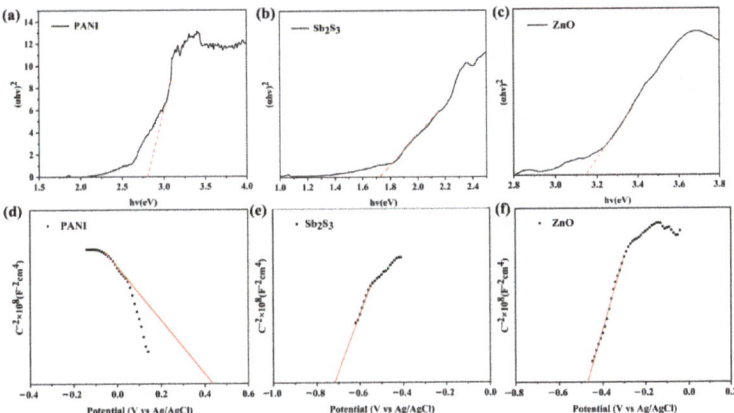

Figure 15. Tauc plots of (**a**) PANI, (**b**) Sb_2S_3, and (**c**) ZnO. Mott–Schottky plots of (**d**) PANI, (**e**) Sb_2S_3, and (**f**) ZnO.

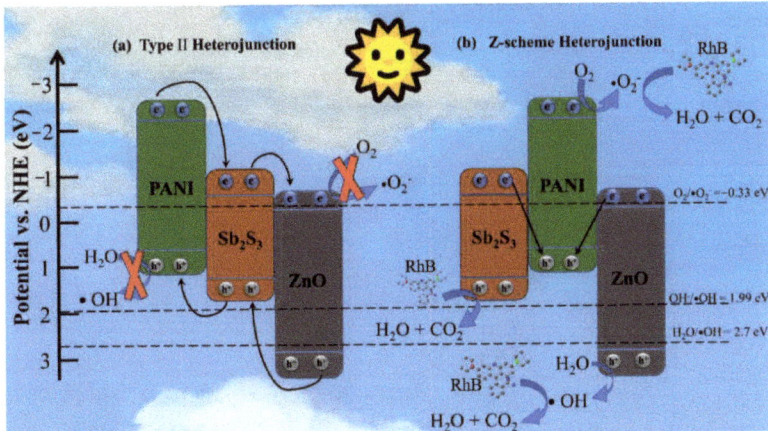

Figure 16. Schematic diagram of the organic dye removal mechanism. (**a**) traditional type-II heterojunction and (**b**) Z-scheme heterojunction charge transfer pathway.

Figure 16b shows the dual Z-type heterojunction mechanism of the composite photocatalyst. The e^- of Sb_2S_3 and ZnO would migrate to the VB of PANI and recombine with the h^+ in PANI; thus, h^+ in the VB of Sb_2S_3/ZnO and e^- in the CB of PANI could be preserved. The LUMO position of PANI was higher than the standard redox potential of $O_2/\bullet O_2^-$ (−0.33 eV), and e^- in the LUMO of PANI could react with O_2 to form $\bullet O_2^-$. The VB position of ZnO (2.88 eV) was higher than the standard redox potential of $H_2O/\bullet OH$ (2.7 eV) and $OH^-/\bullet OH$ (1.99 eV), so h^+ in the VB of ZnO could react with H_2O or OH^- to produce $\bullet OH$. $\bullet OH$ and $\bullet O_2^-$ were the main photocatalytic reactive species in the assumed dual Z-type heterojunction mechanism, which in agreement with the conclusion of the free radical trapping experiment.

2.6. Reusability and Stability of PAN/PANI–Sb_2S_3–ZnO

The degradation efficiency after recycling the photodegradation experiments five times is shown in Figure 17a. It could be seen that the degradation efficiency for RhB dye was only slightly reduced, indicating that the photocatalyst had great reusability. Figure 18b shows the FTIR of the PAN/PANI–Sb_2S_3–ZnO membrane before and after recycling the photodegradation experiments five times. It was observed that the FTIR peak had not significantly changed before and after the degradation reaction, indicating that PAN/PANI–Sb_2S_3–ZnO had good stability.

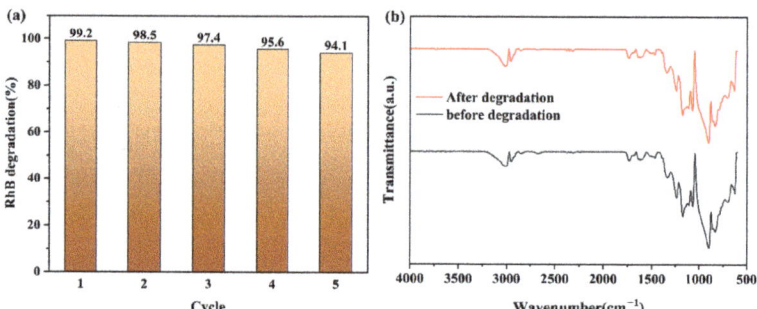

Figure 17. (a) Cyclic degradation efficiency of RhB and (b) FTIR before and after degradation of PAN/PANI–Sb_2S_3–ZnO composite film.

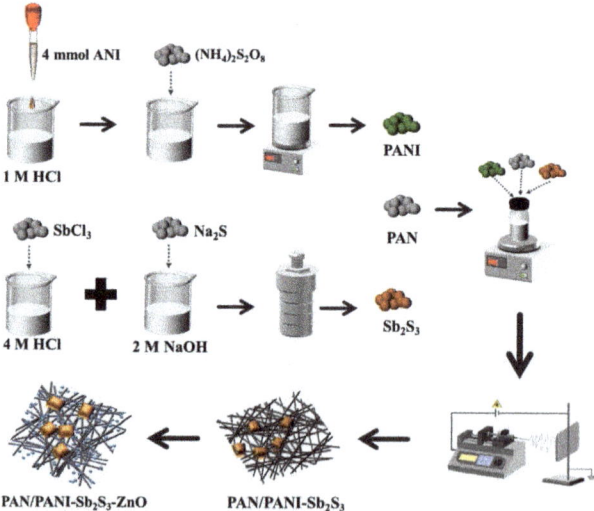

Figure 18. Schematic illustration of the preparation of composite membranes.

3. Experimental

3.1. Materials

Aniline (ANI), ammonium persulfate ($(NH_4)_2S_2O_8$), ZnO, Rhodamine B (RhB) (99%), methylene blue (MB) (99%), Congo red (CR) (99%), methyl orange (MO) (99%), $SbCl_3$, and $Na_2S \cdot 9H_2O$ were all purchased from Aladdin Industrial Corporation, Shanghai, China. Sodium hydroxide (NaOH), hydrochloric acid (HCl, 36~38%), N, N-dimethylformamide (DMF), polyacrylonitrile (PAN, Mw = 150,000 g/mol), and ethyl alcohol were obtained from Beijing Chemical Works. Benzoquinone (BQ), oxalic acid (OA), and tert-Butanol (t-BuOH) were ordered from Sinopharm Chemical Reagent Co. LTD (Shanghai, China). Deionized (DI) water was obtained from the laboratory. All of these chemicals were of analytical grade and utilized without further purification.

3.2. The Preparation of Composite Membranes

3.2.1. The Preparation of PANI

First, 4 mmol ANI was added drop by drop into 200 mL 1 M HCl solution and stirred for 1 h as solution A. Then, 4 mmol $(NH_4)_2S_2O_8$ was dissolved in 200 mL 1 M HCl and stirred for 1 h as solution B. Solution B was slowly added to solution A and stirred at 0 °C in ice bath for 12 h. Finally, the prepared PANI was washed with anhydrous ethanol and deionized water, and dried in oven at 60 °C.

3.2.2. The Preparation of Sb_2S_3

A total of 1.52 g $SbCl_3$ was dissolved in 50 mL solution of 4 M HCl and 2.4 g Na_2S was dissolved in 100 mL solution of 2 M NaOH. After stirring at room temperature for 30 min, $SbCl_3$/HCl solution was slowly added into Na_2S/NaOH solution, and stirring continued for 1 h. Then, the suspension was heated at 100 °C in a 200 mL Teflon-lined autoclave for 12 h. At room temperature, the obtained Sb_2S_3 was separated using filtration, washed with deionized water, and dried at 60 °C for 6 h.

3.2.3. The Preparation of PAN/PANI-Sb_2S_3 Fiber Membrane

PAN was selected to improve spinnability of PANI in this experiment. A total of 1 g PAN and 0.2 g PANI were dissolved in 10 mL DMF and stirred at room temperature for 12 h to obtain PAN/PANI electrospinning precursor solution. Then, 2% w/v Sb_2S_3 was added into PAN/PANI solution. The PAN/PANI-Sb_2S_3 precursor solution was obtained through stirring for 12 h. Before electrospinning, the precursor solution was ultrasonically treated for 1 h.

A 5 mL plastic syringe with an inner diameter of 0.6 mm was used as the spinneret and 10 cm × 10 cm aluminum foil was used as the receiver in the electrospinning process. Electrospinning was performed at a voltage of 16 kV, the distance between the syringe tip and the receiver was 15 cm, and the feeding rate was set to 1 mL/h. During the experiment, the ambient temperature was 20 ± 2 °C and the relative humidity was 10–20%.

3.2.4. The Preparation of PAN/PANI–Sb_2S_3–ZnO Membrane

To obtain the well-dispersed ZnO in PAN/PANI-Sb_2S_3, the surface ultrasonic method was used to prepare the PAN/PANI–Sb_2S_3–ZnO membrane. The PAN/PANI-Sb_2S_3 fiber membrane was immersed in the ZnO suspension (0.5% w/v) and ultrasound treatment for 1 h. The preparation process is shown in Figure 1.

3.3. Characterizations

The membranes' morphologies were observed using FE-SEM (JSM-6700 F, JEOL, Shimadzu Corporation, Kyoto, Japan). TEM (JEM-2100 F, JEOL, Kyoto Japan) was used to detected the internal structures and elements distribution of the membranes. The component structures of membranes were characterized using Fourier transform infrared spectroscopy (FTIR-4100, JASCO, Kyoto, Japan) in the wave number range of 400–4000 cm^{-1}. The crystal structure was analyzed using a Siemens D5000 X-ray diffractometer (XRD) using copper target Kα ray, 40 mA current, 40 kV voltage, and 4° min^{-1} scan speed (Shanghai Metash Instrument Co. Ltd., Shanghai, China). The surface chemical components of the membranes were examined using an X-ray photoelectron spectroscope (XPS, ESCALab220i-XL VG Science, Walser M, MA, USA). The absorption performance for dye solutions of the samples was evaluated using UV-vis spectroscopy (UV-6100S, MAPADA, Shanghai Metash Instrument Co. Ltd., Shanghai, China). The electron spin resonance (ESR) spectra data were obtained using a Bruker EPR A 300-10/12 spectrometer for tracing the active species. The pH values of dyes were adjusted in the range of 2.0–8.0 using HCl or NaOH. Pen-type pH meter (pH8180-0-00, Hong Kong Xima Instrument Technology Co. Ltd., Shanghai, China) was used to measure the pH value of dye solutions. The surface zeta potential of the membranes was determined using zeta potential analyzer (NanoZS90, London, UK). UV-vis diffuse reflectance spectroscopy (UV-vis DRS) was obtained on a Cary 500 spectrometer. The electrochemical properties of the membranes were investigated using electrochemical workstations (CHI660D, Beijing Merry Change Technology Co. Ltd., Beijing, China). The composite membranes, Pt nod, and Ag/AgCl electrode were used as the working, counter, and reference electrodes, respectively. The photoluminescence (PL) spectra and time-resolved photoluminescence spectra (TRPS) were measured with an Edinburgh FLS 980 apparatus at an excitation wavelength of 253 nm (Shimadzu, Kyoto, Japan). A xenon lamp source (PLS-SXE300+/300UV) with a spectral range of 320–780 nm was used to simulate sunlight (Beijing Merry Change Technology Co. Ltd., Beijing, China). Chromatographic separation of the aliquots was performed using an Agilent 1290 infinity LC (Agilent Technologies, Bellevue, WA, USA), an Agilent Eclipse Plus C18 column (2.1 mm × 50 mm, 1.8 µm), and a mobile phase composed of 0.1% FA in water (A) and 0.1% FA in ACN (B). The flow rate was 0.3 mL/min, and the injection volume was 1 µL, which was injected into the column using a thermostated HiP-ALS autosampler. For profiling and identification of by-products formed during the photocatalytic process, the separated peaks were analyzed using an Agilent 6550 QTOF (Agilent Technologies, Beijing, China) providing high-resolution mass measurement.

3.4. Photocatalytic Degradation of Dyes

A total of 12 mg of dyes (RhB, MB, CR, MO) was dissolved in 1 L of distilled water to obtain 12 mg/L dye concentration. The prepared fiber membrane (0.2 g) was immersed in 60 mL organic dye solution (12 mg L^{-1}) under dark condition for 30 min. Then, the entire experimental setup was irradiated under visible light source for 60 min, and the light intensity was approximately 20 mW (cm^2)$^{-1}$. The pH of dyes was adjusted using HCl or NaOH. A UV-vis spectrophotometer was used to measure the concentrations of organic dye solutions before and after degradation. The removal rate (R%) was calculated using the following Equation (9):

$$R(\%) = \frac{C_0 - C_t}{C_0} \times 100\% \tag{9}$$

where C_0 (mg L^{-1}) and C_t (mg L^{-1}) represent the initial and equilibrium concentration of organic dyes, respectively.

3.5. Kinetic Model Analysis

The nonlinear kinetic model of pseudo-first order (Equation (10)) was fitted to experimental data.

$$\frac{C_0}{C} = e^{-k_{app}t} \tag{10}$$

where C_0 is the dyes' initial concentration, C is the dyes' concentration at the time t (min), and k_{app} is the apparent first-order rate constant. From k_{app} the half-life time ($t_{1/2}$) was obtained using Equation (11), as follows:

$$t_{1/2} = \frac{ln2}{k_{app}} \tag{11}$$

4. Conclusions

In this work, the novel dual Z-scheme organic/inorganic PAN/PANI–Sb_2S_3–ZnO photocatalyst was designed using electrospinning and surface ultrasound. PAN/PANI–Sb_2S_3–ZnO exhibited a wide spectral absorption range from ultraviolet to visible light. Under 40 min irradiation, the PAN/PANI–Sb_2S_3–ZnO membrane almost achieved complete degradation for RhB, MB, MO, and CR dyes. The photocatalytic mechanism showed that PANI, Sb_2S_3, and ZnO formed a dual Z-type heterostructure, which not only achieved efficient carrier separation, but also retained photogenerated e^- and h^+ with strong redox capacity. It was also verified that •OH and •O_2^- were the main active species during the photocatalytic reaction. In addition, the PAN/PANI–Sb_2S_3–ZnO composite membrane also had high stability and reusability. This work provided a research strategy for constructing novel dual organic/inorganic photocatalysts with designable heterojunction structures.

Supplementary Materials: The following supporting information can be downloaded at: https://www.mdpi.com/article/10.3390/catal13111391/s1, Figure S1: Corresponding reaction rate diagram of RhB with different initial concentrations by PAN/PANI-Sb_2S_3-ZnO, Figure S2: The degradation of RhB by ZnO with or without TEMPOL. [Experimental conditions: Initial concentration = 12 mg/L, photocatalyst = 0.2 g, TEMPOL concentration = 10 Mm, pH value for RhB solutions = 10, t = 40 min, UV light irradiation, Figure S3: ESR spectra of radical adducts trapped by DMPO in ZnO for superoxide radical, Table S1: The pseudo-first order kinetic model kinetic parameters of RhB photocatalytic experiments, Table S2: The pseudo-first order kinetic model kinetic parameters of RhB, MB, CR and MO photocatalytic experiments by PAN/PANI-Sb_2S_3-ZnO, Table S3: Comparison of degradation rate of PAN/PANI-Sb_2S_3-ZnO with other photocatalysts toward RhB and MB, Table S4: L-H model kinetic parameters for different initial concentrations of RhB by PAN/PANI-Sb_2S_3-ZnO. References [54–59] are cited in the Supplementary Materials.

Author Contributions: T.Y.: data analysis and writing. D.W.: conceptualization, methodology, formal analysis, validation, data curation, and writing—original draft. L.L.: validation, resources, writing—review and editing, visualization, supervision, project administration, and funding acquisition. W.S.: data curation and visualization. C.L.: validation, resources, writing—review and editing, and visualization. X.P.: validation, resources, and writing—review and editing. All authors have read and agreed to the published version of the manuscript.

Funding: This research was supported by the Science and Technology Development Plan of Jilin Province (20230201152GX).

Data Availability Statement: All research data are included in this article.

Conflicts of Interest: The authors declare no conflict of interests.

References

1. Zhou, L.; Xiao, G.Q.; He, Y.; Wu, J.C.; Shi, H.; Yin, X.Y.; He, T.; Li, Z.Y.; Chen, J.Y. Multi-Functional Composite Membrane with Strong Photocatalysis to Effectively Separate Emulsified-Oil/Dyes from Complex Oily Sewage. *Colloids Surf. A-Physicochem. Eng. Asp.* **2022**, *643*, 13. [CrossRef]
2. Du, F.Q.; Yang, D.M.; Kang, T.X.; Ren, Y.X.; Hu, P.; Song, J.M.; Teng, F.; Fan, H.B. SiO_2/Ga_2O_3 Nanocomposite for Highly Efficient Selective Removal of Cationic Organic Pollutant Via Synergistic Electrostatic Adsorption and Photocatalysis. *Sep. Purif. Technol.* **2022**, *295*, 10. [CrossRef]
3. Cionti, C.; Pargoletti, E.; Falletta, E.; Bianchi, C.L.; Meroni, D.; Cappelletti, G. Combining Ph-Triggered Adsorption and Photocatalysis for the Remediation of Complex Water Matrices. *J. Environ. Chem. Eng.* **2022**, *10*, 9. [CrossRef]
4. Huang, X.F.; Xu, X.Y.; Yang, R.Y.; Fu, X.H. Synergetic adsorption and photocatalysis performance of g-C3N4/Ce-doped MgAl-LDH in degradation of organic dye under LED visible light. *Colloid Surf. A-Physicochem. Eng. Asp.* **2022**, *643*, 13. [CrossRef]
5. Lan, F.; Liu, C.S.; Zhou, C.; Huang, X.Z.; Wu, J.Y.; Zhang, X. Developing highly reducing conjugated porous polymer: A metal-free and recyclable approach with superior performance for pinacol C-C coupling under visible light. *J. Mater. Chem. A* **2022**, *10*, 16578–16584. [CrossRef]
6. Yuan, X.J.; Remita, H. Conjugated Polymer Polypyrrole Nanostructures: Synthesis and Photocatalytic Applications. *Top. Curr. Chem.* **2022**, *380*, 34. [CrossRef] [PubMed]
7. Wu, B.; Liu, Y.; Zhang, Y.X.; Fan, L.; Li, Q.Y.; Yu, Z.Y.; Zhao, X.S.; Zheng, Y.C.; Wang, X.J. Molecular engineering of covalent triazine frameworks for highly enhanced photocatalytic aerobic oxidation of sulfides. *J. Mater. Chem. A* **2022**, *10*, 12489–12496. [CrossRef]

8. Li, Z.L.; Fang, H.; Chen, Z.P.; Zou, W.X.; Zhao, C.X.; Yang, X.F. Regulating donor-acceptor interactions in triazine-based conjugated polymers for boosted photocatalytic hydrogen production. *Appl. Catal. B-Environ.* **2022**, *312*, 9. [CrossRef]
9. Kumar, H.; Luthra, M.; Punia, M.; Singh, D. Ag2O@PANI nanocomposites for advanced functional applications: A sustainable experimental and theoretical approach. *Colloid Surf. A-Physicochem. Eng. Asp.* **2022**, *640*, 12. [CrossRef]
10. Celebi, N.; Soysal, F.; Salimi, K. Dual-functional PANI/MIL-125(Ti) photoanodes for enhanced photoelectrochemical H-2 generation and photothermal heating. *Mater. Chem. Phys.* **2022**, *291*, 9. [CrossRef]
11. Lin, A.J.; Ren, M.; Tan, X.; Ma, J.; Zhang, Y.; Yang, T.X.; Pei, Y.S.; Cui, J. Visible-light-driven melamine foam/PANI/N,O,P containing covalent organic polymer for achieving H_2O_2 production and boosting Fe(II)/Fe(III) cycle. *J. Clean. Prod.* **2022**, *345*, 12. [CrossRef]
12. Sboui, M.; Niu, W.K.; Li, D.Z.; Lu, G.; Zhou, N.; Zhang, K.; Pan, J.H. Fabrication of electrically conductive TiO_2/PANI/PVDF composite membranes for simultaneous photoelectrocatalysis and microfiltration of azo dye from wastewater. *Appl. Catal. A-Gen.* **2022**, *644*, 10. [CrossRef]
13. Xu, D.; Huang, Y.; Ma, Q.; Qiao, J.Z.; Guo, X.; Wu, Y.Q. A 3D porous structured cellulose nanofibrils-based hydrogel with carbon dots-enhanced synergetic effects of adsorption and photocatalysis for effective Cr(VI) removal. *Chem. Eng. J.* **2023**, *456*, 11. [CrossRef]
14. Wang, F.; Xu, J.; Wang, Z.P.; Lou, Y.; Pan, C.S.; Zhu, Y.F. Unprecedentedly efficient mineralization performance of photocatalysis-self-Fenton system towards organic pollutants over oxygen-doped porous g-C_3N_4 nanosheets. *Appl. Catal. B-Environ.* **2022**, *312*, 9. [CrossRef]
15. Chen, Q.S.; Zhou, H.Q.; Wang, J.C.; Bi, J.H.; Dong, F. Activating earth-abundant insulator $BaSO_4$ for visible-light induced degradation of tetracycline. *Appl. Catal. B-Environ.* **2022**, *307*, 12. [CrossRef]
16. Xia, L.H.; Sun, Z.L.; Wu, Y.N.; Yu, X.F.; Cheng, J.B.; Zhang, K.S.; Sarina, S.; Zhu, H.Y.; Weerathunga, H.; Zhang, L.X.; et al. Leveraging doping and defect engineering to modulate exciton dissociation in graphitic carbon nitride for photocatalytic elimination of marine oil spill. *Chem. Eng. J.* **2022**, *439*, 15. [CrossRef]
17. Feng, Z.Y.; Zhu, X.W.; Yang, J.M.; Zhong, K.; Jiang, Z.F.; Yu, Q.; Song, Y.H.; Hua, Y.J.; Li, H.M.; Xu, H. Inherent Facet-Dominant effect for cobalt oxide nanosheets to enhance photocatalytic CO_2 reduction. *Appl. Surf. Sci.* **2022**, *578*, 9. [CrossRef]
18. Chen, Y.N.; Yang, L.; Sun, Y.N.; Guan, R.Q.; Liu, D.; Zhao, J.; Shang, Q.K. A high-performance composite CDs@Cu-HQCA/TiO_2 flower photocatalyst: Synergy of complex-sensitization, TiO_2-morphology control and carbon dot-surface modification. *Chem. Eng. J.* **2022**, *436*, 17. [CrossRef]
19. Zhang, Y.T.; Zhang, T.; Jia, J.; Lin, G.; Li, K.K.; Zheng, L.S.; Li, X.; Kong, Z.H. Construction of $Zn_{0.2}Cd_{0.8}S$/g-C_3N_4 nanosheet array heterojunctions toward enhanced photocatalytic reduction of CO_2 in visible light. *Colloid Surf. A-Physicochem. Eng. Asp.* **2022**, *655*, 10. [CrossRef]
20. Peng, Q.; Tang, X.K.; Liu, K.; Zhong, W.L.; Zhang, Y.J.; Xing, J.J. Synthesis of silica nanofibers-supported BiOCl/TiO_2 heterojunction composites with enhanced visible-light photocatalytic performance. *Colloid Surf. A-Physicochem. Eng. Asp.* **2022**, *652*, 11. [CrossRef]
21. Yang, H.; Hao, H.S.; Zhao, Y.R.; Hu, Y.T.; Min, J.K.; Zhang, G.L.; Bi, J.R.; Yan, S.; Hou, H.M. An efficient construction method of S-scheme Ag_2CrO_4/$ZnFe_2O_4$ nanofibers heterojunction toward enhanced photocatalytic and antibacterial activity. *Colloid Surf. A-Physicochem. Eng. Asp.* **2022**, *641*, 10. [CrossRef]
22. Xu, F.; Chai, B.; Liu, Y.Y.; Liu, Y.L.; Fan, G.Z.; Song, G.S. Superior photo-Fenton activity toward tetracycline degradation by 2D?-Fe_2O_3 anchored on 2D g-C_3N_4: S-scheme heterojunction mechanism and accelerated Fe^{3+}/Fe^{2+} cycle. *Colloid Surf. A-Physicochem. Eng. Asp.* **2022**, *652*, 14. [CrossRef]
23. Ge, L.; Han, C.C.; Liu, J. In situ synthesis and enhanced visible light photocatalytic activities of novel PANI-g-C_3N_4 composite photocatalysts. *J. Mater. Chem.* **2012**, *22*, 11843–11850. [CrossRef]
24. Wang, Q.Z.; Hui, J.; Li, J.J.; Cai, Y.X.; Yin, S.Q.; Wang, F.P.; Su, B.T. Photodegradation of methyl orange with PANI-modified BiOCl photocatalyst under visible light irradiation. *Appl. Surf. Sci.* **2013**, *283*, 577–583. [CrossRef]
25. Chen, Y.Q.; Cheng, Y.F.; Zhao, J.F.; Zhang, W.W.; Gao, J.H.; Miao, H.; Hu, X.Y. Construction of Sb_2S_3/CdS/$CdIn_2S_4$ cascaded S-scheme heterojunction for improving photoelectrochemical performance. *J. Colloid Interface Sci.* **2022**, *627*, 1047–1060. [CrossRef] [PubMed]
26. Wang, F.; Yang, C.L.; Wang, M.S.; Ma, X.G. Photocatalytic hydrogen evolution reaction with high solar-to-hydrogen efficiency driven by the Sb_2S_3 monolayer and RuI_2/Sb_2S_3 heterostructure with solar light. *J. Power Sources* **2022**, *532*, 10. [CrossRef]
27. Cheng, Y.F.; Gong, M.; Xu, T.T.; Liu, E.Z.; Fan, J.; Miao, H.; Hu, X.Y. Epitaxial Grown Sb_2Se_3@Sb_2S_3 Core-Shell Nanorod Radial-Axial Hierarchical Heterostructure with Enhanced Photoelectrochemical Water Splitting Performance. *ACS Appl. Mater. Interfaces* **2022**, *14*, 23785–23796. [CrossRef]
28. Xu, W.; Xu, L.H.; Pan, H.; Wang, L.M.; Shen, Y. Superamphiphobic cotton fabric with photocatalysis and ultraviolet shielding property based on hierarchical ZnO/halloysite nanotubes hybrid particles. *Colloid Surf. A-Physicochem. Eng. Asp.* **2022**, *654*, 11. [CrossRef]
29. Meenakshi, G.; Sivasamy, A. Enhanced photocatalytic activities of CeO_2@ZnO core-shell nanostar particles through delayed electron hole recombination process. *Colloid Surf. A-Physicochem. Eng. Asp.* **2022**, *645*, 13. [CrossRef]
30. Ramos, P.G.; Sanchez, L.A.; Rodriguez, J.M. A review on improving the efficiency of photocatalytic water decontamination using ZnO nanorods. *J. Sol-Gel Sci. Technol.* **2022**, *102*, 105–124. [CrossRef]

31. Gao, Y.; Yan, N.; Jiang, C.; Xu, C.; Yu, S.; Liang, P.; Zhang, X.; Liang, S.; Huang, X. Filtration-enhanced highly efficient photocatalytic degradation with a novel electrospun rGO@TiO$_2$ nanofibrous membrane: Implication for improving photocatalytic efficiency. *Appl. Catal. B Environ.* **2020**, *268*, 118737. [CrossRef]
32. Selvinsimpson, S.; Gnanamozhi, P.; Pandiyan, V.; Govindasamy, M.; Habila, M.A.; AlMasoud, N.; Chen, Y. Synergetic effect of Sn doped ZnO nanoparticles synthesized via ultrasonication technique and its photocatalytic and antibacterial activity. *Environ. Res.* **2021**, *197*, 111115. [CrossRef] [PubMed]
33. Qu, C.; Zhao, P.; Wu, C.D.; Zhuang, Y.; Liu, J.M.; Li, W.H.; Liu, Z.; Liu, J.H. Electrospun PAN/PANI fiber film with abundant active sites for ultrasensitive trimethylamine detection. *Sens. Actuator B-Chem.* **2021**, *338*, 9. [CrossRef]
34. Dashairya, L.; Sharma, M.; Basu, S.; Saha, P. Enhanced dye degradation using hydrothermally synthesized nanostructured Sb$_2$S$_3$/rGO under visible light irradiation. *J. Alloys Compd.* **2018**, *735*, 234–245. [CrossRef]
35. Lopez-Lopez, J.; Tejeda-Ochoa, A.; Lopez-Beltran, A.; Herrera-Ramirez, J.; Mendez-Herrera, P. Sunlight Photocatalytic Performance of ZnO Nanoparticles Synthesized by Green Chemistry Using Different Botanical Extracts and Zinc Acetate as a Precursor. *Molecules* **2022**, *27*, 17. [CrossRef] [PubMed]
36. Wang, S.; Zhao, Y.; Zhang, M.Y.; Feng, J.; Wei, T.; Ren, Y.M.; Ma, J. Electrostatic self-assembled layered polymers form supramolecular heterojunction catalyst for photocatalytic reduction of high-stability nitrate in water. *J. Colloid Interface Sci.* **2022**, *622*, 828–839. [CrossRef]
37. Hosny, M.; Fawzy, M.; Eltaweil, A.S. Green synthesis of bimetallic Ag/ZnO@Biohar nanocomposite for photocatalytic degradation of tetracycline, antibacterial and antioxidant activities. *Sci. Rep.* **2022**, *12*, 17. [CrossRef]
38. Zhang, H.L.; Hu, C.G.; Ding, Y.; Lin, Y. Synthesis of 1D Sb$_2$S$_3$ nanostructures and its application in visible-light-driven photodegradation for MO. *J. Alloys Compd.* **2015**, *625*, 90–94. [CrossRef]
39. Yasin, M.; Saeed, M.; Muneer, M.; Usman, M.; ul Haq, A.; Sadia, M.; Altaf, M. Development of Bi$_2$O$_3$-ZnO heterostructure for enhanced photodegradation of rhodamine B and reactive yellow dyes. *Surf. Interfaces* **2022**, *30*, 12. [CrossRef]
40. Wang, C.; Wang, L.; Jin, J.; Liu, J.; Li, Y.; Wu, M.; Chen, L.H.; Wang, B.J.; Yang, X.Y.; Su, B.L. Probing effective photocorrosion inhibition and highly improved photocatalytic hydrogen production on monodisperse PANI@CdS core-shell nanospheres. *Appl. Catal. B-Environ.* **2016**, *188*, 351–359. [CrossRef]
41. Li, F.; Zhang, L.L.; Hu, C.; Xing, X.C.; Yan, B.; Gao, Y.W.; Zhou, L. Enhanced azo dye decolorization through charge transmission by sigma-Sb3+-azo complexes on amorphous Sb$_2$S$_3$ under visible light irradiation. *Appl. Catal. B-Environ.* **2019**, *240*, 132–140. [CrossRef]
42. Sun, J.X.; Yuan, Y.P.; Qiu, L.G.; Jiang, X.; Xie, A.J.; Shen, Y.H.; Zhu, J.F. Fabrication of composite photocatalyst g-C$_3$N$_4$-ZnO and enhancement of photocatalytic activity under visible light. *Dalton Trans.* **2012**, *41*, 6756–6763. [CrossRef] [PubMed]
43. Wang, X.; Zhu, J.Q.; Yu, X.; Fu, X.H.; Zhu, Y.; Zhang, Y.M. Enhanced removal of organic pollutant by separable and recyclable rGH-PANI/BiOI photocatalyst via the synergism of adsorption and photocatalytic degradation under visible light. *J. Mater. Sci. Technol.* **2021**, *77*, 19–27. [CrossRef]
44. Dou, W.Y.; Hu, X.Y.; Kong, L.H.; Peng, X.J. Photo-induced dissolution of Bi2O3 during photocatalysis reactions: Mechanisms and inhibition method. *J. Hazard. Mater.* **2021**, *412*, 10. [CrossRef] [PubMed]
45. Liu, T.T.; Wang, Z.; Wang, X.R.; Yang, G.H.; Liu, Y. Adsorption-photocatalysis performance of polyaniline/dicarboxyl acid cellulose@graphene oxide for dye removal. *Int. J. Biol. Macromol.* **2021**, *182*, 492–501. [CrossRef]
46. Mallappa, M.; Nagaraju, K.; Shivaraj, Y. Enhanced Photocatalytic degradation of methylene blue dye using CuS-CdS nanocomposite under visible light irradiation. *Appl. Surf. Sci.* **2019**, *475*, 828–838. [CrossRef]
47. Dogar, S.; Nayab, S.; Farooq, M.Q.; Said, A.; Kamran, R.; Duran, H.; Yameen, B. Utilization of Biomass Fly Ash for Improving Quality of Organic Dye-Contaminated Water. *ACS Omega* **2020**, *5*, 15850–15864. [CrossRef]
48. Liang, C.; Cui, M.Y.; Zhao, W.; Dong, L.Y.; Ma, S.S.; Liu, X.T.; Wang, D.K.; Jiang, Z.J.; Wang, F. Hybridizing electron-mediated H5PMo10V2O40 with CdS/g-C$_3$N$_4$ for efficient photocatalytic performance of Z-scheme heterojunction in wastewater treatment. *Chemosphere* **2022**, *305*, 12. [CrossRef]
49. Wen, Y.; Wang, Z.W.; Cai, Y.H.; Song, M.X.; Qi, K.M.; Xie, X.Y. S-scheme BiVO4/CQDs/beta-FeOOH photocatalyst for efficient degradation of ofloxacin: Reactive oxygen species transformation mechanism insight. *Chemosphere* **2022**, *295*, 13. [CrossRef]
50. Zhou, X.Y.; Wang, T.Y.; Zhang, L.; Che, S.Y.; Liu, H.; Liu, S.X.; Wang, C.Y.; Su, D.W.; Teng, Z.Y. Highly efficient Ag$_2$O/Na-g-C$_3$N$_4$ heterojunction for photocatalytic desulfurization of thiophene in fuel under ambient air conditions. *Appl. Catal. B-Environ.* **2022**, *316*, 13. [CrossRef]
51. Ma, L.J.; Wu, H.Q.; Chen, B.Y.; Wang, G.; Lei, B.X.; Zhang, D.; Kuang, D.B. 0D/2D CsPbBr$_3$ Nanocrystal/BiOCl Nanoplate Heterostructure with Enhanced Photocatalytic Performance. *Adv. Mater. Interfaces* **2022**, *9*, 9. [CrossRef]
52. Borthakur, P.; Boruah, P.K.; Hussain, N.; Silla, Y.; Das, M.R. Specific ion effect on the surface properties of Ag/reduced graphene oxide nanocomposite and its influence on photocatalytic efficiency towards azo dye degradation. *Appl. Surf. Sci.* **2017**, *423*, 752–761. [CrossRef]
53. Zhou, H.; Qiu, Y.; Yang, C.; Zang, J.; Song, Z.; Yang, T.; Li, J.; Fan, Y.; Dang, F.; Wang, W. Efficient Degradation of Congo Red in Water by UV-Vis Driven CoMoO$_4$/PDS Photo-Fenton System. *Molecules* **2022**, *27*, 8642. [CrossRef] [PubMed]
54. Liang, X.; Liu, J.; Guo, H.; Li, H.; Liu, E.; Zhao, Y.; Ji, Y.; Fan, J. Preparation of a Recyclable and High-Performance Photocatalyst Agins2/Cn/Pan for Rhb and Phenol Degradation. *J. Environ. Chem. Eng.* **2023**, *11*, 3. [CrossRef]

55. Liang, Z.; Cheng, H.; Zhang, X.; Mao, Q. Two Polyoxometalates Based on {P2mo5} Catalysts: Synthesis, Characterization, and Photocatalytic Degradation of Rhb. *J. Mol. Liq.* **2023**, *377*, 121483. [CrossRef]
56. Liu, S.; Ge, Y.; Wang, C.; Li, K.; Mei, Y. Tio2/Bp/G-C3n4 Heterojunction Photocatalyst for the Enhanced Photocatalytic Degradation of Rhb. *Environ. Sci. Pollut. Res.* **2023**, *30*, 84452–84461. [CrossRef] [PubMed]
57. Smok, W.; Zaborowska, M.; Tański, T.; Radoń, A. Novel In2o3/Sno2 Heterojunction 1d Nanostructure Photocatalyst for Mb Degradation. *Opt. Mater.* **2023**, *139*, 113757. [CrossRef]
58. Athavale, S.; Barai, D.P.; Bhanvase, B.A.; Pandharipande, S.L. Investigation on Performance of Eu:Go:Ksrpo4 Nanocomposite for Mb Dye Photocatalytic Degradation: Experimental and Ann Modeling. *Optik* **2023**, *275*, 170561. [CrossRef]
59. Rastgar, S.; Rezaei, H.; Younesi, H.; Abyar, H.; Kordrostami, A. Photocatalytic Degradation of Methylene Blue (Mb) Dye under Uv Light Irradiation by Magnetic Diesel Tank Sludge (Mdts). *Biomass Convers. Biorefin.* **2023**, *in press*. [CrossRef]

Disclaimer/Publisher's Note: The statements, opinions and data contained in all publications are solely those of the individual author(s) and contributor(s) and not of MDPI and/or the editor(s). MDPI and/or the editor(s) disclaim responsibility for any injury to people or property resulting from any ideas, methods, instructions or products referred to in the content.

Review

S-Scheme Heterojunction Photocatalyst for Photocatalytic H₂O₂ Production: A Review

Weili Fang and Liang Wang *

Key Laboratory of Marine Chemistry Theory and Technology, Ministry of Education,
College of Chemistry and Chemical Engineering, Ocean University of China, Qingdao 266100, China;
17854290741@163.com
* Correspondence: wangliangouc@ouc.edu.cn

Abstract: Hydrogen peroxide (H_2O_2) is a clean and mild oxidant that is receiving increasing attention. The photocatalytic H_2O_2 production process utilizes solar energy as an energy source and H_2O and O_2 as material sources, making it a safe and sustainable process. However, the high recombination rate of photogenerated carriers and the low utilization of visible light limit the photocatalytic production of H_2O_2. S-scheme heterojunctions can significantly reduce the recombination rate of photogenerated electron–hole pairs and retain a high reduction and oxidation capacity due to the presence of an internal electric field. Therefore, it is necessary to develop S-scheme heterojunction photocatalysts with simple preparation methods and high performance. After a brief introduction of the basic principles and advantages of photocatalytic H_2O_2 production and S-scheme heterojunctions, this review focuses on the design and application of S-scheme heterojunction photocatalysts in photocatalytic H_2O_2 production. This paper concludes with a challenge and prospect of the application of S-scheme heterojunction photocatalysts in photocatalytic H_2O_2 production.

Keywords: semiconductors; hydrogen peroxide; S-scheme heterojunction; photocatalysis

1. Introduction

Since its first synthesis in 1818 by Thenard [1], hydrogen peroxide (H_2O_2) has been considered a promising liquid fuel and a green oxidizer for a wide range of energy, environmental and chemical synthesis applications [2–4]. Currently, the anthraquinone (AQ) method dominates H_2O_2 production, accounting for about 95% of global H_2O_2 output [3,5]. Despite the maturity of AQ oxidation technology, it suffers from drawbacks such as high energy consumption, dangerous operation and pollution to the environment. In addition, direct synthesis of H_2O_2 using H_2 and O_2 can mitigate environmental concerns [6]. However, this method is cost-prohibitive, lacks selectivity for H_2O_2 and is prone to explosion [7]. Thus, there is a pressing need to discover an environmentally friendly and efficient H_2O_2 production method.

Solar energy is a clean and sustainable source of energy. Since Fujishima and Honda discovered the photo-assisted oxidation of water on TiO_2 electrodes in 1972 [8], semiconductor photocatalysis has been applied in several research fields [9–11]. Photocatalytic H_2O_2 production is a safe and green process using renewable solar energy as an energy source and resource-rich H_2O and O_2 as raw materials. In the long run, photocatalytic H_2O_2 production has great potential in environmental pollution treatment [12]. As shown in Figure 1, a great number of relevant studies have emerged in the field of photocatalytic H_2O_2 production in recent years [2,5,13–21]. However, the low visible light utilization and low solar energy conversion efficiency seriously hinder its commercial feasibility. So far, researchers have adopted various modification methods to enhance the efficiency of photocatalytic H_2O_2 production, such as doping [22,23], vacancy engineering [24], surface engineering [25], nanoparticle deposition [26] and heterojunction construction [17,27,28],

Citation: Fang, W.; Wang, L. S-Scheme Heterojunction Photocatalyst for Photocatalytic H₂O₂ Production: A Review. *Catalysts* **2023**, *13*, 1325. https://doi.org/10.3390/catal13101325

Academic Editors: Weilong Shi, Feng Guo, Xue Lin and Yuanzhi Hong

Received: 6 September 2023
Revised: 23 September 2023
Accepted: 24 September 2023
Published: 27 September 2023

Copyright: © 2023 by the authors. Licensee MDPI, Basel, Switzerland. This article is an open access article distributed under the terms and conditions of the Creative Commons Attribution (CC BY) license (https://creativecommons.org/licenses/by/4.0/).

as well as combinations of two or more of these methods [29]. Thus, there is a pressing need to discover an environmentally friendly and efficient H_2O_2 production method.

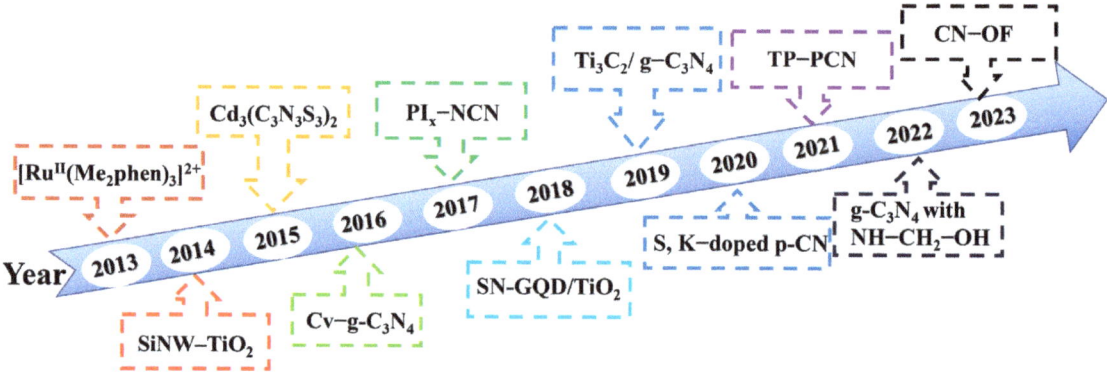

Figure 1. Representative photocatalysts for photocatalytic production of H_2O_2 in the last decade [2,5,13–21].

Mechanisms such as type-II, Z-scheme and S-scheme mechanisms are the most common in the literature used to describe the charge transfer in heterojunction structures. Although type-II heterojunctions can improve the separation efficiency of photogenerated carriers, they also sacrifice the charge of the strong redox potential, resulting in reduced redox capacity. Z-scheme photocatalysts, initially proposed by Bard in 1979, have found application in photocatalytic H_2O_2 production due to their effective charge separation and robust redox capabilities [30]. For instance, Cheng et al. [31] synthesized Z-scheme $Ag/ZnFe_2O_4$–Ag–Ag_3PO_4 composites for photocatalytic H_2O_2 production, which was generated by a continuous two-step one-electron oxygen reduction. Nevertheless, there is still some confusion about the mechanism of Z-scheme heterojunctions. Addressing the limitations inherent in type-II and Z-scheme mechanisms, Yu's team introduced the concept of S-scheme heterojunctions in 2019 [32]. The S-scheme heterojunction is composed of a reduction semiconductor and an oxidation semiconductor, which can be a p-type or n-type semiconductor. Efficient photogenerated carrier migration is achieved by the built-in electric field (IEF) at the interface of the different semiconductors, thus maintaining a high redox capacity [33,34]. In the past few years, S-scheme heterojunctions have attracted unprecedented attention because of their excellent photocatalytic activity. They are widely utilized in the fields of photocatalytic CO_2 reduction [35–39], photocatalytic H_2 production [40–44], photocatalytic H_2O_2 production [45] and other applications [46–49].

In this comprehensive review, we have undertaken a multi-faceted exploration of photocatalytic H_2O_2 production and the pivotal role played by S-scheme heterojunctions. Our journey commenced with an elucidation of the fundamental mechanism governing photocatalytic H_2O_2 production, followed by an in-depth analysis of the latest advancements in S-scheme heterojunctions employed within this context. Notably, recent years have witnessed remarkable progress in S-scheme heterojunction research, a modification strategy that holds immense potential for elevating photocatalyst activity and, consequently, the yield of photocatalytic H_2O_2 production. Our objective is to provide an in-depth reference on the H_2O_2 production system of S-scheme heterojunctions to stimulate new inspirations and promote the industrialization of photocatalytic H_2O_2 production.

2. Mechanism of Photocatalytic H_2O_2 Production Reaction

In general, the process of photocatalytic H_2O_2 production consists of three main steps (Figure 2). In the first step, when the absorbed photon energy of the semiconductor is greater than its band gap (E_g), electrons are excited and jump from the valence band (VB) to the conduction band (CB), while the hole remains in the VB, resulting in photogenerated electron–hole pairs. In the second step, the photogenerated electrons and holes separate and migrate, accompanied by the recombination of photogenerated electrons and holes, only a few of which can migrate to the surface of photocatalyst. In the last step, the electrons and holes migrating to the surface of the photocatalyst are involved in oxidation and reduction reactions, respectively. There are two main pathways for the synthesis of H_2O_2: oxygen reduction reaction (ORR) and water oxidation reaction (WOR).

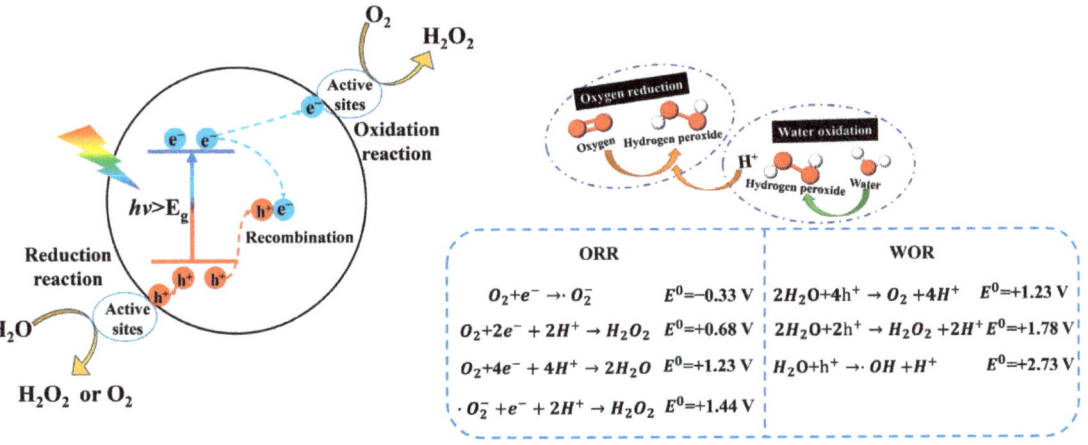

Figure 2. Schematic diagram of photocatalytic H_2O_2 production process.

The reaction potentials of photocatalytic H_2O_2 production are shown in Figure 2. Currently, ORR can be divided into two-step single-electron reduction ($O_2 \rightarrow \cdot O_2^- \rightarrow H_2O_2$) and direct one-step double-electron reduction ($O_2 \rightarrow H_2O_2$) routes, where the protons are mainly derived from the decomposition of H_2O. Since the potential of $O_2/\cdot O_2^-$ (-0.33 V) is much more negative than that of O_2/H_2O_2 (0.68 V), it requires a more negative CB position of the photocatalyst, which unavoidably increases the band gap of the photocatalyst. In general, narrow-band-gap photocatalysts are more utilized to increase their light absorption ability. Therefore, it is necessary to modify the ORR route to a one-step double-electron reaction. However, the presence of the four-electron oxygen reduction reaction makes the photocatalytic production of H_2O_2 less selective.

The WOR pathway is a way to synthesize H_2O_2 by using photogenerated holes (h^+) in the photocatalytic H_2O_2 production process. Similar to the ORR pathway, the WOR pathway can also be divided into two-electron WOR (direct two-electron and indirect two-electron) pathways and a four-electron WOR pathway. As shown in Figure 2, in the direct two-electron WOR pathway, the h^+ can directly oxidize H_2O to H_2O_2 in a one-step two-electron reaction. In addition, in the indirect two-electron reaction, the h^+ can first oxidize H_2O to hydroxyl radicals ($\cdot OH$) and then form H_2O_2 by coupling two $\cdot OH$. Theoretically, the direct two-electron WOR pathway requires a 1.76 V positive valence band (VB) potential of the photocatalyst, while the indirect two-electron WOR pathway requires a 2.73 V positive VB potential. The direct two-electron WOR pathway is thermodynamically more favorable but kinetically unfavorable compared to the indirect two-electron WOR pathway. Similar to the ORR pathway, the WOR pathway also results in low selectivity of H_2O_2 because of the competitive reaction of the four-electron WOR pathway.

In general, photocatalysts can be designed in such a way that H_2O_2 can be produced simultaneously by both two pathways. The dual-channel pathway integrating the ORR and WOR pathways produces H_2O_2 via O_2 and H_2O without the addition of sacrificial agents and achieves 100% atomic utilization. In addition, photocatalytic H_2O_2 production is usually accompanied by the decomposition of H_2O_2. In order to improve the yield and selectivity of H_2O_2 in photocatalytic process, it is essential to prepare photocatalysts with suitable band gaps to provide high redox potential, high separation efficiency of photogenerated charges and excellent visible light absorption performance. To date, the performance of photocatalytic H_2O_2 production has been improved by such modification methods as elemental doping [19,50], morphology modulation [51], deposition of noble metals [52], vacancy engineering [53,54] and construction of heterojunctions [55–57]. Among them, the construction of heterojunctions shows excellent photocatalytic activities because it can induce the maximum separation of photogenerated carriers. Considering this, in the next section, we focus on S-scheme heterojunctions.

3. S-Scheme Heterojunctions

3.1. Mechanism of S-Scheme Heterojunctions

The separation efficiency of photogenerated carriers is an important factor for photocatalysts. In order to avoid the compounding of photogenerated carriers in a single photocatalyst, two photocatalysts were combined to enhance the photocatalytic activities. As shown in Figure 3a, in a type-II heterojunction, photogenerated carriers are generated in each of the two semiconductors under the irradiation of light. The photogenerated electrons and photogenerated holes migrate in opposite directions and aggregate on different semiconductors, thus achieving spatial separation. Although the effective separation of photogenerated carriers is achievable in type-II heterojunctions, this charge transfer reduces the redox ability of the photocatalyst. Moreover, kinetically, the presence of Coulomb repulsion inhibits this charge transfer route.

Z-scheme heterojunctions mainly include traditional Z-scheme, all-solid-state Z-scheme and direct Z-scheme heterojunctions (Figure 3b). Traditional Z-scheme and all-solid-state Z-scheme heterojunctions need to be bonded by an electron acceptor and an electron donor or a metal conductor. Thereby, electron–hole pairs with high redox capacity react with shuttling redox ion pairs or, in all-solid-state Z-scheme heterojunctions, burst each other due to greater thermodynamic driving forces [58]. Direct Z-scheme heterojunctions are derived from traditional Z-scheme and all-solid-state Z-scheme heterojunctions [59]. In a direct Z-scheme heterojunction, when two semiconductors are in contact, due to the Fermi-level difference between them, positive and negative charges collect in the interface region near the two semiconductors, resulting in an internal electric field (IEF). Photogenerated electrons are transferred from the CB of one semiconductor to the VB of the other semiconductor under the action of the IEF, as illustrated in Figure 3b. However, the term "Z-scheme heterojunction" is associated with considerable confusion, theoretical immaturity and problems. In consideration of the above disadvantages, a new charge transfer mechanism needs to be introduced to explain the charge transfer process in heterojunction photocatalysts. Thus, in 2019, Fu et al. [32] presented an S-scheme heterojunction with a similar structure to that of type-II heterojunctions which compensated for the shortcomings of Z-scheme heterojunctions [60]. As shown in Figure 3c, a S-scheme heterojunction is a coupling of an oxidizing photocatalyst (OP) and a reducing photocatalyst (RP) [61]. Like the structure of type-II heterojunctions, the OP and RP exhibit a similar interleaved structure, but the charge transfer routes between them are different. The RP with a small work function and high Fermi energy level and the OP with a large work function and low Fermi energy level form an S-scheme heterojunction by interlocking patterns. When the OP and RP are in close contact, the Fermi energy levels are bent in the interface region until the Fermi energy levels of the two photocatalysts reach equilibrium [62]. A charge accumulation layer and a charge depletion layer are formed at the interface. Energy band bending occurs in the OP and RP, which induces the recombination of electrons on the CB in the OP and holes on the VB in the RP. As a result, the holes on the lower VB in the OP and the electrons on the higher CB in the RP are retained, favoring strong oxidation and reduction reactions, respectively [33,63]. In conclusion, by this mode formation, not only can the separation of photogenerated carriers be achieved, but the strong oxidation and reduction capabilities can also be obtained. The charge transfer path is macroscopically "step-like", so it is termed a step-scheme heterojunction.

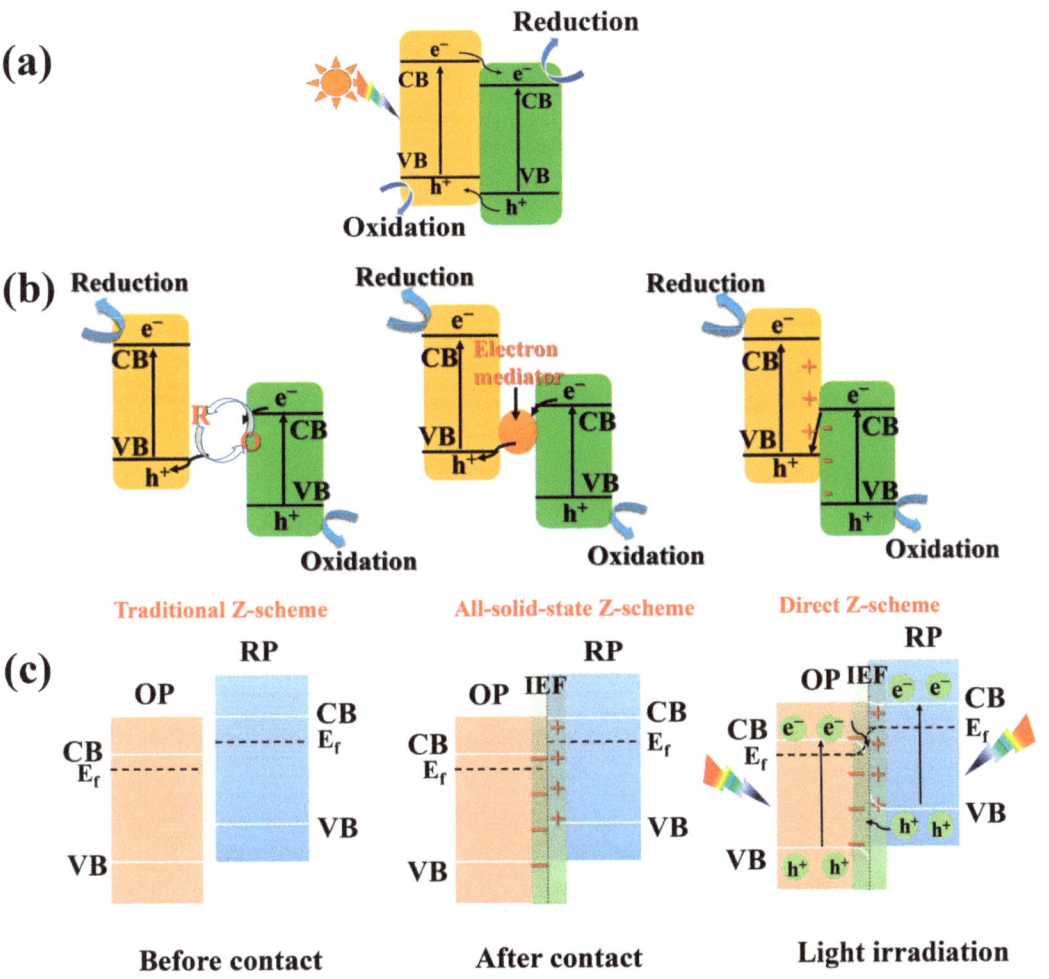

Figure 3. Charge transfer processes in (a) type-II heterojunction, (b) Z-scheme heterojunction, (c) S-scheme heterojunction: before contact; after contact; and under light irradiation.

3.2. Characterization of S-Scheme Heterojunctions

At the moment, the charge transfer pathway in S-scheme heterojunctions can be demonstrated by the characterization of ex situ/in situ irradiated X-ray photoelectron spectroscopy (ISIXPS), Kelvin probe force microscopy (KPFM) and electron paramagnetic resonance spectroscopy (EPR) [62]. The increase or decrease in electron density can be characterized by the shift in binding energy in the in situ XPS spectra under light conditions. The decrease in binding energy represents the increase in electron density and the atom gains electrons. Conversely, the increase in binding energy represents the decrease in electron density and the atom loses electrons [34,64]. Thus, it can be used to determine the direction of charge transfer in heterojunction photocatalysts. For example, Yu et al. synthesized hierarchical TiO$_2$@ZnIn$_2$S$_4$ core–shell hollow spheres and determined the electron transfer paths by XPS. As shown in Figure 4b,c, Ti 2p and O 1s of TiO$_2$@ZnIn$_2$S$_4$ shifted to lower energy levels under dark conditions compared to TiO$_2$, indicating an increase in the electron density of TiO$_2$. The binding energies of Zn 2p, In 3d and S 2p of TiO$_2$@ZnIn$_2$S$_4$ under dark conditions were shifted to higher energy levels compared to those of ZnIn$_2$S$_4$ (Figure 4d–f). This indicates that electrons migrate from ZnIn$_2$S$_4$ to TiO$_2$ when the two photocatalysts are in contact. When light is irradiated, the electron transfer is reversed. That is, the photogenerated electrons migrate from TiO$_2$ to ZnIn$_2$S$_4$. This matches the charge transfer mechanism of the S-scheme heterojunction shown in Figure 4a. In addition, space charge separation in heterojunctions can be revealed by photoirradiated Kelvin probe force microscopy (KPFM) investigation. For example, Cheng et al. [65] prepared a S-scheme heterojunction by growing CdS in situ on the surface of pyrene-alt-triphenylamine conjugated polymer. Figure 5a shows an atomic force microscopy image of the photocatalyst; it can be seen that there is a surface potential difference between the two interfaces. Figure 5b,c shows the surface potential maps of the composites under dark and light conditions. As shown in Figure 5d, the surface potential difference between the PT (A) and CdS (B) is about 100 mV under dark conditions, which proves that an intrinsic electric field is formed between them pointing from the A direction to the B direction. After irradiation, the surface potential of A decreases while the surface potential of B increases. This change in surface potential proves that CdS is an electron donor in the heterojunction (Figure 5e). Furthermore, electron paramagnetic resonance (EPR) and DFT calculations can also indirectly evidence the charge transfer process [66]. EPR can be used to detect the type of radicals contained in the reaction system. Thus, to confirm that the charge transfer path of the synthesized heterojunction follows the S-scheme heterojunction photocatalyst, the presence of •OH and •O$_2$ radicals in the reaction system can be detected by EPR. It is known that the oxidation potential of OH/•OH and the reduction potential of O$_2$/•O$_2$ reach 2.73 V and −0.33 V.

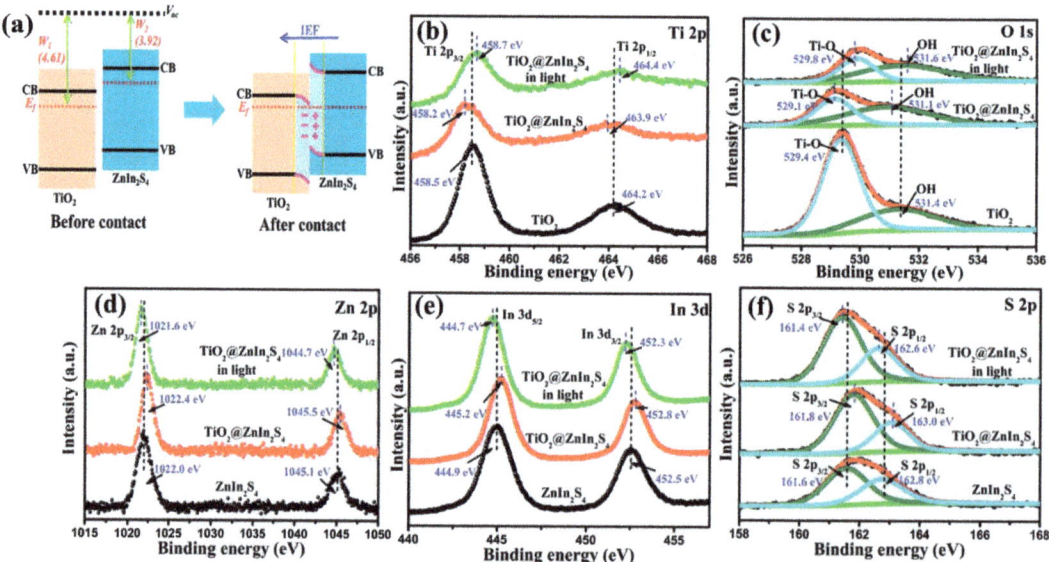

Figure 4. (a) Charge transfer processes in an S-scheme heterojunction: after contact and under light irradiation. High-resolution XPS spectra of (b) Ti 2p, (c) O 1s, (d) Zn 2p, (e) In 3d and (f) S 2p of photocatalysts [36].

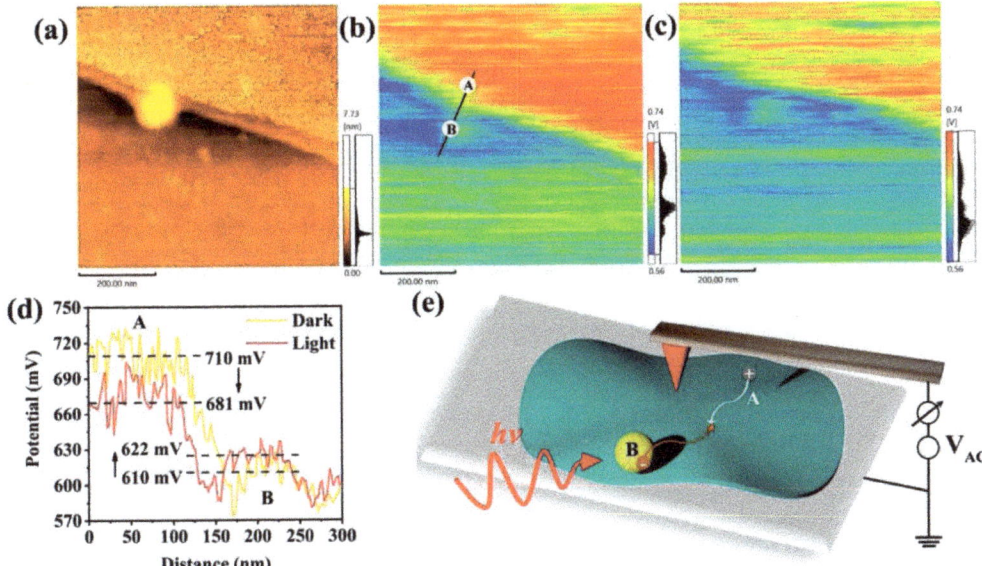

Figure 5. (a) Atomic force microscopy image of photocatalyst. Corresponding surface potential distribution of photocatalyst (b) in dark and (c) under light irradiation. (d) The line-scanning surface potential from point A to B. (e) The schematic illustration of photoirradiation KPFM [65]. (point A: PT; point B: CdS)

3.3. Synthesis Method

Presently, various methods to synthesize S-scheme heterojunctions exist, such as the hydrothermal/solvothermal method [67–69], sol–gel electrostatic spinning method [70,71], self-assembly method [32,72,73] and co-precipitation method [74,75]. For example, Li et al. [76] synthesized a novel S-scheme $TiO_2/ZnIn_2S_4$ heterojunction photocatalyst by the hydrothermal method and evaluated its photocatalytic performance by photocatalytic H_2 production. TiO_2 nanofibers are dispersed in an aqueous ethanol solution containing Zn^{2+} and In^{3+}, which are anchored to the surface of TiO_2 nanofibers by Coulomb electrostatic interactions, while an S source is added. $TiO_2/ZnIn_2S_4$ heterojunctions are obtained by hydrothermal method. It was found the S-scheme mechanism of photogenerated charge transfer made $TiO_2/ZnIn_2S_4$ exhibit the highest H_2 production activity with a H_2 production rate of 6.03 mmol·g^{-1}·h^{-1}.

4. H_2O_2 Production by S-Scheme Heterojunction Photocatalysts

H_2O_2 production by photocatalysis is a safe, sustainable and green process because it requires only water and oxygen from the air as raw materials and sunlight as an energy source [77–79]. In S-scheme heterojunctions, the Fermi energy level difference between semiconductors induces the formation of an intrinsic electric field and energy band bending, which promotes the effective migration and separation of photogenerated electrons and holes. This advantage of S-scheme heterojunctions makes them promising for photocatalytic H_2O_2 production. This review focuses on the application of S-scheme heterojunctions in photocatalytic H_2O_2 production.

4.1. Photocatalytic H_2O_2 Production

As described in Section 2, the two main pathways for photocatalytic H_2O_2 production are the ORR and WOR pathways. Photocatalytic reactions mainly include light absorption, migration and separation of photogenerated charges and redox reactions on surfaces. The most important prerequisite for photocatalytic H_2O_2 production is to satisfy the reaction potential of ORR and WOR pathways. Thus, the band gap position of the photocatalyst is of critical importance in H_2O_2 production. S-scheme heterojunctions have significant advantages in photocatalytic H_2O_2 production because of effective separation of photogenerated carriers and enhanced redox capacity. The oxygen reduction pathway is the most popular photocatalytic H_2O_2 production pathway. For example, Jiang et al. [80] synthesized S-scheme ZnO/WO_3 heterojunction photocatalysts for photocatalytic H_2O_2 production by hydrothermal and calcination methods. FESEM and TEM images show that ZnO/WO_3 exhibits a hierarchical microsphere structure (Figure 6a,b). The prepared ZnO/WO_3 heterojunctions showed superior photocatalytic activity compared to the single component. When the volume of WO_3 was 30%, ZW30 exhibited an H_2O_2 yield of 6788 $\mu mol \cdot L^{-1} \cdot h^{-1}$. In addition, cyclic tests revealed good stability of ZW30, with a small decrease in H_2O_2 yield after four cycles. Figure 6c depicts the mechanism of ZnO/WO_3 for photocatalytic H_2O_2 production. The process is based on a direct $2e^-$ ORR pathway, accompanied by indirect $2e^-$ ORR pathway. The characterization and experimental results demonstrate the formation of a ZnO/WO_3 S-scheme heterojunction with a structure capable of providing more reducing electrons, thus enhancing the driving force of H_2O_2 production by ORR. In another work, Lai et al. [81] developed a $CdS/K_2Ta_2O_6$ S-scheme heterojunction by a two-step hydrothermal method, which exhibits excellent photocatalytic H_2O_2 production activity without using any sacrificial agent and additional O_2. The SEM image shows that the $CdS/K_2Ta_2O_6$ composite exhibits a flower-like structure (Figure 7a). In situ irradiated XPS, EPR and DFT calculations were used to propose the mechanism of an S-scheme heterojunction for H_2O_2 production (Figure 7b). The simultaneous presence of WOR and ORR pathways enables efficient utilization of the redox system. All the above studies provide insights into the design of S-scheme heterojunction photocatalysts for efficient photocatalytic H_2O_2 production. In recent years, there have been a number of S-scheme heterojunctions applied in photocatalytic H_2O_2 production. Table 1 presents the studies of S-scheme heterojunctions for photocatalytic H_2O_2 production.

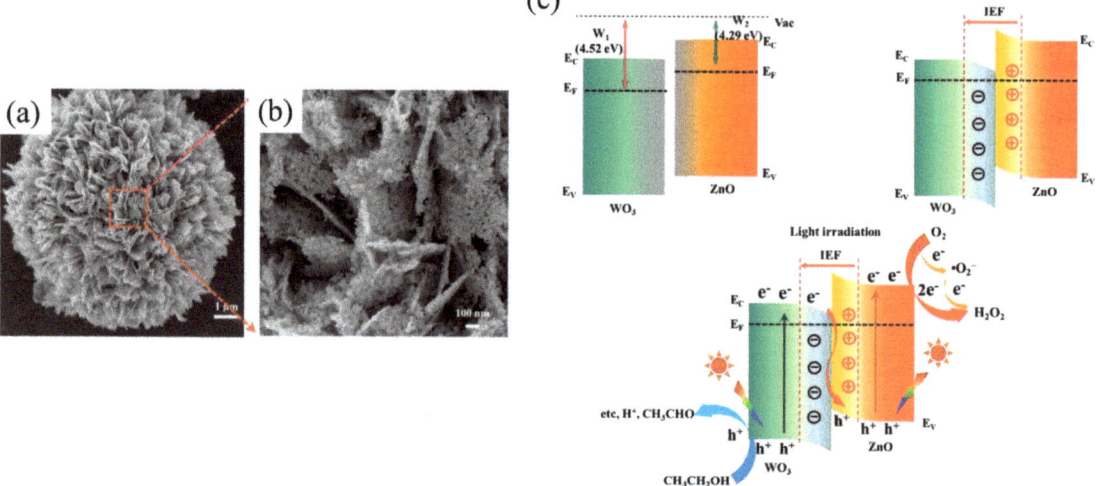

Figure 6. (**a**,**b**) FESEM images of ZnO/WO_3. (**c**) Photocatalytic H_2O_2 production mechanism of ZnO/WO_3 photocatalyst [80].

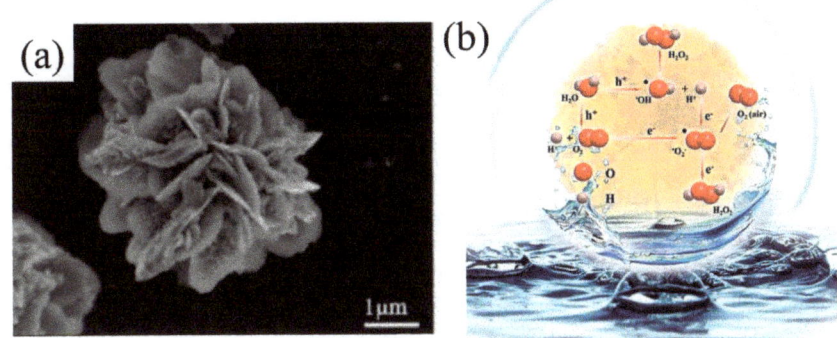

Figure 7. (**a**) Representative SEM image of CdS/K$_2$Ta$_2$O$_6$ photocatalyst. (**b**) Photocatalytic H$_2$O$_2$ production mechanism of CdS/K$_2$Ta$_2$O$_6$ photocatalyst [81].

Table 1. Studies of S-scheme heterojunctions for photocatalytic H$_2$O$_2$ production.

Photocatalyst	Morphology	Light Source	Reaction Solution	Pathway	Concentration of Photocatalyst/g·L^{-1}	H$_2$O$_2$ Yield	Ref.
ZnO/WO$_3$	Hierarchical microsphere structure	300 W Xe lamp	50 mL of 10 vol% ethanol	Direct 2e$^-$ ORR and indirect 2e$^-$ ORR pathways	1.0	6788 µmol·L^{-1}·h^{-1}	[80]
CdS/K$_2$Ta$_2$O$_6$	Flower–like structure	300 W Xe lamp (λ > 420 nm)	Ultra–pure water	2e$^-$ ORR and WOR pathways	0.6	160.89 µmol·L^{-1}·h^{-1}; 346.31 µmol·L^{-1}·h^{-1} with saturated O$_2$	[81]
ZnO/g-C$_3$N$_4$	ZnO NPs dispersed on the CN nanosheet	300 W Xe lamp (λ > 350 nm)	50 mL of 10 vol% ethanol	ORR pathway	0.4	1544 µmol·L^{-1}·h^{-1}	[82]
TiO$_2$/In$_2$S$_3$	Core–shell structure	300 W Xe arc lamp	40 mL of 10 vol% ethanol	Indirect 2e$^-$ ORR pathway	0.5	376 µmol·L^{-1}·h^{-1}	[45]
C$_3$N$_4$/PDA	Nanosheet	300 W Xe arc lamp (λ > 350 nm)	50 mL of 20 vol% ethanol	Indirect 2e$^-$ ORR pathway	0.4	3801.25 µmol·g^{-1}·h^{-1}	[83]
ZnO/COF (TpPa–Cl)	ZnO nanoparticles distributed on the surface of TpPa–Cl	300 W Xe lamp	10 vol% ethanol	Indirect 2e$^-$ ORR pathway	0.5	2443 µmol·g^{-1}·h^{-1}	[84]
TiO$_2$/PDA	Inverse opals	300 W Xe arc lamp	40 mL of 10 vol% ethanol	ORR pathway	0.5	~2.2 mmol·g^{-1}·h^{-1}	[85]
In$_2$O$_3$/ZnIn$_2$S$_4$	Ordered hollow structure	250 W Xe lamp (λ > 420 nm)	50 mL of 5 vol% ethanol	ORR pathway	0.4	5716 µmol·g^{-1}·h^{-1}	[86]
Sv–ZIS/CN	Three–dimensional flower-like structure and agaric shaped with a microporous structure	300 W Xe lamp (λ > 420 nm)	50 mL of 10 vol% isopropanol	Direct 2e$^-$ ORR and indirect 2e$^-$ ORR pathways	0.4	1310.18 µmol·L^{-1}·h^{-1}	[87]
Bi$_2$S$_3$@CdS@RGO	Flaky RGO is wrapped onto the CdS nanoparticles and Bi$_2$S$_3$ rod–aggregate morphology	300 W Xe lamp (λ > 420 nm)	50 mL of 10 vol% isopropanol	Indirect 2e$^-$ ORR pathway	1.0	212.82 µmol·L^{-1} within 180 min	[88]
ZnO@PDA	Inverse Opal	300 W Xe arc lamp	50 mL of 4 vol% glycol	Direct 2e$^-$ ORR and indirect 2e$^-$ ORR pathways	0.4	1011.4 µmol·L^{-1}·h^{-1}	[89]
S-pCN/WO$_{2.72}$	Uniform porous sheet–like two–dimensional structure	300 W Xe lamp (λ > 420 nm)	100 mL water	Direct 2e$^-$ ORR and indirect 2e$^-$ ORR pathways	1.0	90 µmol·L^{-1} within 180 min	[90]
TiO$_2$@RF	Core–shell structure	300 W Xe lamp	15 mL water	2e$^-$ ORR pathway	~0.67	66.6 mmol·g^{-1}·h^{-1}	[91]
sulfur-doped g-C$_3$N$_4$/TiO$_2$	Well-ordered macroporous framework	300 W Xe lamp	50 mL water	2e$^-$ ORR and WOR pathways	0.2	2128 µmol·g^{-1}·h^{-1}	[92]

4.2. Water Splitting

H_2O_2 can also be used as a valuable by-product of photocatalytic overall water splitting to produce H_2. Photocatalytic H_2 production from overall water splitting has been a hot research problem; however, it has the disadvantages of slow kinetics and difficult product separation. The production of H_2 and H_2O_2 from pure water by a two-electron photocatalytic mechanism solves the above problems due to a lower reaction potential than that of the four-electron reaction [93,94]. Two-electron overall water splitting thermodynamically requires a stronger oxidation capacity of the photocatalyst. S-scheme heterojunctions have a strong redox capacity because of their unique step-scheme charge transfer mechanism. For instance, Meng et al. [95] successfully synthesized a g-C_3N_4/$CoTiO_3$ S-scheme heterojunction photocatalyst and applied it in photocatalytic overall water splitting for H_2 production under visible light. The H_2 production efficiency was significantly improved without sacrificial agents, while the presence of H_2O_2 was detected in the photocatalytic process. Based on the results of EPR and DFT calculations, the possible reaction mechanism of the photocatalyst is shown in Figure 8. The difference in the Fermi energy levels of CN and $CoTiO_3$ results in the formation of an intrinsic electric field (IEF) at the contact surface of the two photocatalysts. As a result, energy band bending also occurs in the interface region, forming an S-scheme heterojunction. This means of charge transfer promotes the migration and separation of photogenerated carriers and preserves the strong redox ability of the system, which is beneficial in enhancing the efficiency of photocatalytic overall water splitting.

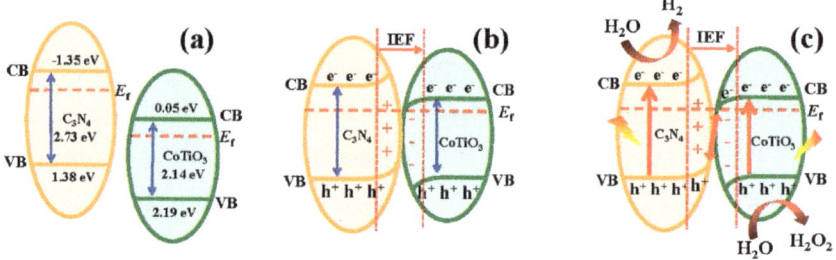

Figure 8. Schematic of the energy levels of CN and $CoTiO_3$ (a) before contact, (b) after contact and (c) under irradiation in S-scheme reaction mechanism of CCT−1.5 [95].

4.3. Coupling of H_2O_2 Production and Organic Synthesis

S-scheme heterojunctions can maximize the redox ability of photocatalysts, effectively utilizing photogenerated electrons and holes and, therefore, having the ability to simultaneously achieve the reduction of O_2 to H_2O_2 and the oxidation of organics [96]. For instance, He et al. [55] synthesized floatable S-scheme TiO_2/Bi_2O_3 photocatalysts by immobilizing hydrophobic TiO_2 and Bi_2O_3 on lightweight polystyrene (PS) spheres by hydrothermal and photodeposition methods. The photocatalysts showed significant H_2O_2 yields and were able to oxidize furfuryl alcohol (FFA) to furoic acid (FA). The mechanism of the photocatalytic reaction was revealed by in situ DRIFT spectroscopy and DFT calculations (Figure 9a,b). In addition, the floatable photocatalyst is able to be in closer contact with O_2 compared to conventional biphasic photocatalytic systems, solving the problem of slow transport of gas reactants from suspended photocatalysts (Figure 9c). Moreover, floatable photocatalysts are less prone to agglomeration, easy to recover and can be recycled. The floatable S-scheme heterojunction photocatalyst not only improves the efficiency of photocatalytic reactions but also provides a new idea for efficient multiphase catalysis. In addition, recently, Yu et al. successfully prepared S-scheme TiO_2@BTTA photocatalysts by synthesizing COF (BTTA) via Schiff-base condensation and by encapsulating TiO_2 NF with BTTA COF. The heterojunction photocatalysts show high H_2O_2 production activity and furoic alcohol (FAL) oxidation activity, with a H_2O_2 production rate of 740 $\mu mol \cdot L^{-1} \cdot h^{-1}$ and a FAL conversion of 96%.

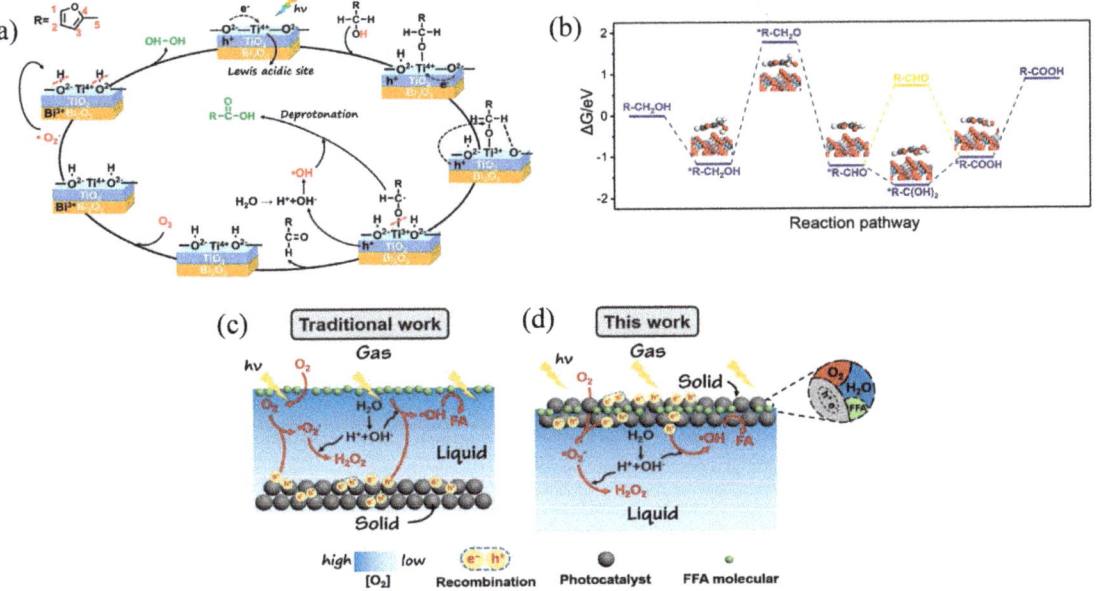

Figure 9. (**a**) The mechanism of photocatalytic FFA oxidation coupling with H_2O_2 production on surface of photocatalyst. (**b**) Free energy diagrams of FFA oxidation steps on active sites of TiO_2. Schematic diagram of O_2 supply for photocatalyst in (**c**) biphase and (**d**) triphase system [55].

4.4. Pollutant Degradation with In Situ H_2O_2 Production

H_2O_2 is usually used in the degradation of contaminants due to its oxidizing ability to improve photocatalytic degradation efficiency. In general, the reactive oxygen species (ROS) used for photocatalytic degradation are mainly H_2O_2, $\cdot O_2^-$ and $\cdot OH$. H_2O_2 is the only stable molecule among them and has a longer lifetime than other active radicals. In situ H_2O_2 production to enhance the degradation of contaminants in photocatalytic processes has proven to be an effective strategy. Recently, S-scheme heterojunction photocatalysts have also been developed for this application (Table 2). Li et al. [97] synthesized a novel layered BP/BiOBr S-scheme heterojunction by self-assembling BiOBr nanosheets on the surface of BP nanosheets by liquid-phase sonication combined with solvothermal methods. The composite exhibited excellent photocatalytic degradation activity of tetracycline (TC) under visible light, which was 7.8 times higher than that of pure BiOBr. The increased activity was attributed to the structure of S-scheme heterojunctions retaining a high redox capacity. The active groups during the experiment were tested by ESR characterization, as shown in Figure 10a,b. After illumination, the signals of both $\cdot O_2^-$ and $\cdot OH$ groups were detected, but the signals of $\cdot O_2^-$ groups became lower with the increase in illumination time, indicating that some $\cdot O_2^-$ and H^+ formed H_2O_2. The results indicate that the main active substances of TC mineralization are in situ generated H_2O_2 and $\cdot OH$. Based on the above results, the photocatalytic mechanism of the S-scheme heterojunction is proposed as shown in Figure 10c.

Table 2. Studies of S-scheme heterojunction photocatalytic H_2O_2 production coupled with pollutant degradation.

Photocatalyst	Morphology	Contaminant or Organics	Light Source	Reaction System	Concentration of Photocatalyst/g·L^{-1}	H_2O_2 Yield	Degradation Efficiency	Ref.
PDI–Urea/BiOBr	BiOBr nanospheres dispersed on the PDI–Urea lamellar layer	Ofloxacin (OFLO), tetracycline (TC)	300 W Xe lamp (λ > 420 nm)	50 mL of TC (50 mg/L) and OFLO (10 mg/L)	1.0	71 μmol·L^{-1}·h^{-1} after 3 h irradiation	93%(~65%) photocatalytic degradation rate for OFLO (TC) after 150 (90) min	[98]
BP/BiOBr	Two-dimensional structure	Tetracycline (TC)	300 W Xe arc lamp (λ > 420 nm)	100 mL of TC (50 mg/L)	1.0	1.62 μmol·L^{-1}·min^{-1}	~85% photocatalytic degradation rate for TC after 90 min	[97]
Graphitic–C_3N_4/ZnCr	Layered structures	Rhodamine B(RhB)	Xe lamp	100 mL of RhB (5 ppm)	1.0	–	99.8% photocatalytic degradation rate for RhB after 210 min	[99]
PDI/g-C_3N_4/TiO_2@Ti_3C_2	Multi-layered 2D frame	Atrazine (ATZ)	300 W Xe lamp (λ > 420 nm)	50 mL of ATZ (10 ppm)	0.8	~160 μmol·L^{-1}·h^{-1}	75% removal rate of ATZ within one hour	[100]
g-C_3N_4/α-MnS	Inhomogeneous morphology with a rough surface	Oxytetracycline (OTC)	300 W Xe lamp (λ > 420 nm)	50 mL of OTC hydrochloride (20 mg·L^{-1})	1.0	111.6 μmol·L^{-1}·h^{-1}	82.2% degradation of OTC in water within 80 min	[101]
Red mud/CdS	RM particles loaded on the surface of CdS nanospheres	Amoxicillin (AMX)	LED lamp (410 < λ < 760 nm)	50 mL of AMX (20 mg·L^{-1})	0.5	1.05 mg·L^{-1}·h^{-1}	73.0% degradation of AMX within 120 min	[102]

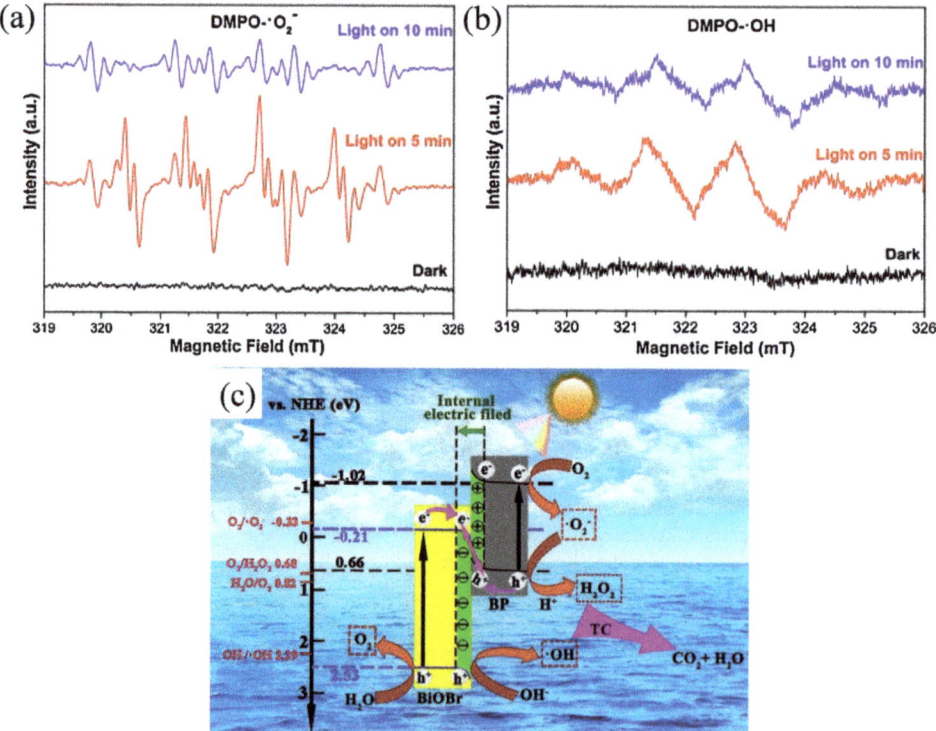

Figure 10. ESR under dark and visible light irradiation: (**a**) DMPO–·O_2^- and (**b**) DMPO–·OH. (**c**) The mechanism of BP/BiOBr S-scheme photocatalyst [97].

Overall, the structure of S-scheme heterojunctions realizes rapid transfer and effective separation of photogenerated carriers and retains the strong redox capability of photocatalysts. This section shows the different S-scheme heterojunction photocatalysts in the literature for H_2O_2 production pathways and provides insights into the synthesis of efficient S-scheme heterojunction photocatalysts.

5. Conclusions and Outlook

Photocatalytic H_2O_2 production is a strategy used to avoid the drawbacks of conventional H_2O_2 production methods and, thus, achieve the conversion from solar energy to chemical energy. However, studies have shown that the efficiency and stability of single-component photocatalysts are not sufficient for practical applications. Therefore, modified photocatalysts obtained by constructing heterojunctions to facilitate the migration and separation of photogenerated carriers have been developed. The novel S-scheme heterojunction proposed by Yu's group overcomes the inherent defects of conventional heterojunctions and obtains a high redox capacity while promoting the effective separation of photogenerated carriers. This paper reviews the mechanism of novel S-scheme heterojunctions and photocatalytic H_2O_2 production and the application of S-scheme heterojunctions in the field of photocatalytic H_2O_2 production.

Up to now, the efficiency of photocatalytic H_2O_2 production has been limited by the energy band position of photocatalysts, the absorption ability of visible light and the migration and separation efficiency of photogenerated carriers. In particular, the inhibition of photogenerated carrier recombination is crucial for photocatalytic efficiency. It is shown that promoting the migration and separation of photogenerated carriers by constructing heterojunctions is most effective. In addition, there are two pathways for photocatalytic H_2O_2 production: two-electron ORR and two-electron WOR pathways. Most of the current studies have focused on the two-electron ORR pathway, which requires the addition of a hole sacrificial agent (isopropyl alcohol, ethanol, etc.) to facilitate the separation of photogenerated carriers. In contrast, the two-electron WOR pathway is rarely realized because it requires a higher oxidation potential than the four-electron WOR pathway to drive the reaction. Therefore, controlling the energy band structure to obtain a sufficient redox potential can improve the selectivity for H_2O_2.

S-scheme heterojunctions are found to be effective in enhancing visible light absorption, promoting the migration and separation of photogenerated charges, extending the lifetime of useful photogenerated charges and keeping a high redox capacity. However, the development of S-scheme heterojunctions in photocatalytic H_2O_2 production is still subject to various limitations. We propose the following aspects to promote the advancement of S-scheme heterojunctions in this field:

1. Modification of the pore size, porosity and particle size of S-scheme heterojunction photocatalysts to increase their surface area, which is conducive to improving the adsorption of reactants (H_2O, O_2) by the photocatalysts;
2. Construction of multiphase catalytic systems. At present, there are few studies on enhancing H_2O_2 yield by constructing multiphase S-scheme heterojunction photocatalytic systems. The disadvantage of slow gas transport kinetics of bi-phase catalysts can be avoided by constructing multiphase catalytic systems, which can promote the adsorption of O_2 by solid photocatalysts and further improve the efficiency of photocatalytic reactions;
3. Combining photocatalysis with electrocatalysis. S-scheme heterojunctions are used to promote the separation of photogenerated charges by using intrinsic electric fields (IEF) at the interface, and other electric fields can be superimposed to further improve their separation efficiency. The introduction of an external electric field by applying a voltage can induce surface charge redistribution of the photocatalyst and can also facilitate the adsorption and activation of O_2 and H_2O;
4. To construct the relationship between the Fermi energy level difference and redox potential. Modulation of redox potential by controlling the Fermi energy level positions

of semiconductors and constructing S-scheme heterojunctions to avoid four-electron competition reactions and improve the selectivity of H_2O_2 products;

5. Optimize the model for theoretical calculations to pre-select semiconductors with suitable Fermi energy levels and energy band structures by theoretical calculations. Meanwhile, theoretical calculations combined with in situ characterization results can also enhance the investigation of the mechanism of photocatalytic H_2O_2 production and contribute to the deeper comprehension of interfacial charge transfer in S-scheme heterojunctions, which is important for the design of efficient S-scheme heterojunction photocatalysts;
6. Considering future commercialization, in addition to the dual-channel pathway of photocatalytic H_2O_2 production, the cost of S-scheme photocatalysts should be controlled and recyclable and reusable photocatalysts should be designed.

Currently, the research of S-scheme heterojunctions in the field of photocatalytic H_2O_2 production is still in the preliminary stage. There are still many challenges on the road to commercialization of photocatalytic H_2O_2 production. We hope that our summary and outlook can facilitate the exploration of S-scheme heterojunctions in photocatalytic H_2O_2 production.

Author Contributions: Conceptualization, L.W. and W.F.; methodology, W.F.; software, W.F.; validation, L.W. and W.F.; formal analysis, W.F.; investigation, W.F.; resources, W.F.; data curation, W.F.; writing—original draft preparation, W.F.; writing—review and editing, L.W. and W.F.; visualization, L.W.; supervision, L.W.; project administration, L.W.; funding acquisition, L.W. All authors have read and agreed to the published version of the manuscript.

Funding: This research was funded by Fundamental Research Funds for the Central Universities, Ocean University of China (grant number 202364004).

Data Availability Statement: Not applicable.

Conflicts of Interest: The authors declare no conflict of interest.

References

1. Campos-Martin, J.M.; Blanco-Brieva, G.; Fierro Jose, L.G. Wasserstoffperoxid-synthese: Perspektiven jenseits des Anthrachinon-Verfahrens. *Angew. Chem.* **2006**, *118*, 7116–7139. [CrossRef]
2. Zhang, P.; Tong, Y.; Liu, Y.; Vequizo, J.J.M.; Sun, H.; Yang, C.; Yamakata, A.; Fan, F.; Lin, W.; Wang, X.; et al. Heteroatom dopants promote two-electron O_2 reduction for photocatalytic production of H_2O_2 on polymeric carbon nitride. *Angew. Chem. Int. Ed. Engl.* **2020**, *59*, 16209–16217. [CrossRef] [PubMed]
3. Keigo Kamata, K.Y. Yasutaka Sumida, Kazuya Yamaguchi, Shiro Hikichi, Noritaka Mizuno. Efficient epoxidation of olefins with >99% selectivity and use of hydrogen peroxide. *Science* **2003**, *300*, 964–966. [CrossRef] [PubMed]
4. Torres-Pinto, A.; Sampaio, M.J.; Silva, C.G.; Faria, J.L.; Silva, A.M.T. Recent strategies for hydrogen peroxide production by metal-free carbon nitride photocatalysts. *Catalysts* **2019**, *9*, 990. [CrossRef]
5. Zheng, L.; Su, H.; Zhang, J.; Walekar, L.S.; Vafaei Molamahmood, H.; Zhou, B.; Long, M.; Hu, Y.H. Highly selective photocatalytic production of H_2O_2 on sulfur and nitrogen co-doped graphene quantum dots tuned TiO_2. *Appl. Catal. B-Environ.* **2018**, *239*, 475–484. [CrossRef]
6. Xia, C.; Xia, Y.; Zhu, P.; Fan, L.; Wang, H. Direct electrosynthesis of pure aqueous H_2O_2 solutions up to 20% by weight using a solid electrolyte. *Science* **2019**, *366*, 226–231. [CrossRef]
7. Lei, J.; Chen, B.; Lv, W.; Zhou, L.; Wang, L.; Liu, Y.; Zhang, J. Robust photocatalytic H_2O_2 production over inverse opal g-C_3N_4 with carbon vacancy under visible light. *ACS Sustain. Chem. Eng.* **2019**, *7*, 16467–16473. [CrossRef]
8. Fujishima, A.; Honda, K. Electrochemical photolysis of water at a semiconductor electrode. *Nature* **1972**, *238*, 37–38. [CrossRef]
9. Zhao, Y.; Liu, Y.; Cao, J.; Wang, H.; Shao, M.; Huang, H.; Liu, Y.; Kang, Z. Efficient production of H_2O_2 via two-channel pathway over ZIF-8/C_3N_4 composite photocatalyst without any sacrificial agent. *Appl. Catal. B-Environ.* **2020**, *278*, 119289. [CrossRef]
10. Feng, C.; Tang, L.; Deng, Y.; Wang, J.; Liu, Y.; Ouyang, X.; Yang, H.; Yu, J.; Wang, J. A novel sulfur-assisted annealing method of g-C_3N_4 nanosheet compensates for the loss of light absorption with further promoted charge transfer for photocatalytic production of H_2 and H_2O_2. *Appl. Catal. B-Environ.* **2021**, *281*, 119539. [CrossRef]
11. Liu, S.; Qi, W.; Adimi, S.; Guo, H.; Weng, B.; Attfield, J.P.; Yang, M. Titanium nitride-supported platinum with metal–support interaction for boosting photocatalytic H_2 evolution of indium sulfide. *ACS Appl. Mater. Inter.* **2021**, *13*, 7238–7247. [CrossRef] [PubMed]

12. Li, Y.; Ma, F.; Zheng, L.; Liu, Y.; Wang, Z.; Wang, P.; Zheng, Z.; Cheng, H.; Dai, Y.; Huang, B. Boron containing metal-organic framework for highly selective photocatalytic production of H_2O_2 by promoting two-electron O_2 reduction. *Mater. Horiz.* **2021**, *8*, 2842–2850. [CrossRef] [PubMed]
13. Kato, S.; Jung, J.; Suenobu, T.; Fukuzumi, S. Production of hydrogen peroxide as a sustainable solar fuel from water and dioxygen. *Energy Environ. Sci.* **2013**, *6*, 3756. [CrossRef]
14. Kaynan, N.; Berke, B.A.; Hazut, O.; Yerushalmi, R. Sustainable photocatalytic production of hydrogen peroxide from water and molecular oxygen. *J. Mater. Chem. A* **2014**, *2*, 13822–13826. [CrossRef]
15. Zhuang, H.; Yang, L.; Xu, J.; Li, F.; Zhang, Z.; Lin, H.; Long, J.; Wang, X. Robust photocatalytic H_2O_2 production by octahedral $Cd_3(C_3N_3S_3)_2$ coordination polymer under visible light. *Sci. Rep.* **2015**, *5*, 16947. [CrossRef] [PubMed]
16. Li, S.; Dong, G.; Hailili, R.; Yang, L.; Li, Y.; Wang, F.; Zeng, Y.; Wang, C. Effective photocatalytic H_2O_2 production under visible light irradiation at g-C_3N_4 modulated by carbon vacancies. *Appl. Catal. B-Environ.* **2016**, *190*, 26–35. [CrossRef]
17. Yang, L.; Dong, G.; Jacobs, D.L.; Wang, Y.; Zang, L.; Wang, C. Two-channel photocatalytic production of H_2O_2 over g-C_3N_4 nanosheets modified with perylene imides. *J. Catal.* **2017**, *352*, 274–281. [CrossRef]
18. Yang, Y.; Zeng, Z.T.; Zeng, G.M.; Huang, D.L.; Xiao, R.; Zhang, C.; Zhou, C.Y.; Xiong, W.P.; Wang, W.J.; Cheng, M.; et al. Ti_3C_2 Mxene/porous g-C_3N_4 interfacial Schottky junction for boosting spatial charge separation in photocatalytic H_2O_2 production. *Appl. Catal. B-Environ.* **2019**, *258*, 117956. [CrossRef]
19. Che, H.; Gao, X.; Chen, J.; Hou, J.; Ao, Y.; Wang, P. Iodide-induced fragmentation of polymerized hydrophilic carbon nitride for high-performance Quasi-Homogeneous photocatalytic H_2O_2 production. *Angew. Chem. Int. Ed. Engl.* **2021**, *60*, 25546–25550. [CrossRef]
20. Liu, B.; Du, J.; Ke, G.; Jia, B.; Huang, Y.; He, H.; Zhou, Y.; Zou, Z. Boosting O_2 reduction and H_2O dehydrogenation kinetics: Surface N-hydroxymethylation of g-C_3N_4 photocatalysts for the efficient production of H_2O_2. *Adv. Funct. Mater.* **2021**, *32*, 2111125. [CrossRef]
21. Wang, P.; Fan, S.; Li, X.; Duan, J.; Zhang, D. Modulating the molecular structure of graphitic carbon nitride for identifying the impact of the piezoelectric effect on photocatalytic H_2O_2 production. *ACS Catal.* **2023**, *13*, 9515–9523. [CrossRef]
22. Xue, L.; Sun, H.; Wu, Q.; Yao, W. P-doped melon-carbon nitride for efficient photocatalytic H_2O_2 production. *J. Colloid Interface Sci.* **2022**, *615*, 87–94. [CrossRef]
23. Che, H.; Wang, J.; Gao, X.; Chen, J.; Wang, P.; Liu, B.; Ao, Y. Regulating directional transfer of electrons on polymeric g-C_3N_5 for highly efficient photocatalytic H_2O_2 production. *J. Colloid Interface Sci.* **2022**, *627*, 739–748. [CrossRef] [PubMed]
24. Zhao, C.; Shi, C.; Li, Q.; Wang, X.; Zeng, G.; Ye, S.; Jiang, B.; Liu, J. Nitrogen vacancy-rich porous carbon nitride nanosheets for efficient photocatalytic H_2O_2 production. *Mater. Today Energy* **2022**, *24*, 100926. [CrossRef]
25. Xu, Y.; Fu, H.; Zhao, L.; Jian, L.; Liang, Q.; Xiao, X. Insight into facet-dependent photocatalytic H_2O_2 production on BiOCl nanosheets. *New J. Chem.* **2021**, *45*, 3335–3342. [CrossRef]
26. Zhu, H.; Xue, Q.; Zhu, G.; Liu, Y.; Dou, X.; Yuan, X. Decorating Pt@cyclodextrin nanoclusters on C_3N_4/MXene for boosting the photocatalytic H_2O_2 production. *J. Mater. Chem. A* **2021**, *9*, 6872–6880. [CrossRef]
27. Zhang, H.; Bai, X. Protonated g-C_3N_4 coated Co_9S_8 heterojunction for photocatalytic H_2O_2 production. *J. Colloid Interface Sci.* **2022**, *627*, 541–553. [CrossRef]
28. Zhao, X.; You, Y.; Huang, S.; Wu, Y.; Ma, Y.; Zhang, G.; Zhang, Z. Z-scheme photocatalytic production of hydrogen peroxide over $Bi_4O_5Br_2$/g-C_3N_4 heterostructure under visible light. *Appl. Catal. B-Environ.* **2020**, *278*, 119251. [CrossRef]
29. Wu, S.; Yu, H.; Chen, S.; Quan, X. Enhanced photocatalytic H_2O_2 production over carbon nitride by doping and defect engineering. *ACS Catal.* **2020**, *10*, 14380–14389. [CrossRef]
30. Xu, Y.; Liao, J.; Zhang, L.; Sun, Z.; Ge, C. Dual sulfur defect engineering of Z-scheme heterojunction on Ag-CdS_{1-x}@$ZnIn_2S_{4-x}$ hollow core-shell for ultra-efficient selective photocatalytic H_2O_2 production. *J. Colloid Interface Sci.* **2023**, *647*, 446–455. [CrossRef]
31. Ma, X.; Cheng, H. Facet-dependent photocatalytic H_2O_2 production of single phase Ag_3PO_4 and Z-scheme Ag/$ZnFe_2O_4$-Ag-Ag_3PO_4 composites. *Chem. Eng. J.* **2022**, *429*, 132373. [CrossRef]
32. Fu, J.; Xu, Q.; Low, J.; Jiang, C.; Yu, J. Ultrathin 2D/2D WO_3/g-C_3N_4 step-scheme H_2-production photocatalyst. *Appl. Catal. B-Environ.* **2019**, *243*, 556–565. [CrossRef]
33. Xu, Q.; Zhang, L.; Cheng, B.; Fan, J.; Yu, J. S-scheme heterojunction photocatalyst. *Chem* **2020**, *6*, 1543–1559. [CrossRef]
34. Xu, F.; Meng, K.; Cheng, B.; Wang, S.; Xu, J.; Yu, J. Unique S-scheme heterojunctions in self-assembled TiO_2/$CsPbBr_3$ hybrids for CO_2 photoreduction. *Nat. Commun.* **2020**, *11*, 4613. [CrossRef] [PubMed]
35. He, F.; Zhu, B.; Cheng, B.; Yu, J.; Ho, W.; Macyk, W. 2D/2D/0D TiO_2/C_3N_4/Ti_3C_2 MXene composite S-scheme photocatalyst with enhanced CO_2 reduction activity. *Appl. Catal. B-Environ.* **2020**, *272*, 119006. [CrossRef]
36. Wang, L.; Cheng, B.; Zhang, L.; Yu, J. In situ irradiated XPS investigation on S-scheme TiO_2@$ZnIn_2S_4$ photocatalyst for efficient photocatalytic CO_2 reduction. *Small* **2021**, *17*, 2103447. [CrossRef]
37. Meng, A.; Cheng, B.; Tan, H.; Fan, J.; Su, C.; Yu, J. TiO_2/polydopamine S-scheme heterojunction photocatalyst with enhanced CO_2 reduction selectivity. *Appl. Catal. B-Environ.* **2021**, *289*, 120039. [CrossRef]
38. Wang, L.; Chen, D.; Miao, S.; Chen, F.; Guo, C.; Ye, P.; Ning, J.; Zhong, Y.; Hu, Y. Nitric acid-assisted growth of $InVO_4$ nanobelts on protonated ultrathin C_3N_4 nanosheets as an S-scheme photocatalyst with tunable oxygen vacancies for boosting CO_2 conversion. *Chem. Eng. J.* **2022**, *434*, 133867. [CrossRef]

39. Han, X.; Lu, B.; Huang, X.; Liu, C.; Chen, S.; Chen, J.; Zeng, Z.; Deng, S.; Wang, J. Novel p- and n-type S-scheme heterojunction photocatalyst for boosted CO_2 photoreduction activity. *Appl. Catal. B-Environ.* **2022**, *316*, 121587. [CrossRef]
40. Jin, Z.; Jiang, X.; Guo, X. Hollow tubular Co_9S_8 grown on In_2O_3 to form S-scheme heterojunction for efficient and stable hydrogen evolution. *Int. J. Hydrogen Energy* **2022**, *47*, 1669–1682. [CrossRef]
41. Hao, X.; Xiang, D.; Jin, Z. Zn-vacancy engineered S-scheme ZnCdS/ZnS photocatalyst for highly efficient photocatalytic H_2 evolution. *ChemCatChem* **2021**, *13*, 4738–4750. [CrossRef]
42. Shen, R.; Lu, X.; Zheng, Q.; Chen, Q.; Ng, Y.H.; Zhang, P.; Li, X. Tracking S-scheme charge transfer pathways in Mo_2C/CdS H_2-evolution photocatalysts. *Solar RRL* **2021**, *5*, 2100177. [CrossRef]
43. Dai, M.; He, Z.; Zhang, P.; Li, X.; Wang, S. $ZnWO_4$-$ZnIn_2S_4$ S-scheme heterojunction for enhanced photocatalytic H_2 evolution. *J. Mater. Sci. Technol.* **2022**, *122*, 231–242. [CrossRef]
44. Feng, K.; Tian, J.; Hu, X.; Fan, J.; Liu, E. Active-center-enriched $Ni_{0.85}Se$/g-C_3N_4 S-scheme heterojunction for efficient photocatalytic H_2 generation. *Int. J. Hydrogen Energy* **2022**, *47*, 4601–4613. [CrossRef]
45. Yang, Y.; Cheng, B.; Yu, J.; Wang, L.; Ho, W. TiO_2/In_2S_3 S-scheme photocatalyst with enhanced H_2O_2-production activity. *Nano Res.* **2021**, *16*, 4506–4514. [CrossRef]
46. Wu, S.; Yu, X.; Zhang, J.; Zhang, Y.; Zhu, Y.; Zhu, M. Construction of BiOCl/$CuBi_2O_4$ S-scheme heterojunction with oxygen vacancy for enhanced photocatalytic diclofenac degradation and nitric oxide removal. *Chem. Eng. J.* **2021**, *411*, 128555. [CrossRef]
47. He, R.; Ou, S.; Liu, Y.; Liu, Y.; Xu, D. In situ fabrication of Bi_2Se_3/g-C_3N_4 S-scheme photocatalyst with improved photocatalytic activity. *Chin. J. Catal.* **2022**, *43*, 370–378. [CrossRef]
48. Wang, Y.; Wang, K.; Wang, J.; Wu, X.; Zhang, G. Sb_2WO_6/BiOBr 2D nanocomposite S-scheme photocatalyst for NO removal. *J. Mater. Sci. Technol.* **2020**, *56*, 236–243. [CrossRef]
49. Le, S.; Ma, Y.; He, D.; Wang, X.; Guo, Y. CdS/$NH_4V_4O_{10}$ S-scheme photocatalyst for sustainable photo-decomposition of amoxicillin. *Chem. Eng. J.* **2021**, *426*, 130354. [CrossRef]
50. Lee, J.H.; Cho, H.; Park, S.O.; Hwang, J.M.; Hong, Y.; Sharma, P.; Jeon, W.C.; Cho, Y.; Yang, C.; Kwak, S.K.; et al. High performance H_2O_2 production achieved by sulfur-doped carbon on CdS photocatalyst via inhibiting reverse H_2O_2 decomposition. *Appl. Catal. B-Environ.* **2021**, *284*, 119690. [CrossRef]
51. Pan, C.; Bian, G.; Zhang, Y.; Lou, Y.; Zhang, Y.; Dong, Y.; Xu, J.; Zhu, Y. Efficient and stable H_2O_2 production from H_2O and O_2 on $BiPO_4$ photocatalyst. *Appl. Catal. B-Environ.* **2022**, *316*, 121675. [CrossRef]
52. Wang, Y.; Wang, Y.; Zhao, J.; Chen, M.; Huang, X.; Xu, Y. Efficient production of H_2O_2 on Au/WO_3 under visible light and the influencing factors. *Appl. Catal. B-Environ.* **2021**, *284*, 119691. [CrossRef]
53. Xie, H.; Zheng, Y.; Guo, X.; Liu, Y.; Zhang, Z.; Zhao, J.; Zhang, W.; Wang, Y.; Huang, Y. Rapid microwave synthesis of mesoporous oxygen-doped g-C_3N_4 with carbon vacancies for efficient photocatalytic H_2O_2 production. *ACS Sustain. Chem. Eng.* **2021**, *9*, 6788–6798. [CrossRef]
54. Luo, J.; Liu, Y.; Fan, C.; Tang, L.; Yang, S.; Liu, M.; Wang, M.; Feng, C.; Ouyang, X.; Wang, L.; et al. Direct attack and indirect transfer mechanisms dominated by reactive oxygen species for photocatalytic H_2O_2 production on g-C_3N_4 possessing nitrogen vacancies. *ACS Catal.* **2021**, *11*, 11440–11450. [CrossRef]
55. He, B.; Wang, Z.; Xiao, P.; Chen, T.; Yu, J.; Zhang, L. Cooperative coupling of H_2O_2 production and organic synthesis over a floatable polystyrene-sphere-supported TiO_2/Bi_2O_3 S-scheme photocatalyst. *Adv. Mater.* **2022**, *34*, e2203225. [CrossRef]
56. Chen, X.; Zhang, W.; Zhang, L.; Feng, L.; Zhang, C.; Jiang, J.; Wang, H. Construction of porous tubular In_2S_3@In_2O_3 with plasma treatment-derived oxygen vacancies for efficient photocatalytic H_2O_2 production in pure water via two-electron reduction. *ACS Appl. Mater. Interfaces* **2021**, *13*, 25868–25878. [CrossRef]
57. Zhao, Y.; Liu, Y.; Wang, Z.; Ma, Y.; Zhou, Y.; Shi, X.; Wu, Q.; Wang, X.; Shao, M.; Huang, H.; et al. Carbon nitride assisted 2D conductive metal-organic frameworks composite photocatalyst for efficient visible light-driven H_2O_2 production. *Appl. Catal. B-Environ.* **2021**, *289*, 120035. [CrossRef]
58. Yuan, Y.; Guo, R.-T.; Hong, L.-F.; Ji, X.-Y.; Lin, Z.-D.; Li, Z.-S.; Pan, W.-G. A review of metal oxide-based Z-scheme heterojunction photocatalysts: Actualities and developments. *Mater. Today Energy* **2021**, *21*, 100829. [CrossRef]
59. Li, X.; Garlisi, C.; Guan, Q.; Anwer, S.; Al-Ali, K.; Palmisano, G.; Zheng, L. A review of material aspects in developing direct Z-scheme photocatalysts. *Mater. Today* **2021**, *47*, 75–107. [CrossRef]
60. Bao, Y.; Song, S.; Yao, G.; Jiang, S. S-Scheme Photocatalytic Systems. *Solar RRL* **2021**, *5*, 2100118. [CrossRef]
61. Hasija, V.; Kumar, A.; Sudhaik, A.; Raizada, P.; Singh, P.; Van Le, Q.; Le, T.T.; Nguyen, V.-H. Step-scheme heterojunction photocatalysts for solar energy, water splitting, CO_2 conversion, and bacterial inactivation: A review. *Environ. Chem. Lett.* **2021**, *19*, 2941–2966. [CrossRef]
62. Zhang, L.; Zhang, J.; Yu, H.; Yu, J. Emerging S-scheme photocatalyst. *Adv. Mater.* **2022**, *34*, e2107668. [CrossRef] [PubMed]
63. Zhu, B.; Tan, H.; Fan, J.; Cheng, B.; Yu, J.; Ho, W. Tuning the strength of built-in electric field in 2D/2D g-C_3N_4/SnS_2 and g-C_3N_4/ZrS_2 S-scheme heterojunctions by nonmetal doping. *J. Mater.* **2021**, *7*, 988–997. [CrossRef]
64. Wang, L.; Zhu, B.; Cheng, B.; Zhang, J.; Zhang, L.; Yu, J. In-situ preparation of TiO_2/N-doped graphene hollow sphere photocatalyst with enhanced photocatalytic CO_2 reduction performance. *Chin. J. Catal.* **2021**, *42*, 1648–1658. [CrossRef]
65. Cheng, C.; He, B.; Fan, J.; Cheng, B.; Cao, S.; Yu, J. An inorganic/organic S-scheme heterojunction H_2-production photocatalyst and its charge transfer mechanism. *Adv. Mater.* **2021**, *33*, 2100317. [CrossRef]

66. Wang, X.; Sayed, M.; Ruzimuradov, O.; Zhang, J.; Fan, Y.; Li, X.; Bai, X.; Low, J. A review of step-scheme photocatalysts. *Appl. Mater. Today* **2022**, *29*, 101609. [CrossRef]
67. Deng, X.; Wang, D.; Li, H.; Jiang, W.; Zhou, T.; Wen, Y.; Yu, B.; Che, G.; Wang, L. Boosting interfacial charge separation and photocatalytic activity of 2D/2D g-C_3N_4/$ZnIn_2S_4$ S-scheme heterojunction under visible light irradiation. *J. Alloys Compd.* **2022**, *894*, 162209. [CrossRef]
68. Dou, L.; Jin, X.; Chen, J.; Zhong, J.; Li, J.; Zeng, Y.; Duan, R. One-pot solvothermal fabrication of S-scheme OVs-Bi_2O_3/Bi_2SiO_5 microsphere heterojunctions with enhanced photocatalytic performance toward decontamination of organic pollutants. *Appl. Surf. Sci.* **2020**, *527*, 146775. [CrossRef]
69. Yang, H.; Zhang, J.F.; Dai, K. Organic amine surface modified one-dimensional $CdSe_{0.8}S_{0.2}$-diethylenetriamine/two-dimensional $SnNb_2O_6$ S-scheme heterojunction with promoted visible-light-driven photocatalytic CO_2 reduction. *Chin. J. Catal.* **2022**, *43*, 255–264. [CrossRef]
70. Liao, X.; Li, T.T.; Ren, H.T.; Zhang, X.; Shen, B.; Lin, J.H.; Lou, C.W. Construction of BiOI/TiO_2 flexible and hierarchical S-scheme heterojunction nanofibers membranes for visible-light-driven photocatalytic pollutants degradation. *Sci. Total Environ.* **2022**, *806*, 150698. [CrossRef]
71. Xia, P.; Cao, S.; Zhu, B.; Liu, M.; Shi, M.; Yu, J.; Zhang, Y. Designing a 0D/2D S-scheme heterojunction over polymeric carbon nitride for visible-light photocatalytic inactivation of bacteria. *Angew. Chem. Int. Ed. Engl.* **2020**, *59*, 5218–5225. [CrossRef] [PubMed]
72. Zhang, L.; Jiang, X.; Jin, Z.; Tsubaki, N. Spatially separated catalytic sites supplied with the CdS–MoS_2–In_2O_3 ternary dumbbell S-scheme heterojunction for enhanced photocatalytic hydrogen production. *J. Mater. Chem. A* **2022**, *10*, 10715–10728. [CrossRef]
73. Wang, K.; Liu, S.; Li, Y.; Wang, G.; Yang, M.; Jin, Z. Phosphorus ZIF-67@NiAl LDH S-scheme heterojunction for efficient photocatalytic hydrogen production. *Appl. Surf. Sci.* **2022**, *601*, 154174. [CrossRef]
74. Zhang, L.; Meng, Y.; Shen, H.; Li, J.; Yang, C.; Xie, B.; Xia, S. Photocatalytic degradation of rhodamine B by Bi_2O_3@LDHs S–scheme heterojunction: Performance, kinetics and mechanism. *Appl. Surf. Sci.* **2021**, *567*, 150760. [CrossRef]
75. Kamali, M.; Sheibani, S.; Ataie, A. Magnetic $MgFe_2O_4$-$CaFe_2O_4$ S-scheme photocatalyst prepared from recycling of electric arc furnace dust. *J. Environ. Manag.* **2021**, *290*, 112609. [CrossRef] [PubMed]
76. Li, J.; Wu, C.; Li, J.; Dong, B.; Zhao, L.; Wang, S. 1D/2D TiO_2/$ZnIn_2S_4$ S-scheme heterojunction photocatalyst for efficient hydrogen evolution. *Chin. J. Catal.* **2022**, *43*, 339–349. [CrossRef]
77. Yu, W.; Hu, C.; Bai, L.; Tian, N.; Zhang, Y.; Huang, H. Photocatalytic hydrogen peroxide evolution: What is the most effective strategy? *Nano Energy* **2022**, *104*, 107906. [CrossRef]
78. Wang, L.; Zhang, J.; Zhang, Y.; Yu, H.; Qu, Y.; Yu, J. Inorganic metal-oxide photocatalyst for H_2O_2 production. *Small* **2022**, *18*, e2104561. [CrossRef]
79. Hou, H.; Zeng, X.; Zhang, X. Production of hydrogen peroxide by photocatalytic processes. *Angew. Chem. Int. Ed. Engl.* **2020**, *59*, 17356–17376. [CrossRef]
80. Jiang, Z.; Cheng, B.; Zhang, Y.; Wageh, S.; Al-Ghamdi, A.A.; Yu, J.; Wang, L. S-scheme ZnO/WO_3 heterojunction photocatalyst for efficient H_2O_2 production. *J. Mater. Sci. Technol.* **2022**, *124*, 193–201. [CrossRef]
81. Lai, C.; Xu, M.; Xu, F.; Li, B.; Ma, D.; Li, Y.; Li, L.; Zhang, M.; Huang, D.; Tang, L.; et al. An S-scheme CdS/$K_2Ta_2O_6$ heterojunction photocatalyst for production of H_2O_2 from water and air. *Chem. Eng. J.* **2023**, *452*, 139070. [CrossRef]
82. Liu, B.; Bie, C.; Zhang, Y.; Wang, L.; Li, Y.; Yu, J. Hierarchically porous ZnO/g-C_3N_4 S-scheme heterojunction photocatalyst for efficient H_2O_2 production. *Langmuir* **2021**, *37*, 14114–14124. [CrossRef] [PubMed]
83. Zhang, X.; Yu, J.; Macyk, W.; Wageh, S.; Al-Ghamdi, A.A.; Wang, L. C_3N_4/PDA S-scheme heterojunction with enhanced photocatalytic H_2O_2 production performance and its mechanism. *Adv. Sustain. Syst.* **2022**, *7*, 2200113. [CrossRef]
84. Zhang, Y.; Qiu, J.; Zhu, B.; Fedin, M.V.; Cheng, B.; Yu, J.; Zhang, L. ZnO/COF S-scheme heterojunction for improved photocatalytic H_2O_2 production performance. *Chem. Eng. J.* **2022**, *444*, 136584. [CrossRef]
85. Wang, L.; Zhang, J.; Yu, H.; Patir, I.H.; Li, Y.; Wageh, S.; Al-Ghamdi, A.A.; Yu, J. Dynamics of photogenerated charge carriers in inorganic/organic S-scheme heterojunctions. *J. Phys. Chem. Lett.* **2022**, *13*, 4695–4700. [CrossRef]
86. Bariki, R.; Das, K.; Pradhan, S.K.; Prusti, B.; Mishra, B.G. MOF-derived hollow tubular In_2O_3/$M^{II}In_2S_4$ (M^{II}: Ca, Mn, and Zn) heterostructures: Synergetic charge-transfer mechanism and excellent photocatalytic performance to boost activation of small atmospheric molecules. *ACS Appl. Energy Mater.* **2022**, *5*, 11002–11017. [CrossRef]
87. Sun, L.; Liu, X.; Jiang, x.; Feng, Y.; Ding, X.-L.; Jiang, N.; Wang, J. Internal electric field and interfacial S-C bonds jointly accelerate S-scheme charge transfer achieving efficient sunlight-driven photocatalysis. *J. Mater. Chem. A* **2022**, *10*, 25279–25294. [CrossRef]
88. Ghoreishian, S.M.; Ranjith, K.S.; Park, B.; Hwang, S.-K.; Hosseini, R.; Behjatmanesh-Ardakani, R.; Pourmortazavi, S.M.; Lee, H.U.; Son, B.; Mirsadeghi, S.; et al. Full-spectrum-responsive Bi_2S_3@CdS S-scheme heterostructure with intimated ultrathin RGO toward photocatalytic Cr(VI) reduction and H_2O_2 production: Experimental and DFT studies. *Chem. Eng. J.* **2021**, *419*, 129530. [CrossRef]
89. Han, G.; Xu, F.; Cheng, B.; Li, Y.; Yu, J.; Zhang, L. Enhanced photocatalytic H_2O_2 production over inverse opal ZnO@Polydopamine S-scheme heterojunctions. *Acta Phys. Chim. Sin.* **2022**, *38*, 2112037. [CrossRef]
90. Li, X.; Kang, B.; Dong, F.; Zhang, Z.; Luo, X.; Han, L.; Huang, J.; Feng, Z.; Chen, Z.; Xu, J.; et al. Enhanced photocatalytic degradation and H_2/H_2O_2 production performance of S-pCN/$WO_{2.72}$ S-scheme heterojunction with appropriate surface oxygen vacancies. *Nano Energy* **2021**, *81*, 105671. [CrossRef]

91. Xia, C.; Yuan, L.; Song, H.; Zhang, C.; Li, Z.; Zou, Y.; Li, J.; Bao, T.; Yu, C.; Liu, C. Spatial specific janus S-scheme photocatalyst with enhanced H_2O_2 production Performance. *Small* **2023**, *19*, 2300292. [CrossRef] [PubMed]
92. Jiang, Z.; Long, Q.; Cheng, B.; He, R.; Wang, L. 3D ordered macroporous sulfur-doped g-C_3N_4/TiO_2 S-scheme photocatalysts for efficient H_2O_2 production in pure water. *J. Mater. Sci. Technol.* **2023**, *162*, 1–10. [CrossRef]
93. Cao, S.; Chan, T.-S.; Lu, Y.-R.; Shi, X.; Fu, B.; Wu, Z.; Li, H.; Liu, K.; Alzuabi, S.; Cheng, P.; et al. Photocatalytic pure water splitting with high efficiency and value by Pt/porous brookite TiO_2 nanoflutes. *Nano Energy* **2020**, *67*, 104287. [CrossRef]
94. Xue, F.; Si, Y.; Wang, M.; Liu, M.; Guo, L. Toward efficient photocatalytic pure water splitting for simultaneous H_2 and H_2O_2 production. *Nano Energy* **2019**, *62*, 823–831. [CrossRef]
95. Meng, A.; Zhou, S.; Wen, D.; Han, P.; Su, Y. g-C_3N_4/$CoTiO_3$ S-scheme heterojunction for enhanced visible light hydrogen production through photocatalytic pure water splitting. *Chin. J. Catal.* **2022**, *43*, 2548–2557. [CrossRef]
96. He, R.; Xu, D.; Li, X. Floatable S-scheme photocatalyst for H_2O_2 production and organic synthesis. *J. Mater. Sci. Technol.* **2023**, *138*, 256–258. [CrossRef]
97. Li, X.; Xiong, J.; Gao, X.; Ma, J.; Chen, Z.; Kang, B.; Liu, J.; Li, H.; Feng, Z.; Huang, J. Novel BP/BiOBr S-scheme nano-heterojunction for enhanced visible-light photocatalytic tetracycline removal and oxygen evolution activity. *J. Hazard. Mater.* **2020**, *387*, 121690. [CrossRef]
98. Wang, W.; Li, X.; Deng, F.; Liu, J.; Gao, X.; Huang, J.; Xu, J.; Feng, Z.; Chen, Z.; Han, L. Novel organic/inorganic PDI-Urea/BiOBr S-scheme heterojunction for improved photocatalytic antibiotic degradation and H_2O_2 production. *Chin. Chem. Lett.* **2022**, *33*, 5200–5207. [CrossRef]
99. Khamesan, A.; Esfahani, M.M.; Ghasemi, J.B.; Farzin, F.; Parsaei-Khomami, A.; Mousavi, M. Graphitic-C_3N_4/ZnCr-layered double hydroxide 2D/2D nanosheet heterojunction: Mesoporous photocatalyst for advanced oxidation of azo dyes with in situ produced H_2O_2. *Adv. Power Technol.* **2022**, *33*, 103777. [CrossRef]
100. Tang, R.; Gong, D.; Deng, Y.; Xiong, S.; Deng, J.; Li, L.; Zhou, Z.; Zheng, J.; Su, L.; Yang, L. π-π Stacked step-scheme PDI/g-C_3N_4/TiO_2@Ti_3C_2 photocatalyst with enhanced visible photocatalytic degradation towards atrazine via peroxymonosulfate activation. *Chem. Eng. J.* **2022**, *427*, 131809. [CrossRef]
101. Wang, Y.; He, Y.; Chi, Y.; Yin, P.; Wei, L.; Liu, W.; Wang, X.; Zhang, H.; Song, H. Construction of S-scheme p-n heterojunction between protonated g-C_3N_4 and α-MnS nanosphere for photocatalytic H_2O_2 production and in situ degradation of oxytetracycline. *J. Environ. Chem. Eng.* **2023**, *11*, 109968. [CrossRef]
102. Sun, X.; He, K.; Chen, Z.; Yuan, H.; Guo, F.; Shi, W. Construction of visible-light-response photocatalysis-self-Fenton system for the efficient degradation of amoxicillin based on industrial waste red mud/CdS S-scheme heterojunction. *Sep. Purif. Technol.* **2023**, *324*, 124600. [CrossRef]

Disclaimer/Publisher's Note: The statements, opinions and data contained in all publications are solely those of the individual author(s) and contributor(s) and not of MDPI and/or the editor(s). MDPI and/or the editor(s) disclaim responsibility for any injury to people or property resulting from any ideas, methods, instructions or products referred to in the content.

Review

A Brief Review on the Latest Developments on Pharmaceutical Compound Degradation Using g-C₃N₄-Based Composite Catalysts

Subhadeep Biswas [1] and Anjali Pal [2,*]

[1] Civil Engineering Department, Swami Vivekananda Institute of Science and Technology, Kolkata 700145, West Bengal, India; sdeep0819@gmail.com
[2] Civil Engineering Department, Indian Institute of Technology Kharagpur, Kharagpur 721302, West Bengal, India
[*] Correspondence: anjalipal@civil.iitkgp.ac.in; Tel.: +91-3222-281920; Fax: +91-3222-282254

Abstract: Pharmaceutical compounds (PCs) are one of the most notable water pollutants of the current age with severe impacts on the ecosystem. Hence, scientists and engineers are continuously working on developing different materials and technologies to eradicate PCs from aqueous media. Among various new-age materials, graphitic carbon nitride (g-C₃N₄) is one of the wonder substances with excellent catalytic property. The current review article describes the latest trend in the application of g-C₃N₄-based catalyst materials towards the degradation of various kinds of drugs and pharmaceutical products present in wastewater. The synthesis procedure of different g-C₃N₄-based catalysts is covered in brief, and this is followed by different PCs degraded as described by different workers. The applicability of these novel catalysts in the real field has been highlighted along with different optimization techniques in practice. Different techniques often explored to characterize the g-C₃N₄-based materials are also described. Finally, existing challenges in this field along with future perspectives are presented before concluding the article.

Keywords: pharmaceutical compounds; g-C₃N₄; catalyst; wastewater; treatment

1. Introduction

Emerging contaminants are the most alarming water pollutants of the present era. Pharmaceutical compounds (PCs) have become one of the notable classes of emerging water pollutants [1]. These compounds have been detected in various concentration ranges (ng/L to µg/L) in natural water bodies. These compounds are reported to be present in natural water bodies, such as Taihu Lake in China and Mississippi River in the USA [2]. Moreover, an estimation has shown that till now at least 3000 species of PCs have been detected in water and wastewater [3]. Among various categories of drug compounds, nonsteroidal anti-inflammatory drugs and sulfonamides are often used for the treatment of various categories of diseases. However, the long-term existence of these compounds in natural water bodies leads to the chronic poisoning of the aquatic lives. Various drugs have been marked by several environmental agencies for their toxic effects. Naproxen (NPX), a drug often used to treat menstrual cramps, has been enlisted by US EPA as Class I in Current Contaminant Candidate List. Diclofenac (DC), a commonly used painkiller, is well known for its bioaccumulation and toxic characteristics. Evidence shows that it causes hemodynamic changes and thyroid tumor in humans [4].

Moreover, due to the COVID-19 pandemic, their concentration has increased drastically in the environmental matrices. Morales-Paredes et al. [5] in their recent review article mentioned the abnormal increase in the concentration range of these compounds in natural water and wastewater. Among several antiviral drugs used during the COVID-19 times, azithromycin and chloroquine are some of the most commonly used PCs. The concentration of azithromycin drug is increased by 217 times in comparison to the normal concentration (WWTP-river-estuary at Wuhan, China). Due to various toxic effects, its removal from water streams is of utmost importance to environmental scientists and engineers. Conventional biological water treatment processes, adsorption, and coagulation are not feasible for the ultimate destruction of these recalcitrant compounds in aqueous media.

Eradication of organic compounds is possible through photocatalytic reaction, which undoubtedly is one of the green waste management options. In this regard, it may be noted that g-C_3N_4 is a prominent photocatalyst of the new age. It provides an attractive option to the research community for synthesizing noble metal-free efficient semiconductor-based photocatalysts. It possesses several distinguishing features such as higher thermal and chemical stability, due to the presence of strong covalent bonds between the C and N atoms in the conjugated g-C_3N_4 framework in it. It is easily prepared and nontoxic in nature [6]. However, the high probability of charge recombination, small surface area, and low reusability are some of the major hindrances behind its large-scale application [7]. Often these hurdles are minimized by means of doping [8], co-doping, co-polymerization via hybridization, exfoliation [9,10], formation of heterojunction structures [11], etc. A significant amount of research work has been performed by scientists all over the globe in recent times regarding the modification of g-C_3N_4 and its application for the treatment of PC-bearing wastewater. The current review article aims to focus on the latest trends in the applications of g-C_3N_4-based composite photocatalysts towards PC wastewater remediation. Firstly, different types of g-C_3N_4-based catalysts and their novel synthesis procedures are discussed. After that, various types of g-C_3N_4-based catalysts applied for PC wastewater treatment in recent times have been elaborated. The next two sections deal with the optimization and applicability of different g-C_3N_4-based catalysts for real wastewater treatment. Several characterization procedures often followed to get better insight into the removal process are then mentioned. The last section mentions the current challenges and future perspectives. This is followed by the concluding remarks. Almost all the papers discussed here have been published within the last five years.

2. Synthesis of Different Types of g-C_3N_4-Based Catalysts for PC Degradation

Different types of g-C_3N_4-based catalysts have been reported in the literature regarding the degradation of PCs. In the current section of this article, three categories have been selected for distinguishing g-C_3N_4-based catalysts in respect of preparation technique. Firstly, they have been differentiated on the basis of the precursor material used for the production of g-C_3N_4. Different starting materials, such as melamine, urea, and dicyanamide, have been reported in the literature for the preparation of g-C_3N_4. Therefore, the first categorization describes the synthesis procedure of different novel catalysts from different precursor materials.

Moreover, it has been stated in the previous section that in comparison to the bare g-C_3N_4, composite formation with other materials and doping improved the catalytic efficiency. Hence, the next two subsections deal with the preparation procedure of several g-C_3N_4 composite and doped catalysts. As the classification is based on different criteria, a particular catalyst may satisfy more than one principle.

2.1. Various Precursors for Preparing g-C_3N_4

In g-C_3N_4, nitrogen is placed in a framework of graphite with a p-conjugated system and the distance between the two layers is 0.326 nm. g-C_3N_4 is often produced from a precursor material such as urea. Upon condensation of urea molecules, NH_3 and CO_2 gases are produced, which ultimately helps in the production of porous g-C_3N_4. However, in comparison to pure g-C_3N_4 catalysts, doped materials are often preferred by the research community for their improved photocatalytic features. Due to the doping with Na or K, the potentials of the valence band and conduction band for absorbing visible light are enhanced, which ultimately leads to increased photocatalytic activity. Guo et al. [12] reported the preparation of g-C_3N_4 nanosheets from urea. Briefly, urea was placed inside an alumina crucible and heated to 550 °C for 3 h at the rate of 5 °C/min. The obtained yellow powder was heated for the second time in a muffle furnace at 52 °C to complete the thermal polymerization reaction and g-C_3N_4 nanosheets were produced.

Many studies reported the synthesis of g-C_3N_4 from melamine as the precursor. Chi et al. [13] prepared a g-C_3N_4-based catalyst from melamine. Guo et al. [14] also prepared a g-C_3N_4-based novel photocatalyst from melamine and urea. Briefly, 7.704 g urea, 5.4 g melamine, and different amount of ammonium chloride (NH_4Cl) were added to 90 mL deionized water. After stirring the solution for 30 min, it was dried at 80 °C to obtain the precursor powder. Then, the precursor powder was calcinated at 550 °C for 3 h at the heating rate of 0.5 °C/min. The whole schematic of the preparation of the Cl-doped g-C_3N_4 nanosheet catalyst is shown in Figure 1. Smykalova et al. [15] produced exfoliated g-C_3N_4 catalyst using melamine as the precursor.

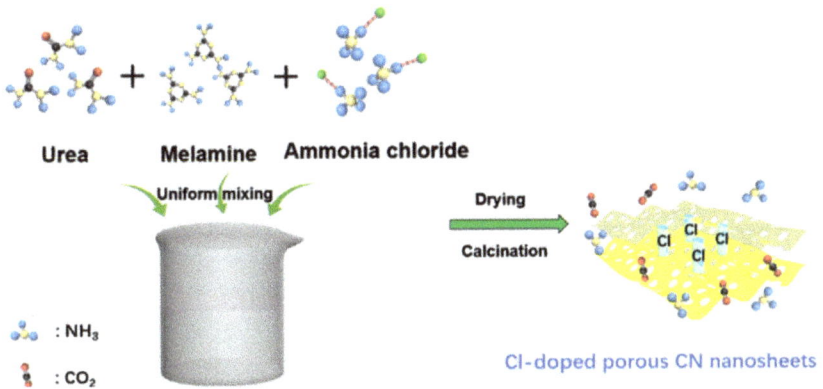

Figure 1. Schematic illustration of the preparation for Cl-doped porous CN nanosheets [14].

He et al. [16] prepared g-C_3N_4 powder from dicyanodiamine. Sixteen grams of dicyanodiamine was heated at the rate of 2.5 °C up to 600 °C and it was maintained for 4 h. After that, the heated powder was cooled and ground for further use. Wang et al. [17] also reported the preparation of g-C_3N_4 from dicyandiamide as the precursor. Sun et al. [18] produced bulk g-C_3N_4 from dicyandiamide as the precursor. To the prepared g-C_3N_4, kaolinite was loaded via the impregnation calcination process and the composite thus produced was named as g-C_3N_4/kaolinite (KCN). To it, Ag was loaded to form the ternary composite Ag/g-C_3N_4/kaolinite composite. The preparation procedure is illustrated in Figure 2.

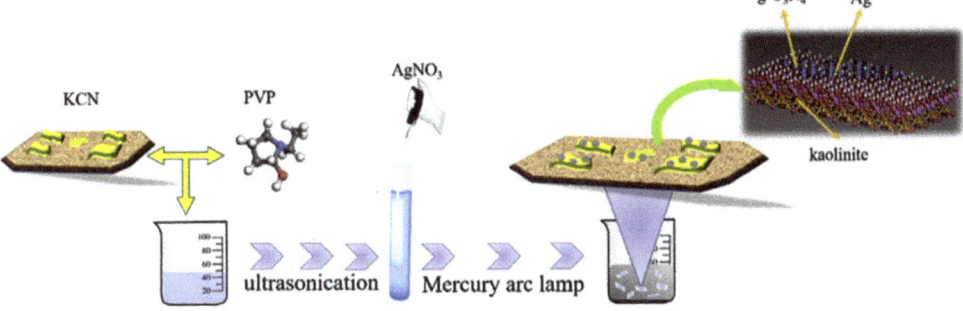

Figure 2. Scheme of the synthetic process for the Ag/KCN-X photocatalysts [18].

2.2. Composite with Other Materials

Often g-C_3N_4 is utilized in making composite catalysts with other suitable materials. ZnO is an excellent semiconductor photocatalyst material and has been proven to be an efficient catalyst for degrading different categories of PCs. Its bandgap is 3.2 eV and there is a huge possibility of electron–hole recombination. Along with this, its optical corrosion is another major hindrance behind its usage for real field purposes. In this regard, composite with g-C_3N_4 offers a sustainable efficient photocatalyst with higher activity. Feyzi et al. [19] prepared ZnO/g-C_3N_4/zeolite P supported photocatalyst for photodegradation of the tetracycline (TC) molecule. The composite catalyst was synthesized in three steps. Firstly, ZnO was prepared by the sol–gel method. Then g-C_3N_4 was synthesized at a large scale by means of pyrolysis of urea. In the final step, the as-synthesized ZnO and g-C_3N_4 were mixed and added to zeolite for the production of the composite catalyst. Mirzaei et al. [20] synthesized a ZnO@g-C_3N_4 catalyst for the mineralization of a sulfamethoxazole (SMX) drug. Baladi et al. [21] reported the preparation and application of g-C_3N_4-$CoFe_2O_4$-ZnO photocatalyst for the degradation of the penicillin G antibiotic compound. Chen et al. [22] applied Ag-$AgVO_3$/g-C_3N_4 photocatalyst for the degradation of TC antibiotic. g-C_3N_4/TiO_2/CFs composite was synthesized by Guo et al. for TC elimination from wastewater [23]. Firstly, different quantity of oxamide and 10 g of urea were blended mechanically. After that, the mixture was subjected to calcination in a muffle furnace at 550 °C to produce g-C_3N_4. In order to synthesize the g-C_3N_4/TiO_2 composite, the as-prepared g-C_3N_4 was dissolved in a previously prepared Ti(SO_4)$_2$ solution and the whole setup was subjected to ultra-sonification. The whole process was repeated in the presence of carbon fibers (CFs) to prepare the composite C_3N_4/TiO_2/CFs. Kumar et al. [24] deployed BiOCl/g-C_3N_4/Cu_2O/Fe_3O_4 ternary composite photocatalyst for oxidative degradation of SMX from wastewater.

2.3. Doping of g-C_3N_4-Based Catalysts

Doping and co-doping with various metals and metal oxides enhanced the catalytic activity of g-C_3N_4-based materials. Wang et al. [25] investigated the co-doping of bimetallic oxides and the effect of oxygen on the catalytic degradation performance of g-C_3N_4 on SMX removal. Liang et al. [26] prepared Ba-atom-embedded g-C_3N_4-based catalyst for the degradation of carbamazepine (CBZ) and DC drug molecules. The Ba atom got anchored onto the surface of the g-C_3N_4 by means of forming an ionic bond with the triazine ring. The catalyst was prepared by the thermal polymerization method. In this method, 2.52 g melamine was mixed with 60 mL of ethylene glycol. This solution was kept in the temperature range of 58–62 °C and was named as solution A. On the other hand, 28% 5 mL HNO_3 was prepared and named as solution B, and 40 mL of Ba(NO_3)$_2$ (in the range of 10–25 mmol) solution as solution C. These three solutions (A, B, and C) were mixed to form a white hydrogel and allowed to stand for 1 h. Then it was filtered and washed thoroughly with ethanol. The solid residue was heated at 550 °C to produce the Ba-embedded g-C_3N_4 catalyst. Tian et al. [27] prepared a Se-doped g-C_3N_4 novel catalyst for SMX detoxification. Due to Se doping, nitrogen vacancy is created in the catalyst matrix to modulate the electron distribution of g-C_3N_4. Additionally, it also helped in exfoliation and creating a large specific surface area of the composite catalyst due to the large radius of the Se atom. Guo et al. [14] reported the preparation and application of a Cl-doped novel porous g-C_3N_4 nanosheet photocatalyst for TC degradation under the irradiation of visible light. For doping purposes, different amounts of Cl were added, and the composite was named accordingly as CN-Cl-0.1, CN-Cl-0.3, CN-Cl-0.5, etc.

3. Different Categories of g-C_3N_4-Based Catalysts Reported in the Literature for PC Degradation

It has been already mentioned in the previous section that although g-C_3N_4 is a promising photocatalytic material, several drawbacks hinder its application. Therefore, researchers look forward to constructing different types of composite materials for enhanced photocatalytic activity. The main purpose of developing a novel g-C_3N_4-based photocatalyst is to reduce the bandgap and prevent the electron–hole recombination. For this purpose, photocatalysts having Z-scheme and S-scheme heterostructures are synthesized. Most of the g-C_3N_4-based catalysts have been found to be efficient under visible light irradiation. However, in some cases, the inclusion of some external agents, such as H_2O_2, peroxymonosulfate (PMS), and ultrasound assistance, facilitates the degradation process.

3.1. g-C_3N_4-Based Z-Scheme Photocatalysts

Z-scheme photocatalysts can be categorized into traditional Z-scheme, direct Z-scheme, and all-solid-state Z-scheme photocatalysts. Traditional Z-scheme photocatalysts were first introduced by Bard in 1979 [28]. It resembles the photosynthetic activity performed by green plants. This type of photocatalysts consists of two semiconductors with suitable intermediate couples. The two semiconductors used in this system have staggered band structure configurations. However, the major disadvantage of this type of catalyst is that it is confined to the solution phase only. Moreover, there is also a possibility of the occurrence of many side reactions.

All-solid-state Z scheme is also known as the indirect Z scheme. The idea of the all-solid-state Z scheme started in the year of 2006. Tada et al. [29] synthesized CdS-Au-TiO_2 ternary composite photocatalyst. In this composite, Au acts as the electron mediator. Noble metals, such as Ag, Au, and Cu nanoparticles, are often explored as the electron mediator. Other than these, carbon quantum dots, graphene, and carbon nanotubes are also used as the electron mediator. As the solid conductor is used in all-solid-state Z scheme, it can be easily utilized in liquid as well as in gas.

However, in all-solid-state Z-scheme-based catalysts, charge transfer is solely based on the conductor. To improve the situation, a direct Z scheme was proposed. In the direct Z-scheme mechanism, no intermediate redox couples exist. Many recent studies on the application of g-C_3N_4-based catalysts are developed based on a direct Z-scheme mechanism. The photodegradation of TC by applying 2D/2D $MnIn_2S_2$/g-C_3N_4 composite is based on the direct Z-scheme mechanism [30]. Photodegradation of DC via S, B-co-doped g-C_3N_4 nanotube@MnO_2 catalyst also proceeded via a direct Z-scheme mechanism [3]. The main reactive species involved in the degradation mechanism were h^+, $O_2^{\cdot -}$, and $SO_4^{\cdot -}$. Moreover, the catalyst showed excellent stability up to 10 cycles. Ghosh and Pal [8], in their recent work, reported the synthesis and application of composite formed by g-C_3N_4 nanosheet, tungsten oxide hydrate nanoplates, and carbon quantum dots towards TC degradation under visible light exposure. It worked on the principle of all-solid-state Z schemes, and the degradation proceeded with a rate constant of 0.044 min^{-1}.

3.2. g-C_3N_4-Based S Scheme Photocatalysts

S scheme is the other name of the step scheme photocatalyst. An S scheme photocatalyst comprises an oxidation photocatalyst and a reduction photocatalyst with a staggered band structure. Its band structure is similar to that of the type II heterojunction but a completely different charge transfer route. Pham et al. [31] prepared S scheme α-Fe_2O_3/g-C_3N_4 nanocomposites as an efficient photocatalyst for the degradation of model antibiotic compounds, such as amoxicillin (AMX) and cefalexin (CFX). Ni et al. [32] prepared a novel g-C_3N_4/TiO_2 catalyst for TCH degradation under UV light irradiation. The S-scheme heterostructure facilitated the formation of $\cdot O_2^-$, h^+, and $OH^\cdot$ reactive species which ultimately helped in degradation performance. Feyzi et al. [19] reported the application of S-scheme ZnO/g-C_3N_4/zeolite P supported catalyst for the degradation of TC molecule. The ternary composite catalyst was loaded on a plasma reactor for degradation purposes. Under optimized reaction conditions, 95.5% degradation efficiency was achieved. Guo et al. [23] synthesized an S-scheme-based novel g-C_3N_4/TiO_2/CFs catalyst for the photocatalytic degradation of TCH. Under 350 W Xe light irradiation, 99.9% degradation was achieved within 90 min.

3.3. g-C$_3$N$_4$-Based Fenton-Type Catalysts

Fenton-type reactions are one of the powerful techniques for the degradation of organic pollutants. It deploys the generation of powerful hydroxyl radicals from H$_2$O$_2$ in the presence of Fe^{2+} as the catalyst. However, there exists a lot of technical problems with the homogeneous Fenton process and therefore, the scientists are continuously developing novel heterogeneous Fenton catalysts for degradation purposes. There are several reported studies on the application of heterogeneous Fenton catalysts for PC wastewater degradation. It is very important to note that g-C$_3$N$_4$ undoubtedly added a new dimension for synthesizing heterogeneous Fenton-type catalysts. In the traditional Fenton process, iron was used as the catalyst, and H$_2$O$_2$ was the oxidant. However, apart from iron, different other transition metals have also been explored as Fenton-type catalysts. Moreover, PMS has replaced H$_2$O$_2$ in many studies. Moreover, visible light irradiation also often enhances degradation and is known as photo-Fenton degradation. This section describes different Fenton, Fenton types, and PMS-mediated g-C$_3$N$_4$-based catalysts for PC degradation.

Zhang et al. [33] reported the application of MnO$_2$/Mn-modified alkalinized g-C$_3$N$_4$ catalyst for TC degradation through the photo-Fenton process. Excellent degradation (96.7%) occurred due to the synergistic effect of the surface-grafted hydroxyl groups, charge transfer via the Z-scheme mechanism, and activation of H$_2$O$_2$ by the redox cycle of Mn(IV)/Mn(III)/Mn(II). He et al. [16] utilized g-C$_3$N$_4$/Fe$_3$O$_4$@MIL-100(Fe) composite for photo-Fenton detoxification of CIP from wastewater. In comparison to the bare g-C$_3$N$_4$ and Fe$_3$O$_4$@MIL-100(Fe), the composite catalyst showed promising performance, exhibiting 94.7% degradation of CIP and having an initial concentration of 200 mg/L in 120 min.

Mei et al. [34] explored a metal-free carboxyl-modified g-C$_3$N$_4$ catalyst for PMS activation in order to degrade model PC, CBZ. As the process did not involve any strong acids or solvents, it has been described as a green low-cost environmentally friendly system. Luo et al. [35] reported the successful application of a Co-MOF-based/g-C$_3$N$_4$ catalyst for degrading antidepressant PC venlafaxine in the presence of PMS. Wang and Wang [36] studied the degradation process of SMX in the presence of γ-Fe$_2$O$_3$/O-g-C$_3$N$_4$/biochar composite via PMS activation. SMX eradication followed a first-order kinetic model with a rate constant value of 0.153 min^{-1}. Wang et al. [25] utilized a Fe-Co-O-co-doped g-C$_3$N$_4$ catalyst for PMS activation in order to degrade SMX molecules. Detailed experimental investigation revealed that both sulfate radical and singlet oxygen were present in the reaction mixture. However, the role of singlet oxygen in the SMX removal process was not clear. Superb degradation efficiency was attributed to the existence of the synergism between the metal oxide and O-g-C$_3$N$_4$.

Liu et al. [37] developed a novel Z-scheme Fe-g-C$_3$N$_4$/Bi$_2$WO$_6$ heterogeneous photo-Fenton catalyst for TC degradation purposes. Experimental investigation revealed that ^{1}O$_2$ and ·O$_2^-$ were the predominant species participating in the degradation phenomenon. Li et al. [38] prepared a g-C$_3$N$_4$/MgO composite which acted as a Fenton-type catalyst for the oxidation of the SMX drug present in wastewater. The composite catalyst showed excellent performance towards the degradation. The H$_2$O$_2$ requirement was also less for oxidative degradation.

Li and Gan [39] applied Cu-doped g-C$_3$N$_4$ composite as the heterogeneous photo-Fenton-type catalyst for the degradation of different PCs at a very low concentration. Due to the doping, the bandgap got reduced from 2.79 eV to 2.17 eV, which proved beneficial for adsorbing visible sunlight for degradation purposes. Cao et al. [40] synthesized novel Fe/g-C$_3$N$_4$/kaolinite as the heterogeneous photo-Fenton catalyst for TCH degradation purposes. High degradation efficiency was achieved due to the large specific surface area of the kaolinite, which can result in the adsorption of the TCH molecule. Moreover, Fe(III) acted as the electron acceptor in the composite matrix, restricted the electron–hole recombination rate, and facilitated the photo-Fenton process. g-C$_3$N$_4$ nanosheets/schwertmannite nanocomposites were explored by Qiao et al. [41] for chlortetracycline eradication from wastewater through the photo-Fenton mechanism.

3.4. g-C$_3$N$_4$-Based Sonocatalysts for PCs Degradation

Ultrasound is often deployed for the destruction of stable organic molecules in wastewater streams. Experimentally it has been proved that sound waves having a frequency greater than 20 kHz have the capability of breaking organic pollutants of higher molecular weight into simpler products. Some g-C$_3$N$_4$-based sonocatalysts have also been reported in the literature related to the degradation of PCs. Zhang et al. [42] utilized CoFe$_2$O$_4$/g-C$_3$N$_4$ composite as the sonocatalyst for the degradation of the TCH molecule. Maximum sonocatalytic efficiency of 26.71% in 10 min was exhibited by the composite catalyst when the amount of CoFe$_2$O$_4$ in the catalyst matrix was 25%. Charge transfer and electron–hole separation mechanism proceeded via S scheme heterojunction. The mechanism is shown in Figure 3.

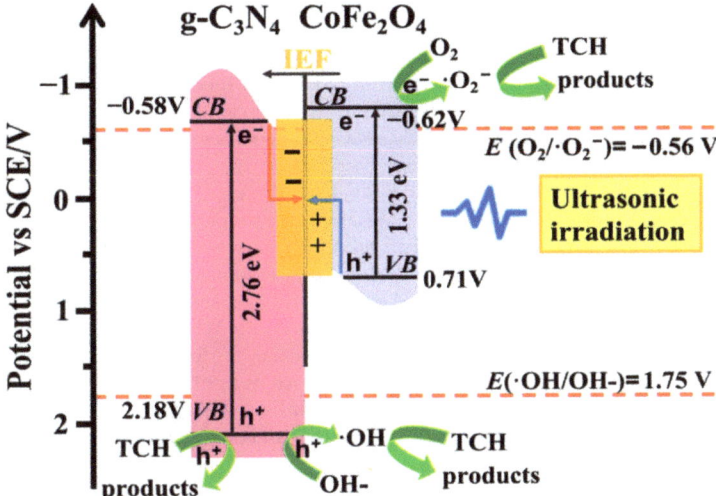

Figure 3. Proposed mechanism for the sonocatalytic degradation of TCH with CFO/CN [42].

He et al. [43] prepared g-C$_3$N$_4$/MoS$_2$ catalyst for levofloxacin (LFX) oxidation through a sonocatalytic mechanism. The as-prepared catalyst showed excellent degradation efficiency towards LFX (75.81%) with an initial concentration of 10 mg/L in 140 min with promising reusability. Experimental investigation revealed that both ˙OH and ˙O$_2^-$ played a major role in the degradation process. The authors validated the sonocatalytic degradation via hotspot and sonoluminescence effect theory.

Vinesh et al. [44] described the application of r-GO supported g-C$_3$N$_4$ nanosheet catalyst for sonophotocatalytic degradation of TC. Influence of ultrasound generated more active sites in the catalyst, which ultimately improved PC degradation. For a TC sample having a 15 mg/L initial concentration, almost 90% degradation was achieved within 60 min of reaction under sonophotocatalysis. Gholami et al. [45] applied Zn-Cu-Mg mixed metal hydroxide/g-C$_3$N$_4$ composite as the effective sonophotocatalyst for the degradation of sulfadiazine (SDZ) drug from wastewater. When the content of mixed metal hydroxide remained at 15 wt%, maximum degradation efficiency of 93% was attained with an initial concentration of SDZ as 0.15 mM, solution pH 6.5, and ultrasonication power of 300 W.

4. Degradation of Different PCs

4.1. Tetracycline (TC)

TC is one of the most common drugs, often explored to test the catalytic efficiency of a newly developed catalyst. It is a broad-spectrum antibiotic. Its major functional groups include phenolic hydroxyl group, dimethylamino group, acylamino group, etc. Owing to the possession of both electron-rich and electron-deficit groups, it has three dissociation constants [1]. TC, OTC (oxytetracycline), and TCH (tetracycline hydrochloride) all belong to the TC group of drugs. Chen et al. [22] explored Ag-AgVO$_3$/g-C$_3$N$_4$ photocatalyst for the degradation of TC in aqueous media. The photocatalytic activity of the composite catalyst was two times higher than that of the pristine g-C$_3$N$_4$. A highly promising MnIn$_2$S$_4$/g-C$_3$N$_4$ photocatalyst was developed by Chen et al. [30] and successively applied for TCH degradation. The composite catalyst exhibited higher degradation efficiency in comparison to the MnIn$_2$S$_4$ nanoflakes and mesoporous g-C$_3$N$_4$ nanosheets. The enhanced removal occurred due to the formation of the Z scheme which results in the transfer and effective separation of the photogenerated charge carriers. Chi et al. [13] reported the exploration of B/Na-co-doped g-C$_3$N$_4$ photocatalyst for the degradation of TC. The synergistic effect between the B/Na co-doping and the porous g-C$_3$N$_4$ nanosheet played a major role in the degradation process. Within a reaction time of 30 min, under visible light irradiation (λ = 430 nm), 78.39% degradation was achieved. Cl-doped g-C$_3$N$_4$-based catalysts were explored by Guo et al. [14] for TC degradation purposes. The optimized removal efficiency of 92% within 120 min of reaction time has been achieved. Due to the Cl doping, the electronic structure of g-C$_3$N$_4$ material was regulated. Moreover, because of the Cl doping, the specific surface area of the composite got increased and the recombination of hole–electrons on the catalyst surface was prevented. Thus ultimately, TC degradation was improved in comparison to the bulk g-C$_3$N$_4$ catalyst.

Jiang et al. [46] developed a nitrogen self-doped g-C$_3$N$_4$ nanosheet catalyst following a self-doping and thermal exfoliation procedure. Thus, the newly developed nitrogen-doped g-C$_3$N$_4$ photocatalyst showed enhanced visible light absorption, high specific surface area, and improved electron–hole separation in comparison to the bulk g-C$_3$N$_4$. Hence, it showed promising efficiency towards TC degradation. The authors proposed an interesting mechanism behind the photocatalytic degradation of TC. The whole degradation process progressed via three steps, such as light harvesting, photogenerated electron–hole pairs separation and transfer, and surface adsorption and redox reaction. Moreover, the catalyst showed excellent repeatability and only a negligible efficiency is lost during the fifth cycle of reuse.

Palanivel et al. [47] reported the synthesis of a novel NiFe$_2$O$_4$-deposited S-doped g-C$_3$N$_4$ nanorod catalyst and deployed it for TC degradation through a photo-Fenton mechanism. Strong chemical interaction between the NiFe$_2$O$_4$ and sulfur-doped g-C$_3$N$_4$ helps in efficient visible light absorption, and electron–hole recombination is also prevented effectively. Preeyangha et al. [48] prepared g-C$_3$N$_4$/BiOBr/Fe$_3$O$_4$ nanocomposite photocatalyst for the degradation of TC under the irradiation of visible light. Complete degradation and 78% TOC removal were achieved within 60 min, where the reaction rate constant was found six times higher than that obtained with the bare g-C$_3$N$_4$ catalyst. Based on the radical scavenging investigations, it was observed that h$^+$ has a major role in the degradation followed by $\cdot$O$_2^-$ and OH$\cdot$. Besides that, the ternary composite showed excellent recyclability also. The possible reaction mechanism is shown in Figure 4.

Various reported g-C$_3$N$_4$-based catalysts for TC removal are presented in Table 1.

Table 1. A list of g-C$_3$N$_4$-based catalysts reported for TC degradation.

Description of the g-C$_3$N$_4$-Based Photocatalyst	Optimized Degradation Efficiency with Reaction Condition	References
B/Na-co-doped porous g-C$_3$N$_4$ nanosheet photocatalyst	TC degradation of 78.39% within 30 min under visible light irradiation (10 W LED lamp)	[13]
Cl-doped porous g-C$_3$N$_4$ nanosheets	At a catalyst dose of 0.5 g/L, TC concentration = 10 mg/L, under visible light irradiation (300 W Xenon lamp, with cut-off filter at 420 nm), 92% degradation within 120 min reaction time	[14]
ZnO/g-C$_3$N$_4$/zeolite P supported catalyst	95.5% TC degradation in plasma reactor (16.5 kV as operating voltage, 300 Hz regulated frequency, airflow rate = 130 mL/min)	[19]

Table 1. *Cont.*

Description of the g-C$_3$N$_4$-Based Photocatalyst	Optimized Degradation Efficiency with Reaction Condition	References
Ag-AgVO$_3$/g-C$_3$N$_4$ composite	83.6% degradation at 120 min (rate constant = 0.0298 min^{-1}) under visible light irradiation (300 W Xenon lamp, with 410 nm filter): TC concentration = 30 mg/L, catalyst dose = 0.2 g/L	[22]
g-C$_3$N$_4$/TiO$_2$/CFs	99.99% TC-HCl degradation (initial concentration = 10 mg/L) with a catalyst dose of 0.5 g/L, under the irradiation of visible light (350 W Xe lamp) for 90 min	[23]
MnIn$_2$S$_4$/g-C$_3$N$_4$ photocatalyst	With TCH concentration of 50 mg/L, catalyst dose = 1 g/L (g-C$_3$N$_4$ kept as 20% mass ratio in the composite), under visible light irradiation (300 W Xenon lamp, with 400 nm filter) almost complete degradation	[30]
g-C$_3$N$_4$/TiO$_2$	In the presence of catalyst (g-C$_3$N$_4$:TiO$_2$ = 1:25) at a dose = 1 g/L, under UV light irradiation (300 W Mercury lamp), maximum degradation efficiency obtained 97.6% for TCH in 90 min	[32]
Porous Z-scheme MnO$_2$/Mn-modified alkalinized g-C$_3$N$_4$ heterojunction	With 0.5 g/L catalyst dose, 96.7% TC removal	[33]
Fe-g-C$_3$N$_4$/Bi$_2$WO$_6$ heterojunctions	98.42% degradation in the presence of 1 mM of H$_2$O$_2$, with TC = 10 mg/L, catalyst dose = 0.4 g/L, solution pH = 6.5	[37]
rGO supported self-assembly of 2D nanosheet of (g-C$_3$N$_4$)	With 0.25 g/L catalyst dose, TC concentration = 15 mg/L, almost complete degradation (under visible light irradiation and exposure to ultrasound)	[44]
Nitrogen self-doped g-C$_3$N$_4$ nanosheets	With 0.5 g/L catalyst dose, 10 mg/L TC concentration, under visible light irradiation, 81.67% degradation in 60 min	[46]
NiFe$_2$O$_4$-deposited S-doped g-C$_3$N$_4$ nanorod	97% degradation in 60 min under visible light irradiation	[47]
g-C$_3$N$_4$/BiOBr/Fe$_3$O$_4$ nanocomposite	With catalyst dose = 0.5 g/L, TC = 15 mg/L, under visible light irradiation (300 W Halogen lamp), complete degradation in 60 min	[48]
Ba-doped g-C$_3$N$_4$ photocatalyst	91.94% TC degradation within 120 min under visible light irradiation at 2% Ba loading at a solution pH 10	[49]
g-C$_3$N$_4$/MoS$_2$ p-n heterojunction photocatalyst	Using photocatalyst dose of 1 g/L, TC concentration of 20 mg/L, irradiation under Xe lamp (300 W) complete degradation achieved within 40 min (rate constant = 547 × 10^{-4} min^{-1})	[50]
PdO/g-C$_3$N$_4$/kaolinite catalyst	At 4% loading of PdO, catalyst dose (0.5 g/L) 94.5% degradation of TCH (40 mg/L) by PMS activation under visible light (300 W Xenon lamp, with 420 nm cut-off filter) irradiation within 20 min	[51]
Nitrogen-doped carbon quantum dots modified g-C$_3$N$_4$ composite	90% TCH degradation under the action of 0.5 g/L catalyst dose, TCH concentration = 20 mg/L, peroxydisulphate (PDS) dose = 0.5 g/L, under visible light irradiation (300 W Xenon lamp, with 420 nm cut-off filter) within 60 min reaction time	[52]
CoO/g-C$_3$N$_4$ p-n heterojunction	Initial concentration of TC = 10 mg/L, catalyst dose = 0.5 g/L (30 wt% CoO), 90% degradation within 60 min under visible light irradiation (300 W Xenon lamp, with cut-off filter at 420 nm)	[53]
CuInS$_2$/g-C$_3$N$_4$ heterojunction photocatalyst	83.7% degradation within 60 min, initial concentration of TC = 20 mg/L, catalyst dose = 0.5 g/L (CuInS$_2$ mass = 50 wt%) under visible light irradiation (300 W Xenon lamp, with cut-off filter at 420 nm)	[54]

Table 1. Cont.

Description of the g-C_3N_4-Based Photocatalyst	Optimized Degradation Efficiency with Reaction Condition	References
CoP nanoparticles anchored on g-C_3N_4 nanosheets	96.7% degradation within 120 min reaction time under visible light irradiation (500 W Xenon lamp with 520 nm cut-off filter)	[55]
g-C_3N_4/RGO/In_2S_3	With initial TCH concentration = 20 mg/L, catalyst dose = 0.5 g/L, 95.6% degradation in 60 min under visible light irradiation	[56]
Metal-free g-C_3N_4-based heterojunction photocatalyst	Catalyst dose of 2 g/L, TC concentration = 20 mg/L, under visible light irradiation (300 W Xe lamp with 420 nm filter) 91% removal in 100 min	[57]
$Bi_xO_yI_z$/g-C_3N_4	TCH concentration = 10 mg/L, under visible light irradiation (500 W Xe lamp with 420 nm filter), 40% degradation in 4 h	[58]
$Bi_2W_2O_9$/g-C_3N_4 heterojunction	2wt% $Bi_2W_2O_9$ in the matrix, with 1 g/L catalyst dose, with initial concentration of TCH = 10 mg/L, under visible light irradiation (35 W Xe lamp), at pH 10.54, 95% degradation occurred	[59]
Carbon-doped g-C_3N_4	More than 95% degradation in 90 min reaction time under visible light irradiation	[60]
Single-atom Fe-g-C_3N_4 catalyst	With TC concentration = 10 mg/L, catalyst dose = 0.1 g/L, PMS = 0.25 mM, 93.29% degradation achieved	[61]
Ag-modified g-C_3N_4 composite	With 8 wt% Ag in the matrix, 1 g/L catalyst dose, 20 mg/L TC concentration, under visible light irradiation (300 W Xe lamp, with 420 nm filter), at pH 11, 90% degradation achieved	[62]
g-C_3N_4/Ag_2CrO_4 photocatalyst	With catalyst dose 1 g/L, TC concentration = 10 mg/L, under visible light irradiation (1000 W halogen lamp), almost complete degradation in 180 min	[63]
FeOOH coupling and nitrogen vacancies functionalized g-C_3N_4 heterojunction	With 4 g/L catalyst dose, initial concentration of OTC = 10 mg/L, under visible light irradiation (300 W Xe lamp, with 420 nm filter), 92.83% degradation in 90 min	[64]
Co-doped KCl/NH_4Cl/g-C_3N_4 catalyst	With catalyst dose = 1 g/L, TC concentration = 10 mg/L, under visible light irradiation (500 W Xe lamp with 420 nm filter), almost complete degradation in 120 min	[65]
Potassium-gluconate-cooperative pore generation based on g-C_3N_4 nanosheets	With catalyst dose = 1 g/L, TC concentration = 20 mg/L, under visible light irradiation (300 W Xe lamp, with 420 nm filter), 82.2% degradation in 30 min	[66]
Sulfur-doped carbon quantum dots loaded hollow tubular g-C_3N_4	With catalyst dose = 1 g/L, TC concentration = 20 mg/L, under visible light irradiation (300 W Xe lamp), about 90% degradation in 60 min	[67]
Nano-confined g-C_3N_4 in mesoporous SiO_2	With 0.33 g/L catalyst dose, TC concentration = 20 mg/L, under visible light irradiation (300 W Xe lamp), complete degradation in 120 min	[68]
Multifunctional 2D porous g-C_3N_4 nanosheets hybridized with 3D hierarchical TiO_2 microflowers	With 0.5 g/L catalyst dose, TC concentration = 20 mg/L, under visible light irradiation, 90% degradation achieved in 60 min	[69]
CuO/g-C_3N_4 2D/2D heterojunction photocatalysts	With catalyst dose of 0.1 g/L, 30 mg/L OTC, under visible light irradiation (300 W Xe lamp, with 420 nm filter), 100% degradation in 10 min	[70]
Sulfur- and tungstate-co-doped porous g-C_3N_4 microrods	With 0.5 g/L catalyst dose, TC concentration of 10 mg/L, under visible light irradiation (300 W Xe lamp, with 420 nm filter), 85.3% degradation in 120 min	[71]
Supramolecular self-assembly synthesis of noble-metal-free (C, Ce) co-doped g-C_3N_4 with porous structure	With 0.5 g/L catalyst dose, 10 mg/L TC concentration, 90% degradation in 60 min	[72]

Table 1. Cont.

Description of the g-C_3N_4-Based Photocatalyst	Optimized Degradation Efficiency with Reaction Condition	References
$Bi_2O_2CO_3$/g-C_3N_4/ Bi_2O_3	With catalyst dose = 0.2 g/L, TC = 10 mg/L, under visible light irradiation (300 W Xe lamp), 95% degradation in 60 min	[73]
Fe-doped surface-alkalinized g-C_3N_4	With catalyst dose = 0.5 g/L, TC concentration = 20 mg/L, under visible light irradiation (300 W Xe lamp), 70% degradation in 80 min	[74]
Donor–acceptor structured g-C_3N_4	With 0.5 g/L catalyst dose, OTC concentration = 20 mg/L, degradation of 93% at 60 min	[75]
C-doped g-C_3N_4/WO_3	With catalyst dose = 1 g/L, TC concentration = 10 mg/L, under visible light irradiation (500 W Xe lamp with 420 nm filter), ~78% degradation in 60 min	[76]
g-C_3N_4/$NiFe_2O_4$ S scheme	79.3% degradation at pH 3	[77]

4.2. Diclofenac (DC)

DC is also another drug like TC which is also investigated by researchers for degradation purposes. It is one of the notable members of the nonsteroidal anti-inflammatory drug group possessing high K_{ow} and bioaccumulating potential power. Due to its wide unrestricted usage during the last four decades, it has become one of the significant emerging contaminants of the current era. Various reports are available in the literature describing the applicability of g-C_3N_4-based photocatalysts towards sustainable eradication of DC from wastewater. Li et al. [78] applied Fe oxide nanoclusters supported on g-C_3N_4 as a robust photocatalyst for DC degradation purposes. He et al. [4] prepared a heterostructure of Ti_3C_2/g-C_3N_4 photocatalyst for the degradation of DC. Polymeric g-C_3N_4 catalyst was explored by Papamichail et al. [79] for degradation of DC. Quantum carbon dots modified reduced ultrathin g-C_3N_4 photocatalyst was designed and applied by Jin et al. [80] for the oxidative degradation of DC from wastewater. Complete degradation of DC took place within 6 min. Hu et al. [81] deployed g-C_3N_4/TiO_2 photocatalyst for the degradation of DC and CBM. The mechanism is shown in Figure 5.

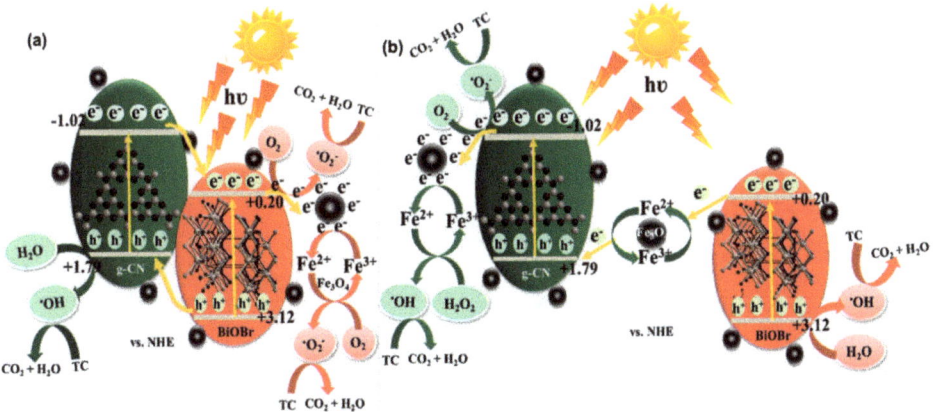

Figure 4. The schematic representation of plausible charge transfer mechanism (**a**) type II and (**b**) Z-scheme heterojunction during the photocatalytic TC degradation over g-CN/BiOBr/Fe_3O_4 nanocomposites [48].

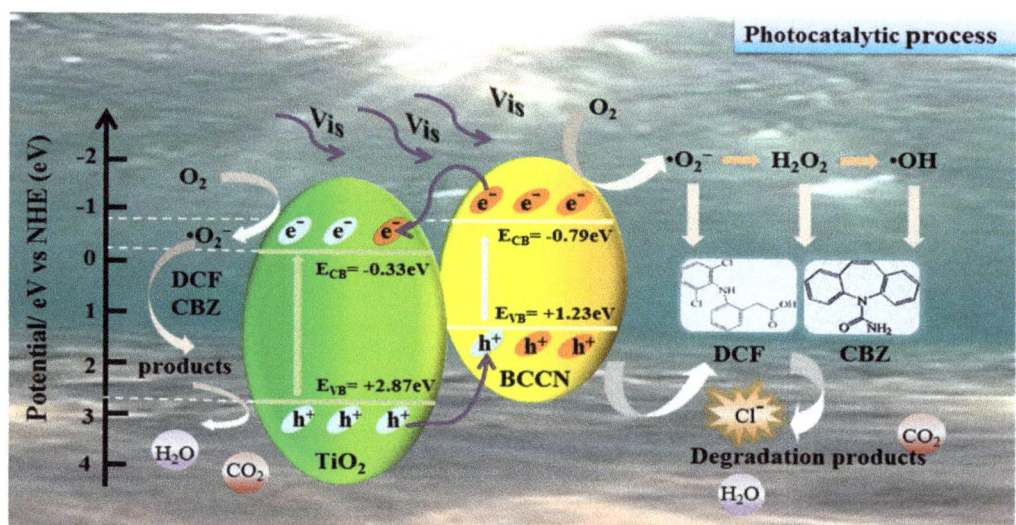

Figure 5. Possible mechanism for the photodegradation of DCF and CBZ under LED lamp irradiation over 30% BCCNT composites [81].

A comprehensive list regarding the application of different g-C_3N_4-based photocatalysts for the degradation of DC is provided in Table 2.

Table 2. A list of g-C_3N_4-based catalysts reported for DC degradation.

Description of the Catalyst	Reaction Conditions for Optimum Degradation Efficiency	Optimized Degradation Efficiency	References
Z-scheme S, B-co-doped g-C_3N_4 nanotube@MnO_2 heterojunction	In the presence of 0.06 mM PMS, photocatalyst dose of 0.5 g/L, DC concentration = 20 mg/L, under the irradiation of 8 × 8 W visible light lamps at a wavelength of 460 nm	99% degradation	[3]
2D/2D heterostructure of Ti_3C_2/g-C_3N_4	Initial concentration of DC = 10 mg/L, catalyst dose = 0.25 g/L, PMS concentration = 0.25 g/L	100% degradation efficiency within 30 min	[4]
Ferric oxide nanoclusters anchored g-C_3N_4 nanorods	Initial concentration of DC = 1 mg/L, dose of catalyst = 0.1 g/L, irradiation under 300 W Xenon arc lamp	Kinetic rate constant of 0.206 min^{-1}	[78]
Polymeric g-C_3N_4 photocatalyst	At a catalyst dose of 1 g/L, initial concentration of DC = 20 mg/L, solution pH = 5	Complete removal within 120 min	[79]
TiO_2/g-C_3N_4	Initial concentration of DC = 5 mg/L, 0.3 g of catalyst loading, pH = 5, irradiation under 1000 W halogen lamp	Maximum degradation efficiency of 93.49%	[82]
Cellulose biochar/g-C_3N_4 composite ($WPBC_{50}$/g-C_3N_4)	At a DC concentration of 0.05 mM, catalyst dose of 1.5 g/L, 3 mM of PMS under visible light irradiation	Complete removal within 25 min	[83]
g-C_3N_4 nanosheets	Initial concentration of DC = 3 mg/L, catalyst dose = 0.65 g/L under solar and LED irradiation	-	[84]
Tunable V_2O_5/boron-doped g-C_3N_4 composite	With 5 wt% B doping, 2 g/L catalyst dose	100% degradation within 105 min under visible light irradiation	[85]
g-C_3N_4/NH_2-MIL-125 photocatalyst	Under the action of the catalyst composed of MOF and g-C_3N_4 in the ratio 50:50. DC concentration kept at 10 mg/L, under UV LED irradiation at 384 nm	Complete eradication within 2 h	[86]

4.3. Sulfamethoxazole (SMX)

SMX is another widely used sulfonamide antibiotic. Due to its overuse, it often comes across in different quantities in different water bodies. In recent years, several studies report the catalytic elimination of sulfamethoxazole from water bodies by deploying g-C_3N_4-based composite catalysts. $ZnIn_2S_4$/g-C_3N_4 photocatalyst was utilized by Reddy et al. [87] to catalytically degrade SMX in a water medium. Tian et al. [27] fabricated a Se-doped g-C_3N_4 catalyst for the degradation of SMX through PMS activation. Experimental findings showed that 93% SMX can be degraded by the synthesized catalyst within 180 min with a reaction rate constant of 0.0149 min^{-1}. The rate constant of the doped catalyst was four times higher in comparison to the bulk g-C_3N_4 catalyst. The inclusion of Se in the composite matrix created nitrogen vacancy to modulate the electron distribution of the g-C_3N_4 catalyst. Peng et al. [88] utilized a "trap-zap" catalyst for the degradative elimination of SMX through PMS activation. β-Cyclodextrin polymer composite with Fe-doped g-C_3N_4 catalyst was used for the degradation. The degradation rate constant by the β-CDPs/Fe-g-C_3N_4 catalyst was found as 0.132 min^{-1} which was 14.7 times and 2.2 times higher than that of the g-C_3N_4 and Fe-g-C_3N_4 catalyst. The inclusion of β-CDPs in the catalyst composition accelerated the electron transfer between the catalyst and PMS. Li et al. [89] applied a $FeCo_2S_4$-modified g-C_3N_4 catalyst for the degradation of the SMX drug. Optimized degradation efficiency has been obtained at the unadjusted pH of 6.5. On the other hand, acidic pH inhibited the degradation. Temperature played a major role in the degradation process. When the reaction temperature was increased from 10 °C to 40 °C, the degradation efficiency got enhanced from 61.2% to 99.9% with a rate constant value of 0.294 min^{-1}. The schematic for the mechanism is shown in Figure 6.

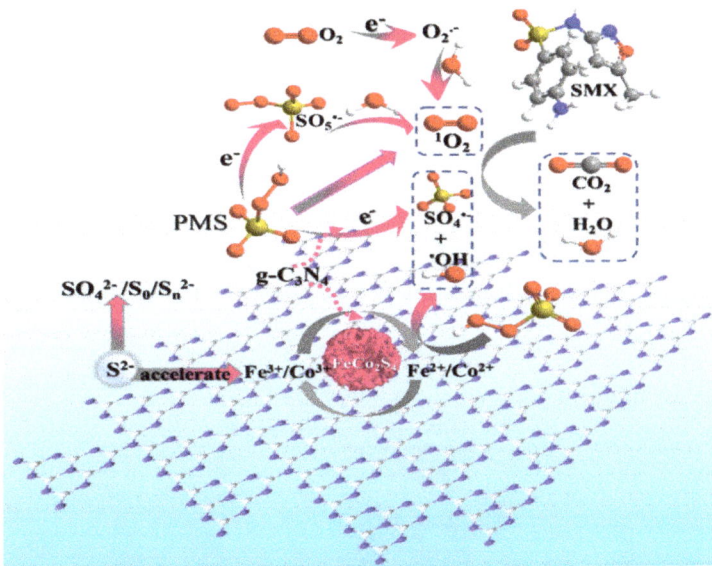

Figure 6. Schematic illustration of the mechanism of PMS activation on $FeCo_2S_4$-CN [89].

A comprehensive list regarding the application of different g-C_3N_4-based photocatalysts for the degradation of SMX is provided in Table 3.

Table 3. A list of g-C_3N_4-based catalysts reported for SMX degradation.

Description of the Catalyst	Reaction Condition	Optimized Degradation Efficiency	References
ZnO@g-C_3N_4	Photocatalyst dose of 0.65 g/L, pH 5.6, airflow rate of 1.89 L/min	90.4% oxidative removal within 60 min of reaction time	[20]
Quaternary magnetic BiOCl/g-C_3N_4/Cu_2O/Fe_3O_4 nano-junction	Degradation reaction with catalyst dose of 0.2 g/L, with initial concentration of SMX = 25.328 mg/L, under visible light irradiation (800 W Xenon lamp, with 400 nm cut-off filter)	99.5% degradation under Xenon lamp for irradiation within 1 h and 92.1% degradation under natural sunlight within 2 h	[24]
Fe-Co-O-co-doped g-C_3N_4	-	Complete degradation of 0.04 mM of sulfamethoxazole within 30 min at a reaction rate of 0.085 min^{-1}	[25]
Se-doped g-C_3N_4	-	93% degradation of SMX in 180 min with a rate constant of 0.0149 min^{-1}	[27]
g-C_3N_4/MgO composite	With the dose of the catalyst = 0.2 g/L, initial concentration of SMX = 20 mg/L	92 ± 3% degraded within 3 h	[38]
Heterostructured 2D/2D $ZnIn_2S_4$/g-C_3N_4 nanohybrids	In the presence of 0.2 g/L catalyst, initial concentration of SMX = 15 mg/L, under visible light irradiation (solar simulation AM 1.5 G, intensity 100 mW/cm^{-2})	89.4% degradation in 2 h	[87]
β-CDPs/Fe-g-C_3N_4 catalyst	In the presence of 0.2 g/L catalyst, 2 mM of PMS	Rate constant value of 0.132 min^{-1}	[88]
$FeCo_2S_4$-modified g-C_3N_4 photocatalyst	At a pH of 6.5, a reaction temperature of 40 °C	91.9% degradation with a rate constant of 0.151 min^{-1}	[89]
Fe-dispersed g-C_3N_4 photocatalyst	Initial concentration of sulfamethoxazole 10 mg/L, dose of catalyst 50 mg/L	98.7% degradation within 6 min	[90]
AgCl/Ag_3PO_4/g-C_3N_4	With 43% AgCl in the matrix	95% removal within 2 h of reaction time	[91]
Porous g-C_3N_4 modified with ammonium bicarbonate	Under the action of 0.05 g/L catalyst dose, initial concentration of SMX = 0.5 mg/L, at pH 9, under visible light irradiation (150 W Xenon lamp)	93.37% degradation within 30 min	[92]
Ag_3PO_4/g-C_3N_4	With a catalyst dose of 1 mg/L, under the irradiation of Xenon lamp	Complete degradation within 90 min of visible light irradiation	[93]
Ag/g-C_3N_4	Under the action of 0.05 g/L catalyst dose, 2.538 mg/L of initial concentration of SMX, under the irradiation of Xe lamp (300 W with 400 nm cut-off filter)	99.5% degradation using 10 wt% Ag in the matrix	[94]
g-C_3N_4 nanotubes	Under the action of 0.4 g/L catalyst dose, 100 mg/L of SMX, under irradiation of Xe lamp (300 W)	Complete degradation within 120 min of reaction time	[95]
Porous loofah-sponge-like ternary heterojunction g-C_3N_4/Bi_2WO_6/MoS_2	-	Under visible light irradiation, over 99% degradation took place in 60 min with a rate constant value of 0.089 min^{-1}	[96]

4.4. Ibuprofen (Ibu)

Ibu is also a nonsteroidal anti-inflammatory drug commonly used for pain relief, fever, and inflammation purposes. In recent years, some studies have been performed on using g-C_3N_4-based catalysts for degrading Ibu from aqueous media. Liu and Tang [97] reported the application of g-C_3N_4/Bi_2WO_6/rGO heterostructured nanocomposite for the degradation of Ibu. Ag/g-C_3N_4/kaolinite composite was applied by Sun et al. [18] for the eradication of Ibu. Mao et al. [98] reported the preparation and application of 1D/2D nanorod FeV$_3$O$_8$/g-C$_3$N$_4$ composite catalyst for the degradation of Ibu. In comparison to the g-C$_3$N$_4$ nanosheets, the catalytic activity of this nanocomposite was nearly four times higher. The authors proposed a Z-scheme mechanism for the composite. Meng et al. [99] utilized layered g-C$_3$N$_4$ and BiOBr photocatalysts for the degradation of Ibu in wastewater. A list is provided in Table 4 regarding the application of various g-C$_3$N$_4$-based catalysts towards catalytic eradication of Ibu from water.

Table 4. A list of g-C$_3$N$_4$-based catalysts reported for Ibu degradation.

Description of the Catalyst	Reaction Conditions	Optimized Degradation Efficiency	References
Ag/g-C$_3$N$_4$/kaolinite composite	7 wt% Ag, with catalyst dose = 1 g/L, Ibu = 5 mg/L, under visible light irradiation (500 W Xe lamp, with 400 nm filter)	Almost complete degradation in 300 min with, rate constant = 0.0113 min^{-1}	[18]
g-C$_3$N$_4$/Bi$_2$WO$_6$/rGO heterostructured composites	At a catalyst dose of 0.2 g/L, Ibu concentration = 5 mg/L, pH = 4.3, under visible light irradiation (300 W Xenon lamp with 420 nm filter)	93% degradation with a rate constant of 0.011 min^{-1} under visible light irradiation and 98.6% degradation under sunlight	[97]
1D/2D FeV$_3$O$_8$/g-C$_3$N$_4$	10 wt% FeV$_3$O$_8$ in the matrix, with catalyst dose = 0.33 g/L, Ibu = 10 mg/L, under visible light irradiation (300 W Xenon lamp with 420 nm filter)	95% degradation in 85 min	[98]
Layered g-C$_3$N$_4$ and BiOBr	With catalyst dose = 0.2 g/L, Ibu = 20 mg/L	Complete degradation in 10 min	[99]
g-C$_3$N$_4$/Ag/AgCl/BiVO$_4$ micro flower composite	Under the catalyst dose of 0.25 g/L, under visible light irradiation (compact fluorescent lamps)	94.7% degradation within 1 h of reaction time	[100]
g-C$_3$N$_4$/MIL-68(In)-NH$_2$ heterojunction composite	Under the action of 0.15 g/L of catalyst, 20 mg/L of Ibu concentration, with visible light irradiation (300 W Xenon lamp, with cut-off filter at 420 nm)	Photocatalytic rate = 0.01739 min^{-1}, 93% degradation, in 120 min	[101]
TiO$_2$/g-C$_3$N$_4$ composite	With catalyst dose of 1 g/L, 5 mg/L of Ibu concentration, under the visible light irradiation (250 W Xe lamp)	Almost complete degradation in 60 min	[102]
TiO$_2$/UV and g-C$_3$N$_4$ visible light	With initial concentration of Ibu as 5 mg/L, catalyst dose of 2.69 g/L, at pH 2.51 under the action of 4–10 W LED lamps	Complete degradation in 120 min	[103]
Au-Ag/g-C$_3$N$_4$ nanohybrids	With initial concentration of Ibu as 5 mg/L, catalyst dose of 2.69 g/L, under the action of natural sunlight and 4–10 W LED lamps	Complete degradation in 120 min under natural sunlight	[104]
g-C$_3$N$_4$/CQDs/CDIn$_2$S$_4$	Initial concentration of Ibu = 80 mg/L, dose of catalyst = 0.1 g/L, under visible light irradiation (300 W Xenon lamp with 420 nm filter)	About 90% degradation in 60 min	[105]
Plasma-treated g-C$_3$N$_4$/TiO$_2$	Using g-C$_3$N$_4$/TiO$_2$ catalyst with 15 min treatment of plasma oxygen	95% degradation within 90 min	[106]
Triple 2D g-C$_3$N$_4$/Bi$_2$WO$_6$/rGO composites	3 wt% rGO in the composite, catalyst dose = 2 g/L, Ibu = 5 mg/L, pH = 4.3	86% degradation under visible light and 98% removal under natural sunlight	[107]
g-C$_3$N$_4$/Bi$_2$WO$_6$ 2D/2D heterojunction	With 0.2 g/L catalyst, initial concentration of Ibu = 103.145 mg/L	96.1% degradation efficiency within 1 h,	[108]
α-SnWO$_4$/UiO-66(NH$_2$)/g-C$_3$N$_4$ ternary heterojunction	With 0.5 g/L catalyst dose, Ibu = 10 mg/L under visible light irradiation (Xe lamp)	More than 90% degraded in 120 min reaction time	[109]

4.5. Other Drugs

Studies have also been conducted on the application of g-C_3N_4-based catalysts to degrade ciprofloxacin (CIP) from water bodies. Deng et al. [110] reported the preparation and application of Ag-modified phosphorus-doped ultrathin g-C_3N_4/BiVO$_4$ photocatalyst for the degradation of CIP drug., In the study, more than 92% degradation efficiency was achieved under visible light and near-infrared light irradiation ($\lambda > 420$ nm, $\lambda > 760$ nm), with an initial concentration of CIP 10 mg/L. Zhang et al. [111] applied Fe_3O_4/CdS/g-C_3N_4 composite for the photocatalytic degradation of CIP under visible light irradiation. CdS itself is a photocatalyst. However, the addition of g-C_3N_4 in the composite improved its optical response and the incorporation of Fe_3O_4 nanoparticles helped in the easy recovery of the catalyst. Triclosan (TCS) is a famous nonionic broad-spectrum antimicrobial pharmaceutical compound. However, US Food and Drug Administration banned its usage in 2016 due to the health risk associated with it [112]. In recent years, some research groups also explored g-C_3N_4-based photocatalysts towards triclosan degradation in water. Wang et al. [113] reported the application of g-C_3N_4/MnFe$_2$O$_4$ catalyst for the degradation of TCS through PMS activation. The as-prepared catalyst showed promising behavior in terms of stability and metal leaching. Dechlorination, hydroxylation, and cyclization along with other bond-breaking mechanisms were attributed to the triclosan degradation purpose. In one of the recent articles, Yu et al. [114] described the highly efficient degradation performance of a novel catalyst g-C_3N_4/Bi$_2$MoO$_6$ towards TCS drug. TCS was converted to 2-phenoxyphenol under visible light irradiation. In 180 min, 95.5% degradation was achieved, which was 3.6 times higher in comparison to that obtained with pure g-C_3N_4 catalyst.

Mafa et al. [115] developed a multi-elemental doped g-C_3N_4 catalyst and tested it towards NPX degradation. Pure g-C_3N_4 catalyst showed poor performance in terms of degradation efficiency (21.5%). With rare earth metals loading (1%), the composite catalyst showed excellent behavior (92.9% efficiency) towards drug removal. A heterojunction was formed between the rare earth metal and the g-C_3N_4 surface which provided the defect for facilitating electron–hole separation. The degradation followed the Z-scheme mechanism with visible light absorption, with the participation of superoxide radicals.

Truong et al. [116] utilized ZnFe$_2$O$_4$/BiVO$_4$/g-C_3N_4 photocatalyst for the efficient removal of lomefloxacin antibiotic degradation. Keeping the amount of ZnFe$_2$O$_4$, BiVO$_4$, and g-C_3N_4 in the ratio 1:8:10, the optimized removal efficiency of 96.1% was obtained after keeping the set-up illuminated for 105 min.

Like the above-mentioned drugs, AMX is another commonly used drug which often occurs in the ecosystem. Mirzaei et al. [117] prepared a magnetic fluorinated mesoporous g-C_3N_4 catalyst for the AMX elimination purpose. Fluorination of the catalyst material provides a facilitating condition for the catalysis by even distribution in the aqueous medium. Due to the inclusion of the iron nanoparticles, the removal efficiency further got enhanced due to the formation of the heterostructure. However, on increasing the iron content, the catalytic efficiency got diminished due to the fact, that the nanoparticles covered the active sites. A list of g-C_3N_4-based catalysts for degrading other PCs is provided in Table 5.

Table 5. A list of g-C_3N_4-based catalysts reported for other PC degradation.

Description of the Catalyst	Target Compound	Optimized Degradation Efficiency	References
g-C_3N_4/Fe$_3$O$_4$@MIL-100(Fe)	CIP	94.7% degradation of CIP having an initial concentration of 200 mg/L within 120 min of visible light irradiation	[16]
Mesoporous g-C_3N_4	CIP	92.3% degradation with an initial concentration of 4 mg/L, catalyst dose = 1 g/L	[17]

Table 5. Cont.

Description of the Catalyst	Target Compound	Optimized Degradation Efficiency	References
Ag-modified phosphorus-doped ultrathin g-C_3N_4 nanosheets/$BiVO_4$ photocatalyst	CIP	92.6% degradation efficiency for CIP with an initial concentration of 10 mg/L	[110]
Fe_3O_4/CdS/g-C_3N_4	CIP	81% degradation with an initial concentration of CIP = 20 mg/L, catalyst dose = 0.5 g/L in 180 min reaction time	[111]
Lignin nanorods/g-C_3N_4 nanocomposite	TCS	99.9% removal with an initial concentration of TCS = 10 mg/L, catalyst dose = 0.5 g/L in 90 min time	[112]
g-C_3N_4/$MnFe_2O_4$	TCS	Almost complete degradation of TCS having initial concentration of 9 mg/L, catalyst dose = 0.2 g/L, in 60 min of reaction time	[113]
g-C_3N_4/Bi_2MoO_6	TCS	95.5% oxidative removal of TCS (initial concentration = 2 mg/L), catalyst dose = 1 g/L	[114]
Multi-elemental doped g-C_3N_4	NPX	92.9% removal with initial concentration of naproxen = 10 mg/L, catalyst dose = 0.3 g/L	[115]
$ZnFe_2O_4$/$BiVO_4$/g-C_3N_4	Lomefloxacin	96.1% removal after 105 min of visible light irradiation, with initial concentration of lomefloxacin = 25 mg/L, with dose of catalyst = 0.5 g/L	[116]
Magnetic fluorinated mesoporous g-C_3N_4	AMX	Initial concentration of AMX 91.35 mg/L, dose of catalyst = 1 g/L, 90% removal	[117]
La/FeO_3/g-C_3N_4/$BiFeO_3$	CIP	Almost complete degradation of CIP at initial concentration of 10 mg/L, catalyst dose = 0.4 g/L in 60 min	[118]
S-Ag/TiO_2@g-C_3N_4	TCS	92.3% degradation with TCS concentration = 10 mg/L, pH = 7.8, catalyst dose = 0.2 g/L in 60 min	[119]
g-C_3N_4/NH_2-MIL-88B(Fe)	Ofloxacin	96.5% removal in 150 min, with ofloxacin concentration = 10 mg/L, catalyst dose = 0.25 g/L	[120]
Carbon-rich g-C_3N_4 nanosheet	AMX	Complete degradation in 150 min under irradiation of simulated solar light and in 300 min under irradiation of visible light	[121]
SnO_2/g-C_3N_4	AMX	92.1% AMX removal in 80 min, with initial concentration of AMX = 10 mg/L, dose of catalyst = 0.25 g/L under 300 W Xe lamp irradiation	[122]

4.6. Application on Multiple Compounds

In some studies, prepared g-C_3N_4-based catalysts have been applied on more than one PC rather than one compound. Dai et al. [123] applied surface hydroxylated g-C_3N_4 nanofibers for the catalytic degradation of TCH, DC, and metaprolol (MT). Within 60 min of reaction time, 97.3%, 88.9%, and 63.2% degradation of TC, DC, and MT, respectively, was achieved. Barium-embedded g-C_3N_4 was tested against CBM and DC degradation [26]. Thang et al. [124] prepared Ag/g-C_3N_4/ZnO nanorods photocatalyst and assigned it for the degradation of commercial drugs, such as paracetamol (PR), cefalexin (CF), and AMX. By applying only a catalyst dose of 0.08 g/L, the degradation of a target compound having a concentration of 40 mg/L was possible. Di et al. [2] applied g-C_3N_4/ZnFeMMO composites for the photocatalytic degradation of Ibu and sulfadiazine (SDZ) drugs. The Z-scheme mechanism was found to be appropriate for the overall process. The degradation of Ibu proceeded via h^+ generation in the process while $OH^·$ was responsible for the elimination of SDZ. A list is presented in Table 6, regarding the application of g-C_3N_4-based catalysts for multiple PC degradation.

Table 6. A list of g-C_3N_4-based catalysts reported for multiple PC degradation.

Catalyst	Target Drugs	Optimized Reaction Condition	Reference
g-C_3N_4/TiO_2 nanomaterials	PR, Ibu, DC	With an initial concentration of paracetamol = 25 mg/L, Ibu = 15 mg/L, DC = 25 mg/L, catalyst dose = 0.9 g/L, complete degradation of paracetamol and Ibu was possible; however, DC did not get fully degraded	[15]
Single barium-atom-embedded g-C_3N_4 catalyst	DC and CBM	Almost complete degradation of CBZ (1 mg/L) and DC (8 mg/L) under visible light irradiation in 60 min	[26]
α-Fe_2O_3/g-C_3N_4	CFX and AMX	Complete degradation of both drugs at initial concentration of 20 mg/L within 180 min	[31]
Carbon quantum dots modified reduced ultrathin g-C_3N_4	DC, TCS, NPX	100% degradation within 6 min	[80]
Carbon-doped supramolecule-based g-C_3N_4/TiO_2 composites	DC and CBM	98.92% and 99.77% degradation of DC and CBZ in 30 min and 6 h illumination under LED	[81]
0D/1D Co_3O_4 quantum dots/surface-hydroxylated g-C_3N_4 nanofibers	TC, DC, MT	97.3% degradation for TC, 88.9% for DC, 63.2% for MT in 60 min of reaction time	[123]
Ag/g-C_3N_4/ZnO nanocomposite	PR, AMX, CFX	With an initial concentration of each drug = 40 mg/L, catalyst dose = 0.08 g/L, 78% degradation for PR, 70% for cefalexin, and 35% for AMX	[124]
g-C_3N_4-supported WO_3/BiOCl heterojunction	LFX and TCH	92.5% degradation of TCH at initial concentration of 20 mg/L	[125]
(2D/3D/2D) rGO/Fe_2O_3/g-C_3N_4 nanostructure	TC and CIP	Complete degradation of both compounds at an initial concentration of 50 mg/L within 60 min	[126]

5. Optimization Techniques

Optimization of the photocatalyst is one of the pertinent areas for proper resource utilization. In the present era, multiparameter optimization has gained more predominance in comparison to single-parameter optimization. Response surface methodology (RSM) and artificial neural network (ANN) are commonly used for this purpose. Mirzaei et al. [20] used the RSM technique for the photodegradation of SMX by taking catalyst dose, solution pH, and airflow rate as the variables. The authors used central composite design (CCD) for RSM analysis and the catalyst dose, pH, and airflow rate were varied in the range of 0.4–0.8 g/L, 3–11, and 0.5–2.5 L/min, respectively. Twenty experiments were run for the purpose and the experimental values showed a good correlation with the predicted values (R^2 = 0.9802). The optimum condition was found as 0.65 g/L of photocatalyst dose, pH of 5.6, and airflow rate of 1.89 L/min and under this condition, the removal efficiency achieved was 94%. Shanavas et al. [126] conducted an optimization study regarding the application of rGO-gC_3N_4-based catalysts for TC and CIP degradation. The reaction time, reaction temperature, and molar concentration were chosen as the variables for the purpose.

John et al. [82] explored RSM for the optimization of DC degradation. Four variables, namely, irradiation time, initial solution pH, initial DC concentration, and g-C_3N_4 loading, were chosen for the study. From the analysis, it was found that the optimum reaction conditions were obtained as irradiation time = 90 min, initial solution pH = 5, initial DC concentration = 5 ppm, and g-C_3N_4 loading = 0.3 g/g TiO_2, and the maximum removal efficiency achieved was 93.49%.

Quarajehdaghi et al. [127] optimized photocatalytic degradation of CIP by application of CdS/g-C_3N_4/rGO/CMC catalyst using RSM and ANN approach. Four parameters were chosen for the RSM study, such as initial concentration of CIP, dose of catalyst, pH, and time of irradiation. Using the CCD model, 30 experiments were run. Predicted removal efficiency shows a good correlation with the experimental values. The quadratic model best fitted the trend with a high R^2 value of 0.9827. From the RSM analysis, the condition for the optimized removal efficiency was determined as initial concentration of CIP = 7.89 mg/L, the dose of catalyst = 0.6 g/L, pH = 6.15, and irradiation time = 33.94 min. The predicted optimized removal was 84.25%, while experimentally, the removal was achieved at 81.93%. From the ANN study, the relative importance of different parameters on the degradation efficiency was determined. It was seen that the influence of the dose of catalyst was the highest (34%), followed by pH (30%), irradiation time (26%), and, lastly, initial concentration of CIP (10%). Furthermore, the values obtained from the ANN model were also compared with those obtained from the RSM model as well as with the experimental values. The values found from RSM, ANN, and experimental investigation were quite close to each other.

AttariKhasraghi et al. [128] also used RSM and ANN models for cefoperazone degradation using a zeolite-supported $CdS/g-C_3N_4$ catalyst. CCD model was used for the purpose using four variables, such as dose of catalyst, initial concentration of cefoperazone, pH, and time. The degradation percentages predicted by both the models (RSM and ANN) showed a good correlation with the experimentally obtained values. Using RSM analysis, the optimized condition was found as the dose of catalyst = 0.4 g/L, initial concentration of cefaperazone = 17 mg/L, pH = 9, and time = 80 min. The predicted maximum removal efficiency was 95.66%, while actually 93.23% was achieved. It proves the accuracy of the model.

6. Real Field Application

It is pertinent to check the performance of the catalyst towards real wastewater. For utilization of the novel $g-C_3N_4$-based catalyst in real wastewater treatment, firstly scaling up the process is mandatory. Plasma reactor is one of the innovative reactors which has the feasibility of being applied in the real field scenario. Feyzi et al. [19] used $ZnO/g-C_3N_4$/zeolite-P-supported catalyst in a dielectric barrier discharge plasma reactor for photodegradation of TC.

Kumar et al. [129] reported the application of novel $g-C_3N_4/TiO_2/Fe_3O_4@SiO_2$ photocatalyst for the degradation of Ibu in sewage. Ibu concentration was kept at 2 mg/L, while the dose of the photocatalyst was maintained at 1 g/L and 2 g/L. With a 1 g/L dose of catalyst, only 13% degradation efficiency was achieved, while with a 2 g/L dose, 92% efficiency was obtained. Nivetha et al. [122] tested the potency of the $SnO_2/g-C_3N_4$ nanocomposite photocatalyst towards the degradation of real pharmaceutical effluent after being successful towards degrading AMX. Untreated pharmaceutical effluent shows a strong absorbance at 280 nm. However, after the application of the novel photocatalyst, the peak started decreasing with the fading of the solution visible to the naked eye.

Rapti et al. [130] investigated the photocatalytic efficiency of $g-C_3N_4$ and 1% $MoS_2/g-C_3N_4$ catalysts towards 10 psychiatric drugs in hospital wastewater effluent. Experiments were conducted in a stainless-steel lamp reactor of volume 46 L provided with 10 UVA lamps and quartz filters connected to a propylene recirculation tank of volume 55–100 L. The composite catalyst (1% $MoS_2/g-C_3N_4$) showed higher catalytic performance compared to that obtained with pure $g-C_3N_4$ catalyst. The degradation efficiency for every pharmaceutical compound varies in the range of 91–100%. Moreover, the reaction was also conducted in a solar simulator under the irradiation of 500 Wm^{-2}, where the degradation efficiency was maintained in the range of 54–100%. Further, the study was escalated to the pilot wastewater treatment plant (parabolic reactor) established at University Hospital, Ioannina City. The degradation efficiency of the targeted pharmaceutical compounds was evaluated using natural sunlight as the source of irradiation. All the samples were analyzed using solid phase extraction followed by chromatographic measurements. Antonopoulou et al. [131] used a $g-C_3N_4$ catalyst under the irradiation of UVA light for the catalytic degradation of amisulpride, a psychiatric drug in distilled water as well as in municipal wastewater. High degradation percentage was maintained in both the distilled water and municipal wastewater matrix. However, a slower reaction rate was observed in the case of the latter due to the complex nature of the real wastewater.

Kumar et al. [132] developed a novel g-C_3N_4 nanorod catalyst by hydrothermal method. Thus, the prepared activated g-C_3N_4 catalyst exhibited higher degradation efficiency in comparison to that of the bare g-C_3N_4 material. The catalytic efficiency was tested towards 17α-ethinylestradiol in real hospital wastewater effluent. Within 45 min of reaction time, high removal percentage was attained.

Jin et al. [80] applied the as-prepared quantum carbon dot modified g-C_3N_4 photocatalyst for the oxidative degradation of DC in distilled water as well as DC spiked in tap water, wastewater treatment plant effluent, Pearl River water, lake water, and South China Sea water. In comparison to the distilled water, degradation efficiency got reduced in the range of 5–13% for different real water matrices. The reduction in removal efficiency might have caused due to the presence of several that act as light filters and electron trappers. Further investigation on the interference study revealed that certain anions such as chloride, sulfate, and nitrate had minimal adverse effects on the degradation efficiency. However, the presence of HCO_3^- caused a moderate hindrance to the reaction as it can quench $OH^.$ to form less reactive CO_3^{2-}. In the presence of Cu^{2+} and Fe^{3+}, removal efficiency got reduced significantly, as it reacted with $O_2^.$ prior to the degradation of DC.

7. Characterization Techniques

7.1. Fourier Transformed Infrared Spectroscopy (FTIR)

FTIR analysis is a common tool for knowing the functional groups involved in the degradation process. In the work of Kumar et al. [132], strong peaks were noticed around 1515 cm^{-1} corresponding to the C-N= stretching vibration. The appearances of a double peak at 1276 and 1350 cm^{-1} are designated to the aromatic C-N stretching. In the photocatalytic degradation study of DC by reduced ultrathin g-C_3N_4 decorated with quantum carbon dots, FTIR spectra of ultrathin carbon nitride (UCN), reduced ultrathin carbon nitride (RUCN), and carbon quantum dots decorated ultrathin g-C_3N_4 were recorded [80]. In all the spectra, peaks appeared at 810 cm^{-1}, 1200–1700 cm^{-1}, and 3000–3700 cm^{-1} corresponding to the bending mode of triazine units, stretching vibration due to C-N heterocycles, and N-H vibrations, respectively.

7.2. Electron Microscopic Analysis

Morphological features of the g-C_3N_4 catalyst are obtained from electron microscopic analysis, such as scanning electron microscopy (SEM) and transmission electron microscopy (TEM). Mafa et al. [115] conducted both SEM and TEM analyses of the different elements (Ce, Er, Gd, and Sm) doped g-C_3N_4 catalysts. The pure g-C_3N_4 catalyst exhibits a tubular structure with small openings. Due to doping, the tubular structure got transformed to flake-like structure. Cerium doping reduces the size of the flake. The SEM image is shown in Figure 7.

In another study, Palanivel et al. found from the SEM image that the structure of S-doped g-C_3N_4 resembled to that of the nanorods [47]. On the other hand, $NiFe_2O_4$ nanoparticles possessed agglomerating structure. The nanorod structure helped in improving charge carrier mobility and facilitated electron channelization which ultimately made the composite a suitable photocatalyst. In the work of Peng et al. [88], the 3D porous structure of βCDPs/Fe-g-C_3N_4 catalyst was confirmed by both SEM and TEM analyses.

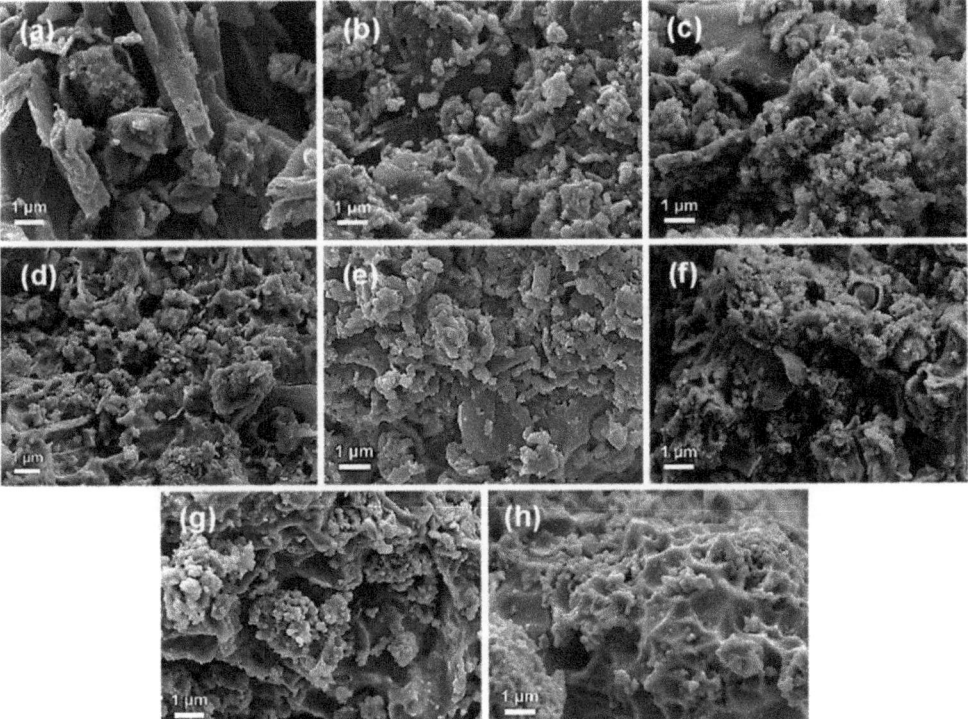

Figure 7. FESEM micrograms of (**a**) CN, (**b**) CeCN, (**c**) ErCN, (**d**) GdCN, (**e**) SmCN, (**f**) 1RECN, (**g**) 3RECN, and (**h**) 5RECN photocatalysts [115].

7.3. BET Surface Area Analysis

BET surface area analysis is important for providing insight into the porosity and surface area of the catalyst material. Mirzaei et al. [117] found the specific surface area of the prepared fluorinated g-C_3N_4 catalyst as ~243 m^2/g. On the other hand, the specific surface area of bare g-C_3N_4 was found to be around 37.85 m^2/g. Enhancement in the specific surface area of the g-C_3N_4 nanosheets occurred due to the acid-assisted hydrothermal treatment. The pore volume of the bare g-C_3N_4 and fluorinated catalyst was 0.082 cm^3/g and 0.427 cm^3/g, respectively. The nitrogen adsorption–desorption isotherm showed a similar trend with the type IV isotherm.

In the study of TC degradation by Ba-doped g-C_3N_4 catalyst, BET surface area analysis for g-C_3N_4 and the composite material was performed [49]. Pure g-C_3N_4 had a specific surface area of 9.98 m^2/g. On the other hand, due to the composite formation, it got increased to 11.41 m^2/g. Similar results were also found in the case of pore volume measurements (0.068 m^3/g for g-C_3N_4 and 0.073 m^3/g for the composite).

7.4. XPS Analysis

The chemical composition of the constructed photocatalyst and the oxidation states of the elements involved are obtained from the XPS analysis. Guo et al. [12] performed the XPS analysis of a 5% Cu_3P-ZSO-CN catalyst. The survey spectra revealed the presence of the Zn 2p, Sn 3d, O 1s, P 2p, Cu 3d, C 1s, and N 1s, and no impurities were found in the catalyst. Furthermore, the high-resolution spectra of Zn 2p displayed strong peaks appearing at 1021.5 eV and 1044.8 eV corresponding to Zn $2p_{3/2}$ and Zn $2p_{1/2}$. In Sn 3d spectrum, peaks appeared at 486.4 eV and 494.7 eV denoting the presence of Sn $3d_{5/2}$ and

Sn $3d_{3/2}$. Chen et al. [102] performed the XPS analysis of TiO_2, g-C_3N_4, and TiO_2/g-C_3N_4 (5% weight) in order to get an idea regarding the surface chemical properties.

Liang et al. [96] conducted the XPS analysis in order to get a clear idea regarding the elemental composition as well as the surface chemical properties of the composite catalyst. From the survey spectrum, prominent peaks of the elements C, N, O, Mo, and Bi were visible. No signal corresponding to S and W was noticeable, which might be due to their presence in very low concentrations. In the O1s spectrum, peaks corresponding to W-O and Bi-O at 232.7 eV and 229.9 eV were visible. On the other hand, Mo 3d peaks at 232.7 eV and 229.9 eV, corresponding to Mo $3d_{3/2}$ and Mo $3d_{5/2}$, implying that Mo is in a +4 state, were observed. The XPS spectra are shown in Figure 8.

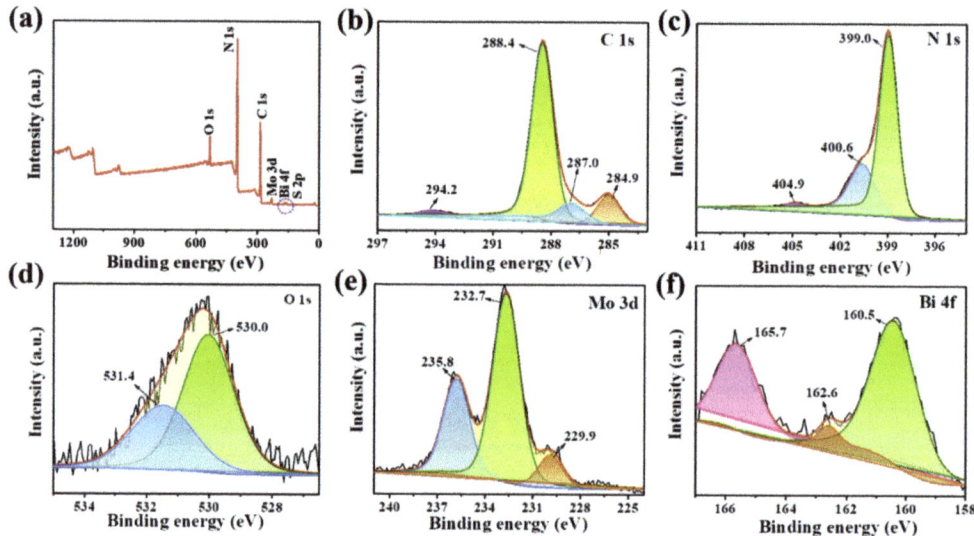

Figure 8. XPS of (**a**) survey spectra, (**b**) C 1s, (**c**) N 1s, (**d**) O 1s, (**e**) Mo 3d, and (**f**) Bi 4f of CN-BM2 [96].

Kumar et al. [129] conducted an XPS analysis of the g-C_3N_4/TiO_2/Fe_3O_4@SiO_2 (gCTFS) catalyst. The wide-scan spectra indicated that the nano photocatalyst was composed of Fe, Si, Ti, C, N, and O. Among all these elements, the signal corresponding to N was very weak indicating that only a few nitrogen atoms were retained after calcination.

7.5. Diffuse Reflectance Spectra (DRS) Analysis

Optical features of the as-synthesized photocatalysts are often explored by conducting diffuse reflectance spectral (DRS) analysis. Baladi et al. [21] performed the DRS analysis of the pure g-C_3N_4, ZnO, and the composite prepared from both. The absorption edges for g-C_3N_4 and ZnO appeared at 445 nm and 405 nm corresponding to the bandgap of 2.78 eV and 3.1 eV. However, after forming the composite, a redshift in comparison to the ZnO appeared indicating a better electron–hole separation efficiency, and an enhanced visible light absorption by the composite photocatalyst. Moreover, the absorption intensity for the composite got increased with respect to the pure g-C_3N_4 indicating the presence of ZnO in the matrix.

Jin et al. [80] carried out the UV-vis DRS analysis of the as-prepared ultrathin g-C_3N_4 photocatalysts. Ultrathin g-C_3N_4 (UCN) showed a strong absorption edge at 460 nm while in the case of reduced UCN, the absorption edge showed a red shift due to the nitrogen defects. Moreover, due to the carbon quantum dot modification, it was shifted to 480 nm which indicates that the incorporation of CQD can enhance the spectral response.

Palanivel et al. [47] found that the absorption of bare g-C_3N_4 occurred at 467 nm with a bandgap of 2.65 eV. On the other hand, carbon nitride nanorod (CNNR) material shows strong absorption at 452 nm. It happened due to the quantum confinement effect. During the formation of the nanocomposite, its electronic state became discrete and hence a blue shift is observed in comparison to the bulk parent material. Moreover, due to sulfur doping, SCNNR showed a strong absorption at 479 nm with a bandgap of 2.58 eV. Due to the incorporation of the S atom, the band gap is shortened.

7.6. X-ray Diffraction (XRD) Analysis

XRD analysis is often explored by researchers for finding out the size of the photocatalyst material. It is based on the Debye–Scherer equation as follows:

$$d = \frac{0.9\lambda}{\beta \cos \theta} \tag{1}$$

Di et al. [2] performed the XRD analysis of g-C_3N_4/ZnMMO composite catalyst. In the XRD spectra of pure g-C_3N_4, two prominent peaks were observed at 13.2° and 27.4°. It refers to the (100) and (002) planes, respectively, which indicates interplanar repeated packing of tri-s-triazine rings and interlayer stacking of graphitic-like structure. In the XRD spectrum of ZnFeMMO, diffraction peaks corresponding to hexagonal wurtzite ZnO and $ZnFe_2O_4$ were clearly observed. On the other hand, after the formation of the composite between the two, the peaks got diminished in magnitude indicating a strong interaction between the two.

Kumar et al. [129] carried out the XRD analysis of Fe_3O_4, Fe_3O_4@SiO_2, g-C_3N_4 nanosheets, and the composite catalyst with TiO_2. In the spectrum of Fe_3O_4, five peaks at 30.1°, 35.5°, 43.1°, 57°, and 62.6° were observed corresponding to (220), (311), (400), (511), and (440) planes. However, in Fe_3O_4@SiO_2, after forming a composite with SiO_2, no characteristic peak due to SiO_2 was observed. Finally, in the spectrum of the composite catalyst (g-C_3N_4/TiO_2/Fe_3O_4@SiO_2), peaks were found at 25.4°, 38°, 48°, 53.9°, 55.19°, 62.72°, 69°, 70.25°, and 75.3° corresponding to the TiO_2.

7.7. Photoluminescence (PL) Spectroscopy

PL spectra are often explored by research groups to have more insight into the photocatalytic property of as-synthesized composite materials. Yan et al. [133] recorded the PL spectra of the pristine g-C_3N_4 material as well as the metal (Na, K, Ca, Mg) doped g-C_3N_4 matrix. A strong absorption peak around 440 nm is observed for the pure g-C_3N_4 while similar spectra with reduced intensity are being observed for the doped materials (as shown in Figure 9). Reduction in the PL intensity indicates that the electron–hole recombination is reduced as well as the lifetime of the charge carrier is increased. Mirzaei et al. [117] reported that due to the fluorination of the g-C_3N_4 catalyst, the intensity of the photoluminescence spectra got reduced.

7.8. Identification of the Intermediate Products

In pharmaceutical degradation, the identification of the intermediate products constitutes an important study. In the photocatalytic degradation of Ibu by g-C_3N_4/TiO_2/Fe_3O_4@SiO_2 photocatalyst, the intermediate compounds were identified [129]. Peaks were obtained at m/z 237, 253, 241, 257, 165, and 163 as a result of the hydroxylation, decarboxylation, and demethylation. Moreover, according to the mass balance equation, 6.5 μmol of CO_2 was obtained after the complete mineralization of 10 μmol of Ibu.

Huang et al. [134] identified the intermediate products by LC-MS analysis while degrading CBZ by a g-C_3N_4 heterogeneous catalyst through the Fenton process. A prominent peak was identified corresponding to an m/z ratio of 253 due to hydroxyl substitution reaction. During the reaction process, a loss of $CONH_2$ occurred which was reflected by the strong signal at the m/z value of 193.

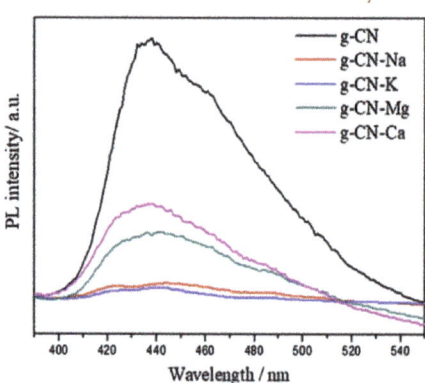

Figure 9. PL spectra of pristine and doped g-C_3N_4 samples [133].

7.9. Photoelectrochemical Tests

Photoelectrochemical tests constitute an important characterization of the photocatalyst material as it gives insight into the charge transfer mechanism of the synthesized material. Guo et al. [14] showed that in comparison to the bare g-C_3N_4 catalyst, the Cl-doped catalyst revealed higher photocurrent density which implied that the electron–hole recombination could be prevented more effectively due to Cl doping. He et al. [4] carried out the photoelectrochemical test of the Ti_3C_2/g-C_3N_4 photocatalyst using Na_2SO_4 as the electrolyte. It was seen that the photocurrent response of the Ti_3C_2/g-C_3N_4 catalyst was higher in comparison to the bare g-C_3N_4 material. Due to the high conductivity of Ti_3C_2, photo-generated electrons from the surface of g-C_3N_4 got transported to the Ti_3C_2 surface resulting in high charge separation and greater photocurrent response. Furthermore, the Nyquist plot revealed that the impedance value of Ti_3C_2/g-C_3N_4 is lower than that of g-C_3N_4, which indicates the efficient generation of photoelectrons under visible light irradiation. Moreover, it also confirmed the 2D/2D nanostructure of the newly synthesized catalyst.

7.10. Electron Spin Resonance (ESR) Tests

The ESR technique is often explored by researchers to investigate the reactive oxygen species (ROS) involved in the degradation process. Chen et al. [30] used 5,5-dimethyl-1-pyrroline N-oxide (DMPO) as a spin-trap chemical in the aqueous dispersion of methanol. Results showed peaks of both DMPO-·O_2^- and DMPO-·OH adducts appeared in the ESR spectrum. Hence, it can be concluded that under visible light irradiation, the MnISCN-20 photocatalyst was able to generate both superoxide and hydroxyl radicals to facilitate TC degradation. Cao et al. [51] utilized DMPO for trapping and detecting radicals such as ·OH, ·O_2^-, and $SO_4^{·-}$ generated during the catalytic degradation of TCH, while 2,2,6,6-tetramethylpiperidine-1-oxyl for the detection of $^1O_2$ radical. The results showed that the 4%P/CNK catalyst was efficient to generate ·OH, ·O_2^-, $SO_4^{·-}$, and $^1O_2$ radicals for TCH degradation purposes under the combined action of visible light irradiation and PMS.

8. Future Perspective and Current Challenges

g-C_3N_4-based photocatalysts are undoubtedly promising materials for pharmaceutical wastewater treatment. Scientists are rigorously working in this field for further development of newer catalysts and their applicability for real field applications. In spite of various achievements, there still exist several challenges and hurdles which need to be overcome. Production of a highly stable g-C_3N_4-based photocatalyst with a narrow band gap is still a challenging task. More control over the surface defects and other properties has to be attained through rigorous research. Another aspect is the synthesis cost of g-C_3N_4-based catalysts causing the major hurdle behind their large-scale applications.

Balakrishnan et al. [7] in their recent review article highlighted serious challenges associated with the usage of g-C_3N_4 catalysts for environmental remediation purposes. One of the major problems associated with the g-C_3N_4-based catalysts is the high electron–hole recombination and significant loss of weight while reusing the catalyst as already mentioned in the introduction part. To improve light absorption, scientists are continuously developing novel techniques for reducing bandgap. Several reports on g-C_3N_4 showed a significant reduction in band gap that occurred due to the modification or immobilization. However, a detailed explanation behind this bandgap reduction is often not dealt with. Hayat et al. [135] also commented that the charge transfer mechanism of the g-C_3N_4-based catalysts is not always fully understood. Hence, a more rigorous effort is recommended.

Many g-C_3N_4-based catalysts have been proven to be toxic to the aquatic ecosystem. Hence, toxicity analysis should also be dealt with while applying these materials for pharmaceutical wastewater treatment.

Often magnetic catalysts are designed for various types of pollutant elimination. However, using a simple bar magnet may lead to severe weight loss during recyclability. Hence, proper design in this respect should be adopted.

g-C_3N_4-based photocatalytic membranes are often explored by researchers for pollutant degradation purposes. Fouling and low mass transfer issues often hinder their use in large-scale industrial purposes. In this respect, uncommon ceramic materials can be explored for the optimization of the cost.

Doping is often tried out to enhance the photocatalytic activity of the g-C_3N_4-based catalysts. Patnaik et al. [136], in their recent review article, mentioned that in many cases doped g-C_3N_4-based catalysts showed poor stability under extreme thermal conditions. Moreover, dopant atoms often become the new area for recombination. Hence, co-doping has been suggested by the authors rather than single-element doping. Although many studies are reported in the literature, still more research is required for controlled doping in order to produce the optimized catalyst.

Yuda and Kumar [137] highlighted some of the important limitations of using g-C_3N_4-based catalysts for real-field wastewater systems. Firstly, their stability under different environmental conditions is not often explored. Hence, stability against acidic conditions, temperature variation, etc., should be checked. Moreover, electron–hole separation can be easily understood in two-component heterostructure systems. However, it becomes complicated in the case of multicomponent systems.

Author Contributions: S.B.: Writing, original draft preparation, A.P.: Writing, reviewing. All authors have read and agreed to the published version of the manuscript.

Funding: This research received no external funding.

Data Availability Statement: No new data is created.

Conflicts of Interest: The authors declare no conflict of interest.

Abbreviations

AMX	amoxicillin
ANN	artificial neural network
CBZ	carbamazepine
CFX	cefalexin
CIP	ciprofloxacin
DC	diclofenac
LFX	levofloxacin
NPX	naproxen
OTC	oxytetracycline
PC	pharmaceutical compound
PMS	peroxymonosulfate

PRRSM	paracetamolresponse surface methodology
SDZ	sulfadiazine
SMX	sulfamethoxazole
TC	tetracycline
TCH	tetracycline hydrochloride
TCS	triclosan

References

1. Mahamallik, P.; Saha, S.; Pal, A. Tetracycline degradation in aquatic environment by highly porous MnO_2 nanosheet assembly. *Chem. Eng. J.* **2015**, *276*, 155–165. [CrossRef]
2. Di, G.; Zhu, Z.; Huang, Q.; Zhang, H.; Zhu, J.; Qiu, Y.; Yin, D.; Zhao, J. Targeted modulation of g-C_3N_4 photocatalytic performance for pharmaceutical pollutants in water using ZnFe-LDH derived mixed metal oxides: Structure-activity and mechanism. *Sci. Total Environ.* **2019**, *650*, 1112–1121. [CrossRef] [PubMed]
3. Nguyen, M.D.; Nguyen, T.B.; Tran, L.H.; Nguyen, T.G.; Fatimah, I.; Kuncoro, E.P.; Doong, R. Z-scheme S, B co-doped g-C_3N_4 nanotube@MnO_2 heterojunction with visible-light-responsive for enhanced photodegradation of diclofenac by peroxymonosulfate activation. *Chem. Eng. J.* **2023**, *452*, 139249. [CrossRef]
4. He, J.; Yang, J.; Jiang, F.; Liu, P.; Zhu, M. Photo-assisted peroxymonosulfate activation via 2D/2D heterostructure of Ti_3C_2/g-C_3N_4 for degradation of diclofenac. *Chemosphere* **2020**, *258*, 127339. [CrossRef]
5. Morales-Paredes, C.A.; Rodriguez-Diaz, J.M.; Boluda-Botella, N. Pharmaceutical compounds used in the COVID-19 pandemic: A review of their presence in water and treatment techniques for their elimination. *Sci. Total Environ.* **2022**, *814*, 152691. [CrossRef]
6. Pattanayak, D.S.; Pal, D.; Mishra, J.; Thakur, C. Noble metal-free doped graphitic carbon nitride (g-C_3N_4) for efficient photodegradation of antibiotics: Progress, limitations, and future directions. *Environ. Sci. Pollut. Res.* **2023**, *30*, 25546–25558. [CrossRef]
7. Balakrishnan, A.; Chinthala, M.; Polagani, R.K.; Vo, D.N. Removal of tetracycline from wastewater using g-C_3N_4 based photocatalysts: A review. *Environ. Res.* **2023**, *216*, 114660. [CrossRef]
8. Ghosh, U.; Pal, A. Insight into the multiple roles of nitrogen doped carbon quantum dots in an ultrathin 2D-0D-2D all-solid-state Z scheme heterostructure and its performance in tetracycline degradation under LED illumination. *Chem. Eng. J.* **2022**, *431*, 133914. [CrossRef]
9. Ghosh, U.; Majumdar, A.; Pal, A. 3D macroporous architecture of self-assembled defect engineered ultrathin g-C_3N_4 nanosheets for tetracycline degradation under LED light irradiation. *Mater. Res. Bull.* **2021**, *133*, 111074. [CrossRef]
10. Ghosh, U.; Pal, A. Drastically enhanced tetracycline degradation performance of a porous 2D g-C_3N_4 nanosheet photocatalyst in real water matrix: Influencing factors and mechanism insight. *J. Water Process Eng.* **2022**, *50*, 103315. [CrossRef]
11. Majumdar, A.; Ghosh, U.; Pal, A. Novel 2D/2D g-C_3N_4/Bi_4NbO_8Cl nano-composite for enhanced photocatalytic degradation of oxytetracycline under visible LED light irradiation. *J. Colloid Interface Sci.* **2021**, *584*, 320–331. [CrossRef]
12. Guo, F.; Huang, X.; Chen, Z.; Cao, L.; Cheng, X.; Chen, L.; Shi, W. Construction of Cu_3P-$ZnSnO_3$-g-C_3N_4 p-n-n heterojunction with multiple built-in electric fields for effectively boosting visible-light photocatalytic degradation of broad-spectrum antibiotics. *Sep. Purif. Technol.* **2021**, *265*, 118477. [CrossRef]
13. Chi, X.; Liu, F.; Gao, Y.; Song, J.; Guan, R.; Yuan, H. An efficient B/Na co-doped porous g-C_3N_4 nanosheets photocatalyst with enhanced photocatalytic hydrogen evolution and degradation of tetracycline under visible light. *Appl. Surf. Sci.* **2022**, *576*, 151837. [CrossRef]
14. Guo, F.; Li, M.; Ren, H.; Huang, X.; Shu, K.; Shi, W.; Lu, C. Facile bottom-up preparation of Cl-doped porous g-C_3N_4 nanosheets for enhanced photocatalytic degradation of tetracycline under visible light. *Sep. Purif. Technol.* **2019**, *228*, 115770. [CrossRef]
15. Smykalova, A.; Sokolova, B.; Foniok, K.; Matejka, V.; Praus, P. Photocatalytic degradation of selected pharmaceuticals using g-C_3N_4 and TiO_2 nanomaterials. *Nanomaterials* **2019**, *9*, 1194. [CrossRef] [PubMed]
16. He, W.; Jia, H.; Li, Z.; Miao, C.; Lu, R.; Zhang, S.; Zhang, Z. Magnetic recyclable g-C_3N_4/Fe_3O_4@MIL-100(Fe) ternary catalyst for photo-Fenton degradation of ciprofloxacin. *J. Environ. Chem. Eng.* **2022**, *10*, 108698. [CrossRef]
17. Wang, F.; Feng, Y.; Chen, P.; Wang, Y.; Su, Y.; Zhang, Q.; Zeng, Y.; Xie, Z.; Liu, H.; Liu, Y.; et al. Photocatalytic degradation of fluoroquinolone antibiotics using ordered mesoporous g-C_3N_4 under simulated sunlight irradiation: Kinetics, mechanism, and antibacterial activity elimination. *Appl. Catal. B Environ.* **2018**, *227*, 114–122. [CrossRef]
18. Sun, Z.; Zhang, X.; Dong, X.; Liu, X.; Tan, Y.; Yuan, F.; Zheng, S.; Li, C. Hierarchical assembly of highly efficient visible-light-driven Ag/g-C_3N_4/kaolinite composite photocatalyst for the degradation of ibuprofen. *J. Mater.* **2020**, *6*, 582–592. [CrossRef]
19. Feyzi, L.; Rahemi, N.; Allahyari, S. Efficient degradation of tetracycline in aqueous solution using a coupled S-scheme ZnO/g-C_3N_4/zeolite P supported catalyst with water falling film plasma reactor. *Process Saf. Environ. Prot.* **2022**, *161*, 827–847. [CrossRef]
20. Mirzaei, A.; Yerushalmi, L.; Chen, Z.; Haghighat, F. Photocatalytic degradation of sulfamethoxazole by hierarchical magnetic ZnO@g-C_3N_4: RSM optimization, kinetic study, reaction pathway and toxicity evaluation. *J. Hazard. Mater.* **2018**, *359*, 516–526. [CrossRef]

21. Baladi, E.; Davar, F.; Hojjati-Najafabadi, A. Synthesis and characterization of g-C$_3$N$_4$-CoFe$_2$O$_4$-ZnO magnetic nanocomposites for enhancing photocatalytic activity with visible light for degradation of penicillin G antibiotic. *Environ. Res.* **2022**, *215*, 114270. [CrossRef] [PubMed]
22. Chen, D.; Li, B.; Pu, Q.; Chen, X.; Wen, G.; Li, Z. Preparation of Ag-AgVO$_3$/g-C$_3$N$_4$ composite photo-catalyst and degradation characteristics of antibiotics. *J. Hazard. Mater.* **2019**, *373*, 303–312. [CrossRef] [PubMed]
23. Guo, Y.; Li, M.; Huang, X.; Wu, Y.; Li, L. S-scheme g-C$_3$N$_4$/TiO$_2$/CFs heterojunction composites with multi-dimensional through-holes and enhanced visible-light photocatalytic activity. *Ceram. Int.* **2022**, *48*, 8196–8208. [CrossRef]
24. Kumar, A.; Kumar, A.; Sharma, G.; Al-Muhtaseb, A.H.; Naushad, M.; Ghfar, A.A.; Stadler, F.J. Quaternary magnetic BiOCl/g-C$_3$N$_4$/Cu$_2$O/Fe$_3$O$_4$ nano-junction for visible light and solar powered degradation of sulfamethoxazole from aqueous environment. *Chem. Eng. J.* **2018**, *334*, 462–478. [CrossRef]
25. Wang, S.; Liu, Y.; Wang, J. Peroxymonosulfate activation by Fe−Co−O−Codoped graphite carbon nitride for degradation of sulfamethoxazole. *Environ. Sci. Technol.* **2020**, *54*, 10361–10369. [CrossRef] [PubMed]
26. Liang, J.; Zhang, W.; Zhao, Z.; Liu, W.; Ye, J.; Tong, M.; Li, Y. Different degradation mechanisms of carbamazepine and diclofenac by single-atom Barium embedded g-C$_3$N$_4$: The role of photosensitation-like mechanism. *J. Hazard. Mater.* **2021**, *416*, 125936. [CrossRef]
27. Tian, Y.; Tian, X.; Zeng, W.; Nie, Y.; Yang, C.; Dai, C.; Li, Y.; Lu, L. Enhanced peroxymonosulfate decomposition into ·OH and ^{1}O$_2$ for sulfamethoxazole degradation over Se doped g-C$_3$N$_4$ due to induced exfoliation and N vacancies formation. *Sep. Purif. Technol.* **2021**, *267*, 118664. [CrossRef]
28. Xu, Q.; Zhang, L.; Cheng, B.; Fan, J.; Yu, J. S-scheme heterojunction photocatalyst. *Chem* **2020**, *6*, 1543–1559. [CrossRef]
29. Tada, H.; Mitsui, T.; Kiyonaga, T.; Akita, T.; Tanaka, K. All solid-state Z-scheme in Cds-Au-TiO$_2$ three component nanojunction system. *Nat. Mater.* **2006**, *5*, 782–786. [CrossRef]
30. Chen, W.; He, Z.; Huang, G.; Wu, C.; Chen, W.; Liu, X. Direct Z scheme 2D/2D MnIn$_2$S$_4$/g-C$_3$N$_4$ architectures with highly efficient photocatalytic activities towards treatment of pharmaceutical wastewater and hydrogen evolution. *Chem. Eng. J.* **2019**, *359*, 244–253. [CrossRef]
31. Pham, V.V.; Truong, T.K.; Hai, L.V.; La, H.P.P.; Nguyen, H.T.; Lam, V.Q.; Tong, H.D.; Nguyen, T.Q.; Sabbah, A.; Chen, K.; et al. S-Scheme α-Fe$_2$O$_3$/g-C$_3$N$_4$ nanocomposites as heterojunction photocatalysts for antibiotic degradation. *Appl. Nanomater.* **2022**, *5*, 4506–4514. [CrossRef]
32. Ni, S.; Fu, Z.; Li, L.; Ma, M.; Liu, Y. Step-scheme heterojunction g-C$_3$N$_4$/TiO$_2$ for efficient photocatalytic degradation of tetracycline hydrochloride under UV light. *Colloids Surf. A Physicochem. Eng. Asp.* **2022**, *649*, 129475. [CrossRef]
33. Zhang, Q.; Peng, Y.; Deng, F.; Wang, M.; Chen, D. Porous Z-scheme MnO$_2$/Mn-modified alkalinized g-C$_3$N$_4$ heterojunction with excellent Fenton-like photocatalytic activity for efficient degradation of pharmaceutical pollutants. *Sep. Purif. Technol.* **2020**, *246*, 116890. [CrossRef]
34. Mei, X.; Chen, S.; Wang, G.; Chen, W.; Lu, W.; Zhang, B.; Fang, Y.; Qi, C. Metal-free carboxyl modified g-C$_3$N$_4$ for enhancing photocatalytic degradation activity of organic pollutants through peroxymonosulfate activation in wastewater under solar irradiation. *J. Solid State Chem.* **2022**, *310*, 123053. [CrossRef]
35. Luo, J.; Dai, Y.; Xu, X.; Liu, Y.; Yang, S.; He, H.; Sun, C.; Xia, Q. Green and efficient synthesis of Co-MOF-based/g-C$_3$N$_4$ composite catalysts to activate peroxymonosulfate for degradation of the antidepressant venlafaxine. *J. Colloid Interf. Sci.* **2022**, *610*, 280–294. [CrossRef] [PubMed]
36. Wang, S.; Wang, J. Magnetic 2D/2D oxygen doped g-C$_3$N$_4$/biochar composite to activate peroxymonosulfate for degradation of emerging organic pollutants. *J. Hazard. Mater.* **2022**, *423*, 127207. [CrossRef] [PubMed]
37. Liu, C.; Dai, H.; Tan, C.; Pan, Q.; Hu, F.; Peng, X. Photo-Fenton degradation of tetracycline over Z-scheme Fe-g-C$_3$N$_4$/Bi$_2$WO$_6$ heterojunctions: Mechanism insight, degradation pathways and DFT calculation. *Appl. Catal. B Environ.* **2022**, *310*, 121326. [CrossRef]
38. Li, T.; Ge, L.; Peng, X.; Wang, W.; Zhang, W. Enhanced degradation of sulfamethoxazole by a novel Fenton-like system with significantly reduced consumption of H$_2$O$_2$ activated by g-C$_3$N$_4$/MgO composite. *Water Res.* **2021**, *190*, 116777. [CrossRef]
39. Li, X.; Gan, X. Photo-Fenton degradation of multiple pharmaceuticals at low concentrations via Cu-doped-graphitic carbon nitride (g-C$_3$N$_4$) under simulated solar irradiation at a wide pH range. *J. Environ. Chem. Eng.* **2022**, *10*, 108290. [CrossRef]
40. Cao, Z.; Jia, Y.; Wang, Q.; Cheng, H. High-efficiency photo-Fenton Fe/g-C$_3$N$_4$/kaolinite catalyst for tetracycline hydrochloride degradation. *Appl. Clay Sci.* **2021**, *212*, 106213. [CrossRef]
41. Qiao, X.; Liu, X.; Zhang, W.; Cai, Y.; Zhong, Z.; Li, Y.; Lu, J. Superior photo-Fenton activity towards chlortetracycline degradation over novel g-C$_3$N$_4$ nanosheets/schwertmannite nanocomposites with accelerated Fe(III)/Fe(II) cycling. *Sep. Purif. Technol.* **2021**, *279*, 119760. [CrossRef]
42. Zhang, J.; Zhao, Y.; Zhang, K.; Zada, A.; Qi, K. Sonocatalytic degradation of tetracycline hydrochloride with CoFe$_2$O$_4$/g-C$_3$N$_4$ composite. *Ultrason. Sonochem.* **2023**, *94*, 106325. [CrossRef]
43. He, Y.; Ma, Z.; Junior, L.B. Distinctive binary g-C$_3$N$_4$/MoS$_2$ heterojunctions with highly efficient ultrasonic catalytic degradation for levofloxacin and methylene blue. *Ceram. Int.* **2020**, *46*, 12364–12372. [CrossRef]
44. Vinesh, V.; Ashokkumar, M.; Neppolian, B. rGO supported self-assembly of 2D nano sheet of (g-C$_3$N$_4$) into rod-like nano structure and its application in sonophotocatalytic degradation of an antibiotic. *Ultrason. Sonochem.* **2020**, *68*, 105218. [CrossRef]

45. Gholami, P.; Khataee, A.; Vahid, B.; Karimi, A.; Golizadeh, M.; Ritala, M. Sonophotocatalytic degradation of sulfadiazine by integration of microfibrillated carboxymethyl cellulose with Zn-Cu-Mg mixed metal hydroxide/g-C_3N_4 composite. *Sep. Purif. Technol.* **2020**, *245*, 116866. [CrossRef]
46. Jiang, L.; Yuan, X.; Zeng, G.; Liang, J.; Wu, Z.; Yu, H.; Mo, D.; Wang, H.; Xiao, Z.; Zhou, C. Nitrogen self-doped g-C_3N_4 nanosheets with tunable band structures for enhanced photocatalytic tetracycline degradation. *J. Colloid Interf. Sci.* **2019**, *536*, 17–29. [CrossRef]
47. Palanivel, B.; Shkir, M.; Alshahrani, T.; Mani, A. Novel $NiFe_2O_4$ deposited S-doped g-C_3N_4 nanorod: Visible-light-driven heterojunction for photo-Fenton like tetracycline degradation. *Diam. Relat. Mater.* **2021**, *112*, 108148. [CrossRef]
48. Preeyanghaa, M.; Dhileepan, M.D.; Madhavan, J.; Neppolian, B. Revealing the charge transfer mechanism in magnetically recyclable ternary g-C_3N_4/BiOBr/Fe_3O_4 nanocomposite for efficient photocatalytic degradation of tetracycline antibiotics. *Chemosphere* **2022**, *303*, 135070. [CrossRef]
49. Bui, T.S.; Bansal, P.; Lee, B.; Mahvelati-Shamsabadi, T. Facile fabrication of novel Ba-doped g-C_3N_4 photocatalyst with remarkably enhanced photocatalytic activity towards tetracycline elimination under visible-light irradiation. *Appl. Surf. Sci.* **2020**, *506*, 144184. [CrossRef]
50. Cao, Y.; Alsharif, S.; El-Shafay, A.S. Preparation, suppressed the charge carriers recombination, and improved photocatalytic performance of g-C_3N_4/MoS_2 p-n heterojunction photocatalyst for tetracycline and dyes degradation upon visible light. *Mater. Sci. Semicond. Process.* **2022**, *144*, 106569. [CrossRef]
51. Cao, Z.; Zhao, Y.; Li, J.; Wang, Q.; Mei, Q.; Cheng, H. Rapid electron transfer-promoted tetracycline hydrochloride degradation: Enhanced activity in visible light-coupled peroxymonosulfate with PdO/g-C_3N_4/kaolinite catalyst. *Chem. Eng. J.* **2023**, *457*, 141191. [CrossRef]
52. Chen, H.; Zhang, X.; Jiang, L.; Yuan, X.; Liang, J.; Zhang, J.; Yu, H.; Chu, W.; Wu, Z.; Li, H.; et al. Strategic combination of nitrogen-doped carbon quantum dots and g-C_3N_4: Efficient photocatalytic peroxydisulfate for the degradation of tetracycline hydrochloride and mechanism insight. *Sep. Purif. Technol.* **2021**, *272*, 118947. [CrossRef]
53. Guo, F.; Shi, W.; Wang, H.; Han, M.; Li, H.; Huang, H.; Liu, Y.; Kang, Z. Facile fabrication of a CoO/g-C_3N_4 p-n heterojunction with enhanced photocatalytic activity and stability for tetracycline degradation under visible light. *Catal. Sci. Technol.* **2017**, *7*, 3325–3331. [CrossRef]
54. Guo, F.; Shi, W.; Li, M.; Shi, Y.; Wen, H. 2D/2D Z-scheme heterojunction of $CuInS_2$/g-C_3N_4 for enhanced visible-light-driven photocatalytic activity towards the degradation of tetracycline. *Sep. Purif. Technol.* **2019**, *210*, 608–615. [CrossRef]
55. Guo, F.; Huang, X.; Chen, Z.; Sun, H.; Chen, L. Prominent co-catalytic effect of CoP nanoparticles anchored on high-crystalline g-C_3N_4 nanosheets for enhanced visible-light photocatalytic degradation of tetracycline in wastewater. *Chem. Eng. J.* **2020**, *395*, 125118. [CrossRef]
56. Guo, B.; Liu, B.; Wang, C.; Lu, J.; Wang, Y.; Yin, S.; Javed, M.S.; Han, W. Boosting photocharge separation in Z-schemed g-C_3N_4/RGO/In_2S_3 photocatalyst for H_2 evolution and antibiotic degradation. *J. Ind. Eng. Chem.* **2022**, *110*, 217–224. [CrossRef]
57. He, Y.; Ma, B.; Yang, Q.; Tong, Y.; Ma, Z.; Junior, L.B.; Yao, B. Surface construction of a novel metal-free g-C_3N_4-based heterojunction photocatalyst for the efficient removal of bio-toxic antibiotic residues. *Appl. Surf. Sci.* **2022**, *571*, 151299. [CrossRef]
58. Huang, H.; Liu, C.; Ou, H.; Ma, T.; Zhang, Y. Self-sacrifice transformation for fabrication of type-I and type-II heterojunctions in hierarchical $Bi_xO_yI_z$/g-C_3N_4 for efficient visible-light photocatalysis. *Appl. Surf. Sci.* **2019**, *470*, 1101–1110. [CrossRef]
59. Obregon, S.; Ruiz-Gomez, M.A.; Rodriguez-Gonzalez, V.; Vaquez, A.; Hernandez-Uresti, D.B. A novel type-II $Bi_2W_2O_9$/g-C_3N_4 heterojunction with enhanced photocatalytic performance under simulated solar irradiation. *Mater. Sci. Semicond. Process.* **2020**, *113*, 105056. [CrossRef]
60. Panneri, S.; Ganguly, P.; Mohan, M.; Nair, B.N.; Mohamed, A.A.P.; Warrier, K.G.; Hareesh, U.S. Photoregenerable, bifunctional granules of carbon-doped g-C_3N_4 as adsorptive photocatalyst for the efficient removal of tetracycline antibiotic. *ACS Sustain. Chem. Eng.* **2017**, *5*, 1610–1618. [CrossRef]
61. Peng, X.; Wu, J.; Zhao, Z.; Wang, X.; Dai, H.; Xu, L.; Xu, G.; Jian, Y.; Hu, F. Activation of peroxymonosulfate by single-atom Fe-g-C_3N_4 catalysts for high efficiency degradation of tetracycline via nonradical pathways: Role of high-valent iron-oxo species and Fe-N_x sites. *Chem. Eng. J.* **2022**, *427*, 130803. [CrossRef]
62. Ren, Z.; Chen, F.; Wen, K.; Lu, J. Enhanced photocatalytic activity for tetracyclines degradation with Ag modified g-C_3N_4 composite under visible light. *J. Photochem. Photobiol. A Chem.* **2020**, *389*, 112217. [CrossRef]
63. Ren, X.; Zhang, X.; Guo, R.; Li, X.; Peng, Y.; Zhao, X.; Pu, X. Hollow mesoporous g-C_3N_4/Ag_2CrO_4 photocatalysis with direct Z-scheme: Excellent degradation performance for antibiotics and dyes. *Sep. Purif. Technol.* **2021**, *270*, 118797. [CrossRef]
64. Shi, Y.; Li, J.; Sun, Y.; Wan, D.; Wan, H.; Wang, Y. FeOOH coupling and nitrogen vacancies functionalized g-C_3N_4 heterojunction for efficient degradation of antibiotics: Performance evaluation, active species evolution and mechanism insight. *J. Alloy. Compd.* **2022**, *903*, 163898. [CrossRef]
65. Song, H.; Liu, L.; Wang, H.; Feng, B.; Xiao, M.; Tang, Y.; Qu, X.; Gai, H.; Huang, T. Adjustment of the band gap of co-doped KCl/NH_4Cl/g-C_3N_4 for enhanced photocatalytic performance under visible light. *Mater. Sci. Semicond. Process.* **2021**, *128*, 105757. [CrossRef]
66. Tian, Y.; Gao, Y.; Wang, Q.; Wang, Z.; Guan, R.; Shi, W. Potassium gluconate-cooperative pore generation based on g-C_3N_4 nanosheets for highly efficient photocatalytic hydrogen production and antibiotic degradation. *J. Environ. Chem. Eng.* **2022**, *10*, 107986. [CrossRef]

67. Wang, W.; Zeng, Z.; Zeng, G.; Zhang, C.; Xiao, R.; Zhou, C.; Xiong, W.; Yang, Y.; Lei, L.; Liu, Y.; et al. Sulfur doped carbon quantum dots loaded hollow tubular g-C_3N_4 as novel photocatalyst for destruction of *Escherichia coli* and tetracycline degradation under visible light. *Chem. Eng. J.* **2019**, *378*, 122132. [CrossRef]
68. Wang, W.; Fang, J.; Chen, H. Nano-confined g-C_3N_4 in mesoporous SiO_2 with improved quantum size effect and tunable structure for photocatalytic tetracycline antibiotic degradation. *J. Alloy. Compd.* **2020**, *819*, 153064. [CrossRef]
69. Wang, Q.; Zhang, L.; Guo, Y.; Shen, M.; Wang, M.; Li, B.; Shi, J. Multifunctional 2D porous g-C_3N_4 nanosheets hybridized with 3D hierarchical TiO_2 microflowers for selective dye adsorption, antibiotic degradation and CO_2 reduction. *Chem. Eng. J.* **2020**, *396*, 125347. [CrossRef]
70. Wang, M.; Jin, C.; Kang, J.; Liu, J.; Tang, Y.; Li, Z.; Li, S. CuO/g-C_3N_4 2D/2D heterojunction photocatalysts as efficient peroxymonosulfate activators under visible light for oxytetracycline degradation: Characterization, efficiency and mechanism. *Chem. Eng. J.* **2021**, *416*, 128118. [CrossRef]
71. Wu, K.; Chen, D.; Fang, J.; Wu, S.; Yang, F.; Zhu, X.; Fang, Z. One-step synthesis of sulfur and tungstate co-doped porous g-C_3N_4 microrods with remarkably enhanced visible-light photocatalytic performances. *Appl. Surf. Sci.* **2018**, *462*, 991–1001. [CrossRef]
72. Wu, K.; Chen, D.; Lu, S.; Fang, J.; Zhu, X.; Yang, F.; Pan, T.; Fang, Z. Supramolecular self-assembly synthesis of noble-metal-free (C, Ce) co-doped g-C_3N_4 with porous structure for highly efficient photocatalytic degradation of organic pollutants. *J. Hazard. Mater.* **2020**, *382*, 121027. [CrossRef] [PubMed]
73. Wu, X.; Zuo, H.; Du, H.; Zhang, S.; Wang, L.; Yan, Q. Construction of layered embedding dual Z-Scheme $Bi_2O_2CO_3$/g-C_3N_4/Bi_2O_3: Tetracycline degradation pathway, toxicity analysis and mechanism insight. *Sep. Purif. Technol.* **2022**, *282*, 120096. [CrossRef]
74. Xu, Y.; Ge, F.; Chen, Z.; Huang, S.; Wei, W.; Xie, M.; Xu, H.; Li, H. One-step synthesis of Fe-doped surface-alkalinized g-C_3N_4 and their improved visible-light photocatalytic performance. *Appl. Surf. Sci.* **2019**, *469*, 739–746. [CrossRef]
75. Zhang, C.; Ouyang, Z.; Yang, Y.; Long, X.; Qin, L.; Wang, W.; Zhou, Y.; Qin, D.; Qin, F.; Lai, C. Molecular engineering of donor-acceptor structured g-C_3N_4 for superior photocatalytic oxytetracycline degradation. *Chem. Eng. J.* **2022**, *448*, 137370. [CrossRef]
76. Zhao, C.; Ran, F.; Dai, L.; Li, C.; Zheng, C.; Si, C. Cellulose-assisted construction of high surface area Z-scheme C-doped g-C_3N_4/WO_3 for improved tetracycline degradation. *Carbohydr. Polym.* **2021**, *255*, 117343. [CrossRef]
77. Lu, C.; Wang, J.; Cao, D.; Guo, F.; Hao, X.; Li, D.; Shi, W. Synthesis of magnetically recyclable g-C_3N_4/$NiFe_2O_4$ S-scheme heterojunction photocatalyst with promoted visible-light-response photo-Fenton degradation of tetracycline. *Mater. Res. Bull.* **2023**, *158*, 112064. [CrossRef]
78. Li, F.; Huang, T.; Sun, F.; Chen, L.; Li, P.; Shao, F.; Yang, X.; Liu, W. Ferric oxide nanoclusters with low-spin Fe^{III} anchored g-C_3N_4 rod for boosting photocatalytic activity and degradation of diclofenac in water under solar light. *Appl. Catal. B Environ.* **2022**, *317*, 121725. [CrossRef]
79. Papamichail, P.; Nannou, C.; Giannakoudakis, D.A.; Bikiaris, N.D.; Papoulia, C.; Pavlidou, E.; Lambropoulou, D.; Samanidou, V.; Deliyanni, E. Maximization of the photocatalytic degradation of diclofenac using polymeric g-C_3N_4 by tuning the precursor and the synthetic protocol. *Catal. Today* **2023**, *418*, 114075. [CrossRef]
80. Jin, X.; Wu, Y.; Wang, Y.; Lin, Z.; Liang, D.; Zheng, X.; Wei, D.; Liu, H.; Lv, W.; Liu, G. Carbon quantum dots-modified reduced ultrathin g-C_3N_4 with strong photoredox capacity for broad spectrum-driven PPCPs remediation in natural water matrices. *Chem. Eng. J.* **2021**, *420*, 129935. [CrossRef]
81. Hu, Z.; Cai, X.; Wang, Z.; Li, S.; Wang, Z.; Xie, X. Construction of carbon-doped supramolecule-based g-C_3N_4/TiO_2 composites for removal of diclofenac and carbamazepine: A comparative study of operating parameters, mechanisms, degradation pathways. *J. Hazard. Mater.* **2019**, *380*, 120812. [CrossRef] [PubMed]
82. John, P.; Johari, K.; Gnanasundaram, N.; Appusamy, A.; Thanabalan, M. Enhanced photocatalytic performance of visible light driven TiO_2/g-C_3N_4 for degradation of diclofenac in aqueous solution. *Environ. Technol. Innov.* **2021**, *22*, 101412. [CrossRef]
83. Han, Y.; Gan, L.; Gong, H.; Han, J.; Qiao, W.; Xu, L. Photoactivation of peroxymonosulfate by wood pulp cellulose biochar/g-C_3N_4 composite for diclofenac degradation: The radical and nonradical pathways. *Biochar* **2022**, *4*, 35. [CrossRef]
84. Jimenez-Salcedo, M.; Monge, M.; Tena, M.T. The photocatalytic degradation of sodium diclofenac in different water matrices using g-C_3N_4 nanosheets: A study of the intermediate by-products and mechanism. *J. Environ. Chem. Eng.* **2021**, *9*, 105827. [CrossRef]
85. Oliveros, A.N.; Pimentel, J.A.I.; Luna, M.D.G.; Garcia-Segura, S.; Abarca, R.R.M.; Doong, R. Visible-light photocatalytic diclofenac removal by tunable vanadium pentoxide/boron-doped graphitic carbon nitride composite. *Chem. Eng. J.* **2021**, *403*, 126213. [CrossRef]
86. Muelas-Ramos, V.; Sampaio, M.J.; Silva, C.G.; Bedia, J.; Rodriguez, J.J.; Faria, J.L. Degradation of diclofenac in water under LED irradiation using combined g-C_3N_4/NH_2-MIL-125 photocatalysts. *J. Hazard. Mater.* **2021**, *416*, 126199. [CrossRef]
87. Reddy, C.V.; Kakarla, R.R.; Cheolho, B.; Shim, J.; Aminabhavi, T.M. Heterostructured 2D/2D $ZnIn_2S_4$/g-C_3N_4 nanohybrids for photocatalytic degradation of antibiotic sulfamethoxazole and photoelectrochemical properties. *Environ. Res.* **2023**, *225*, 115585. [CrossRef] [PubMed]
88. Peng, W.; Liao, J.; Chen, L.; Wu, X.; Zhang, X.; Sun, W.; Ge, C. Constructing a 3D interconnected "trap-zap" β-CDPs/Fe-g-C_3N_4 catalyst for efficient sulfamethoxazole degradation via peroxymonosulfate activation: Performance, mechanism, intermediates and toxicity. *Chemosphere* **2022**, *294*, 133780. [CrossRef]

89. Li, Y.; Li, J.; Pan, Y.; Xiong, Z.; Yao, G.; Xie, R.; Lai, B. Peroxymonosulfate activation on FeCo$_2$S$_4$ modified g-C$_3$N$_4$ (FeCo$_2$S$_4$-CN): Mechanism of singlet oxygen evolution for nonradical efficient degradation of sulfamethoxazole. *Chem. Eng. J.* **2020**, *384*, 123361. [CrossRef]
90. Zhao, G.; Li, W.; Zhang, H.; Wang, W.; Ren, Y. Single atom Fe-dispersed graphitic carbon nitride (g-C$_3$N$_4$) as a highly efficient peroxymonosulfate photocatalytic activator for sulfamethoxazole degradation. *Chem. Eng. J.* **2022**, *430*, 132937. [CrossRef]
91. Zhou, L.; Zhang, W.; Chen, L.; Deng, H.; Wan, J. A novel ternary visible-light-driven photocatalyst AgCl/Ag$_3$PO$_4$/g-C$_3$N$_4$: Synthesis, characterization, photocatalytic activity for antibiotic degradation and mechanism analysis. *Catal. Commun.* **2017**, *100*, 191–195. [CrossRef]
92. Liu, G.; Zhang, Z.; Lv, M.; Wang, H.; Chen, D.; Feng, Y. Photodegradation performance and transformation mechanisms of sulfamethoxazole by porous g-C$_3$N$_4$ modified with ammonia bicarbonate. *Sep. Purif. Technol.* **2020**, *235*, 116172. [CrossRef]
93. Zhou, L.; Zhang, W.; Chen, L.; Deng, H. Z-scheme mechanism of photogenerated carriers for hybrid photocatalyst Ag$_3$PO$_4$/g-C$_3$N$_4$ in degradation of sulfamethoxazole. *J. Colloid Interface Sci.* **2017**, *487*, 410–417. [CrossRef] [PubMed]
94. Song, Y.; Qi, J.; Tian, J.; Gao, S.; Cui, F. Construction of Ag/g-C$_3$N$_4$ photocatalysts with visible-light photocatalytic activity for sulfamethoxazole degradation. *Chem. Eng. J.* **2018**, *341*, 547–555. [CrossRef]
95. Zhang, H.; Li, W.; Yan, Y.; Wang, W.; Ren, Y.; Li, X. Synthesis of highly porous g-C$_3$N$_4$ nanotubes for efficient photocatalytic degradation of sulfamethoxazole. *Mater. Today Commun.* **2021**, *27*, 102288. [CrossRef]
96. Liang, H.; Guo, J.; Yu, M.; Zhou, Y.; Zhan, R.; Liu, C.; Niu, J. Porous loofah-sponge-like ternary heterojunction g-C$_3$N$_4$/Bi$_2$WO$_6$/MoS$_2$ for highly efficient photocatalytic degradation of sulfamethoxazole under visible light irradiation. *Chemosphere* **2021**, *279*, 130552. [CrossRef]
97. Liu, S.; Tang, W. Photodecomposition of ibuprofen over g-C$_3$N$_4$/Bi$_2$WO$_6$/rGO heterostructured composites under visible/solar light. *Sci. Total Environ.* **2020**, *731*, 139172. [CrossRef]
98. Mao, S.; Liu, C.; Xia, M.; Wang, F.; Ju, X. Construction of a Z-scheme 1D/2D FeV$_3$O$_8$/g-C$_3$N$_4$ composite for ibuprofen degradation: Mechanism insight, theoretical calculation and degradation pathway. *Catal. Sci. Technol.* **2021**, *11*, 3466–3480. [CrossRef]
99. Meng, F.; Wang, J.; Tian, W.; Zhang, H.; Liu, S.; Tan, X.; Wang, S. Effects of inter/intralayer adsorption and direct/indirect reaction on photo-removal of pollutants by layered g-C$_3$N$_4$ and BiOBr. *J. Clean. Prod.* **2021**, *322*, 129025. [CrossRef]
100. Akbarzadeh, R.; Fung, C.S.L.; Rather, R.A.; Lo, I.M.C. One-pot hydrothermal synthesis of g-C$_3$N$_4$/Ag/AgCl/BiVO$_4$ micro-flower composite for the visible light degradation of ibuprofen. *Chem. Eng. J.* **2018**, *341*, 248–261. [CrossRef]
101. Cao, W.; Yuan, Y.; Yang, C.; Wu, S.; Cheng, J. In-situ fabrication of g-C$_3$N$_4$/MIL-68(In)-NH$_2$ heterojunction composites with enhanced visible-light photocatalytic activity for degradation of ibuprofen. *Chem. Eng. J.* **2020**, *391*, 123608. [CrossRef]
102. Chen, X.; Li, X.; Yang, J.; Sun, Q.; Yang, Y.; Wu, X. Multiphase TiO$_2$ surface coating g-C$_3$N$_4$ formed a sea urchin like structure with interface effects and improved visible-light photocatalytic performance for the degradation of ibuprofen. *Int. J. Hydrog. Energy* **2018**, *43*, 13284–13293. [CrossRef]
103. Jimenez-Salcedo, M.; Monge, M.; Tena, M.T. Photocatalytic degradation of ibuprofen in water using TiO$_2$/UV and g-C$_3$N$_4$/visible light: Study of intermediate degradation products by liquid chromatography coupled to high-resolution mass spectrometry. *Chemosphere* **2019**, *215*, 605–618. [CrossRef] [PubMed]
104. Jimenez-Salcedo, M.; Monge, M.; Tena, M.T. Combination of Au-Ag plasmonic nanoparticles of varied compositions with carbon nitride for enhanced photocatalytic degradation of Ibuprofen under visible light. *Materials* **2021**, *14*, 3912. [CrossRef]
105. Liang, M.; Zhang, Z.; Long, R.; Wang, Y.; Yu, Y.; Pei, Y. Design of a Z-scheme g-C$_3$N$_4$/CQDs/CdIn$_2$S$_4$ composite for efficient visible-light-driven photocatalytic degradation of ibuprofen. *Environ. Pollut.* **2020**, *259*, 113770. [CrossRef]
106. Liu, R.; Sun, L.; Qiao, Y.; Bie, Y.; Wang, P.; Zhang, X.; Zhang, Q. Efficient photocatalytic degradation of pharmaceutical pollutants using plasma-treated g-C$_3$N$_4$/TiO. *Energy Technol.* **2020**, *8*, 2000095. [CrossRef]
107. Liu, S.; Tang, W.; Chou, P. Microwave-assisted synthesis of triple 2D g-C$_3$N$_4$/Bi$_2$WO$_6$/rGO composites for ibuprofen photodegradation: Kinetics, mechanism and toxicity evaluation of degradation products. *Chem. Eng. J.* **2020**, *387*, 124098. [CrossRef]
108. Wang, J.; Tang, L.; Zeng, G.; Deng, Y.; Liu, Y.; Wang, L.; Zhou, Y.; Guo, Z.; Wang, J.; Zhang, C. Atomic scale g-C$_3$N$_4$/Bi$_2$WO$_6$ 2D/2D heterojunction with enhanced photocatalytic degradation of ibuprofen under visible light irradiation. *Appl. Catal. B Environ.* **2017**, *209*, 285–294. [CrossRef]
109. Wei, Q.; Xiong, S.; Li, W.; Jin, C.; Chen, Y.; Hou, L.; Wu, Z.; Pan, Z.; He, Q.; Wang, Y.; et al. Double Z-scheme system of α-SnWO$_4$/UiO-66(NH$_2$)/g-C$_3$N$_4$ ternary heterojunction with enhanced photocatalytic performance for ibuprofen degradation and H$_2$ evolution. *J. Alloy. Compd.* **2021**, *885*, 160984. [CrossRef]
110. Deng, Y.; Tang, L.; Feng, C.; Zeng, G.; Wang, J.; Zhou, Y.; Liu, Y.; Peng, B.; Feng, H. Construction of plasmonic Ag modified phosphorous-doped ultrathin g-C$_3$N$_4$ nanosheets/BiVO$_4$ photocatalyst with enhanced visible-near-infrared response ability for ciprofloxacin degradation. *J. Hazard. Mater.* **2018**, *344*, 758–769. [CrossRef]
111. Zhang, N.; Li, X.; Wang, Y.; Zhu, B.; Yang, J. Fabrication of magnetically recoverable Fe$_3$O$_4$/CdS/g-C$_3$N$_4$ photocatalysts for effective degradation of ciprofloxacin under visible light. *Ceram. Int.* **2020**, *46*, 20974–20984. [CrossRef]
112. Savunthari, K.V.; Arunagiri, D.; Shanmugam, S.; Ganesan, S.; Arasu, M.V.; Al-Dhabi, N.A.; Chi, N.T.L.; Ponnusamy, V.K. Green synthesis of lignin nanorods/g-C$_3$N$_4$ nanocomposite materials for efficient photocatalytic degradation of triclosan in environmental water. *Chemosphere* **2021**, *272*, 129801. [CrossRef] [PubMed]
113. Wang, J.; Yue, M.; Han, Y.; Xu, X.; Yue, Q.; Xu, S. Highly-efficient degradation of triclosan attributed to peroxymonosulfate activation by heterogeneous catalyst g-C$_3$N$_4$/MnFe$_2$O. *Chem. Eng. J.* **2020**, *391*, 123554. [CrossRef]

114. Yu, B.; Yan, W.; Meng, Y.; Zhang, Y.; Li, X.; Zhong, Y.; Ding, J.; Zhang, H. Selected dechlorination of triclosan by high-performance g-C_3N_4/Bi_2MoO_6 composites: Mechanisms and pathways. *Chemosphere* **2023**, *312*, 137247. [CrossRef]
115. Mafa, P.J.; Malfane, M.E.; Idris, A.O.; Liu, D.; Gui, J.; Mamba, B.B.; Kuvarega, A.T. Multi-elemental doped g-C_3N_4 with enhanced visible light photocatalytic Activity: Insight into naproxen Degradation, Kinetics, effect of Electrolytes, and mechanism. *Sep. Purif. Technol.* **2022**, *282*, 120089. [CrossRef]
116. Truong, H.B.; Huy, B.T.; Ray, S.K.; Gyawali, G.; Lee, Y.; Cho, J.; Hur, J. Magnetic visible-light activated photocatalyst $ZnFe_2O_4$/$BiVO_4$/g-C_3N_4 for decomposition of antibiotic lomefloxacin: Photocatalytic mechanism, degradation pathway, and toxicity assessment. *Chemosphere* **2022**, *299*, 134320. [CrossRef]
117. Mirzaei, A.; Chen, Z.; Haghighat, F.; Yerushalmi, L. Magnetic fluorinated mesoporous g-C_3N_4 for photocatalytic degradation of amoxicillin: Transformation mechanism and toxicity assessment. *Appl. Catal. B Environ.* **2019**, *242*, 337–348. [CrossRef]
118. Saravanakumar, K.; Park, C.M. Rational design of a novel $LaFeO_3$/g-C_3N_4/$BiFeO_3$ double Z-scheme structure: Photocatalytic performance for antibiotic degradation and mechanistic insight. *Chem. Eng. J.* **2021**, *423*, 130076. [CrossRef]
119. Xie, X.; Chen, C.; Wang, X.; Li, J.; Naraginti, S. Efficient detoxification of triclosan by a S–Ag/TiO_2@g-C_3N_4 hybrid photocatalyst: Process optimization and bio-toxicity assessment. *RSC Adv.* **2019**, *9*, 20439–20449. [CrossRef]
120. Su, Q.; Li, J.; Yuan, H.; Wang, B.; Wang, Y.; Li, Y.; Xing, Y. Visible-light-driven photocatalytic degradation of ofloxacin by g-C_3N_4/NH_2-MIL-88B(Fe) heterostructure: Mechanisms, DFT calculation, degradation pathway and toxicity evolution. *Chem. Eng. J.* **2022**, *427*, 131594. [CrossRef]
121. Huang, D.; Sun, X.; Liu, Y.; Ji, H.; Liu, W.; Wang, C.; Ma, W.; Cai, Z. A carbon-rich g-C_3N_4 with promoted charge separation for highly efficient photocatalytic degradation of amoxicillin. *Chin. Chem. Lett.* **2021**, *32*, 2787–2791. [CrossRef]
122. Nivetha, M.S.; Kumar, J.V.; Ajarem, J.S.; Allam, A.A.; Manikandan, V.; Arulmozhi, R.; Abirami, N. Construction of SnO_2/g-C_3N_4 an effective nanocomposite for photocatalytic degradation of amoxicillin and pharmaceutical effluent. *Environ. Res.* **2022**, *209*, 112809. [CrossRef] [PubMed]
123. Dai, S.; Xiao, L.; Li, Q.; Hao, G.; Hu, Y.; Jiang, W. 0D/1D Co_3O_4 quantum dots/surface hydroxylated g-C_3N_4 nanofibers heterojunction with enhanced photocatalytic removal of pharmaceuticals and personal care products. *Sep. Purif. Technol.* **2022**, *297*, 121481. [CrossRef]
124. Thang, N.Q.; Sabbah, A.; Chen, L.; Chen, K.; Thi, C.M.; Viet, P.V. High-efficient photocatalytic degradation of commercial drugs for pharmaceutical wastewater treatment prospects: A case study of Ag/g-C_3N_4/ZnO nanocomposite materials. *Chemosphere* **2021**, *282*, 130971. [CrossRef] [PubMed]
125. Liu, Z.; Zhang, A.; Liu, Y.; Fu, Y.; Du, Y. Local surface plasmon resonance (LSPR)-coupled charge separation over g-C_3N_4-supported WO_3/BiOCl heterojunction for photocatalytic degradation of antibiotics. *Colloids Surf. A Physicochem. Eng. Asp.* **2022**, *643*, 128818. [CrossRef]
126. Shanavas, S.; Roopan, S.M.; Priyadharshan, A.; Devipriya, D.; Jayapandi, S.; Acevedo, R.; Anbarasan, P.M. Computationally guided synthesis of (2D/3D/2D) rGO/Fe_2O_3/g-C_3N_4 nanostructure with improved charge separation and transportation efficiency for degradation of pharmaceutical molecules. *Appl. Catal. B Environ.* **2019**, *255*, 117758. [CrossRef]
127. Qarajehdaghi, M.; Mehrizad, A.; Gharbani, P.; Shahverdizadeh, G.H. Quaternary composite of CdS/g-C_3N_4/rGO/CMC as a susceptible visible-light photocatalyst for effective abatement of ciprofloxacin: Optimization and modeling of the process by RSM and ANN. *Process Saf. Environ. Prot.* **2023**, *169*, 352–362. [CrossRef]
128. AttariKhasraghi, N.; Zare, K.; Mehrizad, A.; Modirshahla, N.; Behnajady, M.A. Zeolite 4A supported CdS/g-C_3N_4 type-II heterojunction: A novel visible-light-active ternary nanocomposite for potential photocatalytic degradation of cefoperazone. *J. Mol. Liq.* **2021**, *342*, 117479. [CrossRef]
129. Kumar, A.; Khan, M.; Zeng, X.; Lo, I.M.C. Development of g-C_3N_4/TiO_2/Fe_3O_4@SiO_2 heterojunction via sol-gel route: A magnetically recyclable direct contact Z-scheme nanophotocatalyst for enhanced photocatalytic removal of ibuprofen from real sewage effluent under visible light. *Chem. Eng. J.* **2018**, *353*, 645–656. [CrossRef]
130. Rapti, I.; Boti, V.; Albanis, T.; Konstantinou, I. Photocatalytic degradation of psychiatric pharmaceuticals in hospital WWTP secondary effluents using g-C_3N_4 and g-C_3N_4/MoS_2 catalysts in laboratory-scale pilot. *Catalysts* **2023**, *13*, 252. [CrossRef]
131. Antonopoulou, M.; Papadaki, M.; Rapti, I.; Konstantinou, I. Photocatalytic degradation of pharmaceutical amisulpride using g-C_3N_4 catalyst and UV-A irradiation. *Catalysts* **2023**, *13*, 226. [CrossRef]
132. Kumar, V.V.; Avisar, D.; Prasanna, L.V.; Betzalel, Y.; Mamane, H. Rapid visible-light degradation of EE2 and its estrogenicity in hospital wastewater by crystalline promoted g-C_3N. *J. Hazard. Mater.* **2020**, *398*, 122880.
133. Yan, W.; Yan, L.; Jing, C. Impact of doped metals on urea-derived g-C_3N_4 for photocatalytic degradation of antibiotics: Structure, photoactivity and degradation mechanisms. *Appl. Catal. B Environ.* **2019**, *244*, 475–485. [CrossRef]
134. Huang, Y.; Luo, X.; Du, Y.; Fu, Y.; Guo, C.; Zou, Y. The role of iron-doped g-C_3N_4 heterogeneous catalysts in Fenton-like process investigated by experiment and theoretical simulation. *Chem. Eng. J.* **2022**, *446*, 137252. [CrossRef]
135. Hayat, A.; Al-Sehemi, A.G.; El-Nasser, K.S.; Taha, T.A.; Al-Ghamdi, A.A.; Syed, J.A.S.; Amin, M.A.; Ali, T.; Bashir, T.; Palamanit, A.; et al. Graphitic carbon nitride (g-C_3N_4)-based semiconductor as a beneficial candidate in photocatalysis diversity. *Int. J. Hydrog. Energy* **2022**, *47*, 5142–5191. [CrossRef]

136. Patnaik, S.; Sahoo, D.P.; Parida, K. Recent advances in anion doped g-C_3N_4 photocatalysts: A review. *Carbon* **2021**, *172*, 682–711. [CrossRef]
137. Yuda, A.; Kumar, A. A review of g-C_3N_4 based catalysts for direct methanol fuel cells. *Int. J. Hydrog. Energy* **2022**, *47*, 3371–3395. [CrossRef]

Disclaimer/Publisher's Note: The statements, opinions and data contained in all publications are solely those of the individual author(s) and contributor(s) and not of MDPI and/or the editor(s). MDPI and/or the editor(s) disclaim responsibility for any injury to people or property resulting from any ideas, methods, instructions or products referred to in the content.

Article

Visible-Light-Driven GO/Rh-SrTiO$_3$ Photocatalyst for Efficient Overall Water Splitting

Shuai Zhang [1], Enhui Jiang [2], Ji Wu [1], Zhonghuan Liu [1], Yan Yan [1], Pengwei Huo [1,*] and Yongsheng Yan [1,*]

[1] School of Chemistry and Chemical Engineering, Jiangsu University, Zhenjiang 212013, China
[2] School of Materials Science and Engineering, Jiangsu University, Zhenjiang 212013, China
* Correspondence: huopw@ujs.edu.cn (P.H.); yys@ujs.edu.cn (Y.Y.)

Abstract: The combining of the heterostructure construction and active sites modification to remodel the traditional wide-band-gap semiconductor SrTiO$_3$ for improving visible light absorption capacity and enhancing photocatalytic performance is greatly desired. Herein, we research a novel GO/Rh-SrTiO$_3$ nanocomposite via a facile hydrothermal method. The champion GO/Rh-SrTiO$_3$ nanocomposite exhibits the superior photocatalytic overall water splitting performance with an H$_2$ evolution rate of 55.83 µmol·g^{-1}·h^{-1} and O$_2$ production rate of 23.26 µmol·g^{-1}·h^{-1}, realizing a breakthrough from zero with respect to the single-phased STO under visible light ($\lambda \geq 420$ nm). More importantly, a series of characterizations results showed that significantly improving photocatalytic performance originated mainly from the construction of heterostructure and more active sites rooted in Rh metal. In addition, the possible photocatalytic reaction mechanisms and the transport behavior of photogenerated carriers have been revealed in deeper detail. This work provides an effective strategy for heterostructure construction to improve solar utilization through vastly expanding visible light response ranges from traditional UV photocatalysts.

Keywords: SrTiO$_3$; GO; Rh active sites modification; photocatalytic overall water splitting; energy conversion

1. Introduction

The excessive consumption of fossil energy has led to serious energy and environmental problems [1] while promoting the research of green alternative energy sources [2]. Hydrogen is considered to be an optimal clean energy source because of its low density, high calorific value, easy storage, and non-toxicity [3]. At the same time, its combustion product is only water, which effectively avoids greenhouse gas emissions, so achieving efficient, clean, and low-energy hydrogen production is significant in the current development. Unlike fossil fuels, hydrogen is not readily available directly in nature. Fortunately, hydrogen can be produced from primary energy sources such as coal, crude oil, natural gas, biomass, and solar energy. Many process technologies for hydrogen production have been developed, such as pyrolysis, fermentation, electrocatalysis, and photocatalysis [4]. Currently, the largest hydrogen production comes from coal gasification, methane steam reforming, and methane partial oxidation technologies [5]. However, the excessive depletion of fossil fuels and significant environmental issues have promoted the development of viable alternatives. Among them, photocatalytic water splitting, which uses solar energy and water, is a promising option [6]. In recent years, photocatalytic technology has received widespread attention because of its clean, safe, sustainable, and longevous advantages. In the photocatalytic system, a decisive role in the overall water splitting efficiency is played by the photocatalyst, which is the core of the whole system [7,8]. It is well known that suitable band structure, abundant reactive sites, and efficient photogenerated carrier separation are the criteria for excellent photocatalysts [9,10]. However, there are very few photocatalysts that can meet these criteria. Therefore, it is essential to study an efficient photocatalyst.

In recent years, SrTiO$_3$ (abbreviated as STO) has been widely used in photocatalysis [11], sensors [12], dye-sensitized solar cells (DSCC) [13], and supercapacitors [14] because of its excellent properties, including high dielectric constant, resistance to photochemical corrosion, relatively stable properties, and non-toxicity [15,16]. Nevertheless, the excessively wide energy band (~3.2 eV) of STO semiconductors leads to poor photogenerated electron–hole pair separation and low utilization of visible light, which results in poor photocatalytic performance [17]. Therefore, improving the visible light utilization of STO is a major challenge. Currently, the modification of photocatalysts involves heterostructure construction [18,19], defect generation [20,21], elemental doping [22,23], and active sites modification [24,25]. Ha's group synthesized SrTiO$_3$/TiO$_2$ heterostructures with different morphologies by a simple hydrothermal method. The SrTiO$_3$/TiO$_2$ heterostructures not only facilitate the rapid separation of photogenerated electron–hole pairs but also allow more light absorption, which has a synergistic effect that enhances the photocatalytic activity of SrTiO$_3$/TiO$_2$ heterostructures [26]. Deng's group synthesized SrTiO$_3$-SrCO$_3$ heterostructures by a simple hydrothermal method. The SrTiO$_3$-SrCO$_3$ heterostructures facilitate the rapid separation and transmission of photogenerated carriers. Under simulated sunlight irradiation, the champion SrTiO$_3$-SrCO$_3$ showed a great photocatalytic H$_2$ production rate of 4.73 mmol·h^{-1}·g^{-1}, which was 21 times higher than that of pure SrTiO$_3$ [27]. The strategy of constructing heterostructures was applied to STO and other semiconductors, significantly improving the utilization of visible light and the separation and migration of charge carriers [28].

Graphene oxide (GO) is widely used for the synthesis of efficient photocatalysts because of its big specific surface area, outstanding electrical conductivity, and powerful light absorption ability [29,30]. For example, Jung's group combined GO with TiO$_2$ and ZnO to increase light utilization and reduce electron–hole pairs recombination in the composite [31]. Gao's group synthesized GO/CdS composites by a novel two-phase hybrid method. The efficient electron transfer from CdS to GO decreased the recombination of photogenerated carriers which improved the catalytic activity [32]. Hunge's group has synthesized GO/TiO$_2$ composite using a hydrothermal process. A series of characterizations showed that the occurrence of GO not only accelerates electron transfer and thus inhibits the recombination of photogenerated electron–hole pairs but also increased light absorption and utilization, which improved the photocatalytic activity [33]. Herein, GO was applied to improve STO light utilization and conductivity. Meanwhile, the active sites are also important for overall water splitting, and metal elements are usually chosen as the electron acceptor and as the active sites for the reaction. For example, Cai's group successfully constructed metal/phosphide heterostructures by photo-deposition which drove a dual active site mechanism thereby overcoming the low efficiency of precious metal water splitting and providing an effective strategy for the preparation of high performance water splitting catalysts [34]. The use of the Rh metal as the electron acceptors greatly increases the exposed active sites and promotes the electron–hole separation.

In this work, heterostructure construction and active site modification strategies have been applied to fabricate GO/Rh-STO composites by a hydrothermal process. Notably, the prepared GO/Rh-STO composites show excellent photocatalytic overall water splitting performance under visible light. In addition, the influencing factors to enhance the photocatalytic activity and the rational photocatalytic reaction mechanism were also investigated in depth. This study combines heterostructure construction techniques (GO improved light absorption) and active sites modification (Rh metal increased active sites) to improve photocatalytic overall water splitting performance, which provides a new strategy to transform traditional UV catalysts into visible-light-responsive catalysts with abundant active sites applied to energy conversion.

2. Results and Discussions

2.1. Characterization of GO/Rh-STO Photocatalysts

The microstructure and morphological characteristics of the prepared catalysts were analyzed by SEM, TEM, HRTEM, and EDS mapping. Figure 1a,d show the SEM and TEM images of STO which show a smooth surface of hexagonal nanosheets without any other impurities, thereby confirming the formation of STO nanosheets. Figure 1b shows the SEM image of GO with a large specific surface area, clearly illustrating the disorderly stacked, and folded sheet-like accumulation. Figure 1c,e show SEM and TEM images of GO/Rh-STO heterostructures, showing that STO nanoparticles grown on the GO flakes. In Figure 1f, the HRTEM image of the GO/Rh-STO heterostructure shows the heterostructure interface between GO and STO, thereby confirming the formation of the GO/Rh-STO heterostructure. Moreover, the clean lattice spacing of 0.224 nm matches well with the (111) plane of STO, while GO is amorphous with no clear lattice. In addition, Figure 1g shows the HAADF and EDS mapping images which show all elements of Sr, Ti, O, Rh, and C in the GO/Rh-STO composite catalysts, which further confirms the formation of GO/Rh-STO composite catalysts. The above results confirm the formation of GO/Rh-STO heterostructures.

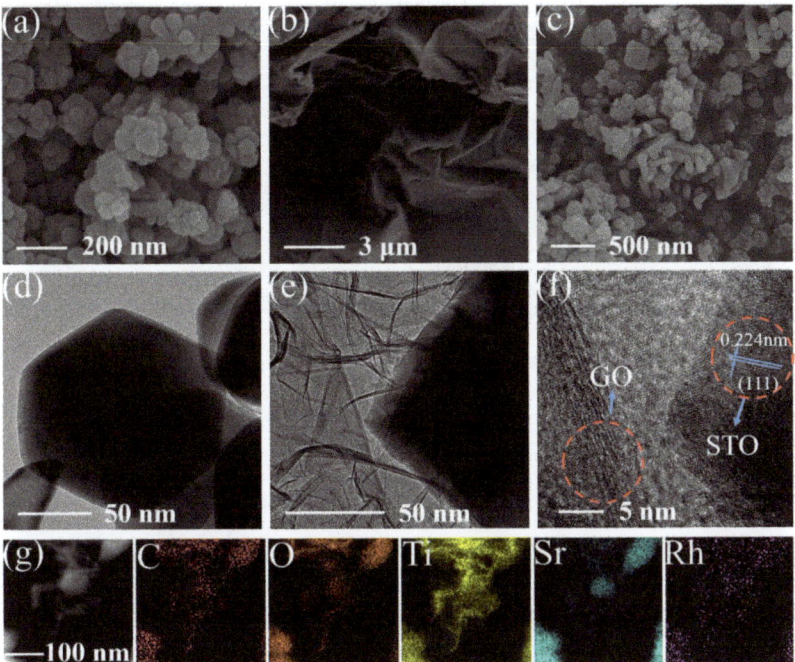

Figure 1. SEM images of (**a**) STO, (**b**) GO and (**c**) GO/Rh-STO; (**d**) TEM image of STO, (**e**) TEM, (**f**) HRTEM, (**g**) HADDF and EDS mapping images of GO/Rh-STO.

2.2. The Crystal Structure and Chemical States Analysis

To better understand the structure and chemical state of pure STO and GO/Rh-STO, XRD patterns and XPS analysis were investigated. As shown in Figure 2a, the STO nanosheets show good crystallinity with diffraction peaks at 22.75°, 32.40°, 39.96°, 46.47°, 52.35°, 57.76°, and 67.83°, which corresponds to (100), (110), (111), (200), (012), (211), and (220) crystal planes of the STO structure, respectively. These peaks are characteristic peaks of STO, indexing to JCPDS card number 35-0734 [35]. The STO crystalline structure is well-developed based on the well-defined and very sharp peaks in the XRD patterns of the STO nanosheets. As shown in Figure 2b, the XRD pattern of GO shows an intense and sharp peak at 2θ = 10.40° that correspond to the (001) plane of GO, which is usually determined by the synthesis process and the number of water layers in the interplanar space of GO [36,37]. The XRD patterns of GO/Rh-STO nanocomposites clearly show GO and STO characteristic diffraction peaks, which proves the presence of GO and STO in the nanocomposites, while the decrease of GO characteristic peak intensity indicates that the aggregation of GO sheets has been greatly decreased. Therefore, the XRD patterns confirmed the successful synthesis of GO, STO, and GO/Rh-STO nanocomposites. Figure 3a shows that there are Ti, Sr, Rh, C, and O elements in GO/Rh-STO. As shown in Figure 3b, two symmetric diffraction peaks (309.4 eV and 314.4 eV) appeared in the Rh 3d XPS spectrum which corresponded to Rh $3d_{5/2}$ and $3d_{3/2}$, representing Rh^{4+} rather than Rh^{3+} or the oxide form of Rh [38,39]. This suggests that Rh is successfully modified into the STO lattice by substitution at the Ti position. Figure 3c shows the XPS patterns of O elements in GO and GO/Rh-STO samples. The diffraction peak of lattice oxygen in GO/Rh-STO sample shifts from 529.8 eV to lower binding energy, which is due to the interference generated by Rh atoms entering the STO crystal structure, and further confirms the successful modification of Rh in STO. The XPS spectra of Ti 2p and Sr 3d of STO and GO/Rh-STO are shown in Figure 3d,e, respectively, which show that the integration of Rh metal active sites modification and GO heterostructure construction in the STO nanosheets does not change the XPS of Ti 2p and Sr 3d too much. This result suggests that the incorporation of GO and the Rh metal does not change the core structure of STO.

The XPS atomic percentage analysis of xGO/Rh-STO as shown in Table 1. The percentages of C atomic were 3.8%, 7.2%, 12.2%, and 16.9%, respectively, and named 3.8% GO/Rh-STO, 7.2% GO/Rh-STO, 12.2% GO/Rh-STO, and 16.9% GO/Rh-STO, according to the atomic percentage analysis of XPS.

Table 1. The XPS atomic percentage analysis of xGO/Rh-STO.

GO Addition Amount (mg)	Atomic %C	Atomic %O	Atomic %Rh	Atomic %Sr	Atomic %Ti
10	3.8	53.9	0.7	25.1	16.5
20	7.2	53.7	0.6	23.1	15.4
40	12.2	51.4	0.5	21.7	14.2
60	16.9	48.1	0.4	18.4	12.5

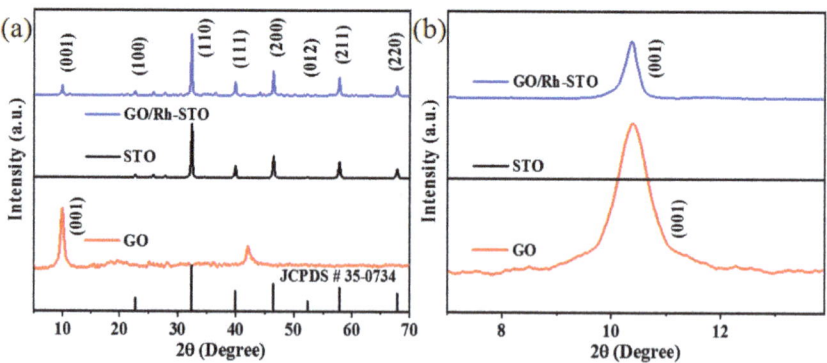

Figure 2. (a,b) XRD patterns of the GO, STO, and GO/Rh-STO.

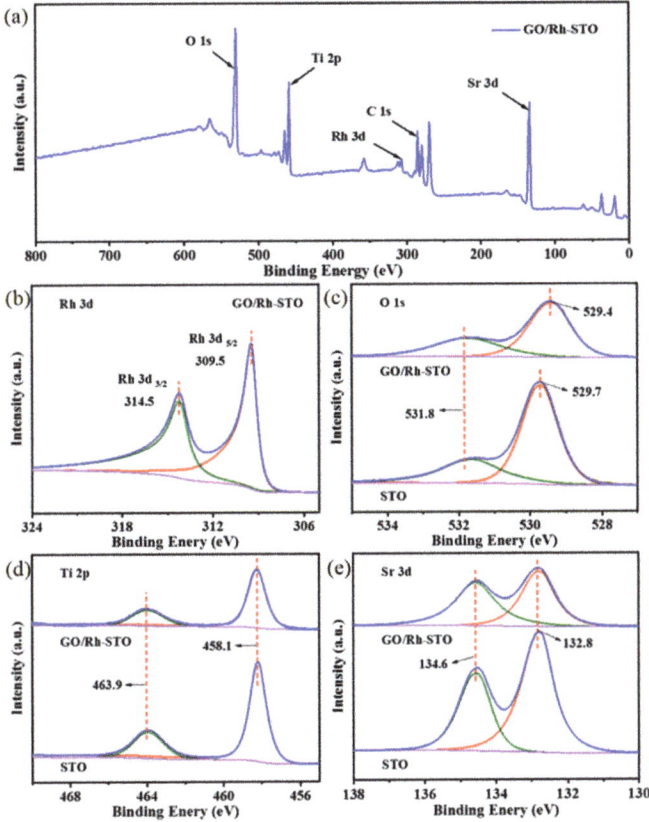

Figure 3. XPS spectra of STO and GO/Rh-STO: (**a**) the total survey spectra, (**b**) Rh 3d, (**c**) O 1s, (**d**) Ti 2p, (**e**) Sr 3d.

2.3. Photocatalytic Performance Evaluation

The photocatalytic performance of the prepared photocatalysts was investigated by overall water splitting for H_2 and O_2 production. As shown in Figure 4a, pure STO cannot overall water splitting under visible light ($\lambda \geq 420$ nm) due to the fact that it can only absorb and utilize UV light. However, the overall water splitting of GO/STO heterostructure can work under visible light. The champion GO/Rh-SrTiO$_3$ nanocomposite exhibits the superior photocatalytic overall water splitting performance with an H_2 evolution rate of 55.83 µmol·g^{-1}·h^{-1} and O_2 production rate of 23.26 µmol·g^{-1}·h^{-1} under visible light ($\lambda \geq 420$ nm), illustrating that the heterostructure construction and Rh metal active sites modification enhanced the utilization of visible light and photocatalytic performance of STO. As shown in Figure 4b, the H_2 and O_2 production rates over the xGO/Rh-STO samples are increased and then decreased with the increasing amount of GO, the possible reason for which can be ascribed to the fact that the excess GO reduces the exposed active sites of STO, resulting in a reduction in the overall water splitting rate. Notably, the 7.2% GO/Rh-STO photocatalyst displays the highest H_2 production rate (55.83 µmol·g^{-1}·h^{-1}) and O_2 production rate (23.26 µmol·g^{-1}·h^{-1}) under visible light ($\lambda \geq 420$ nm).

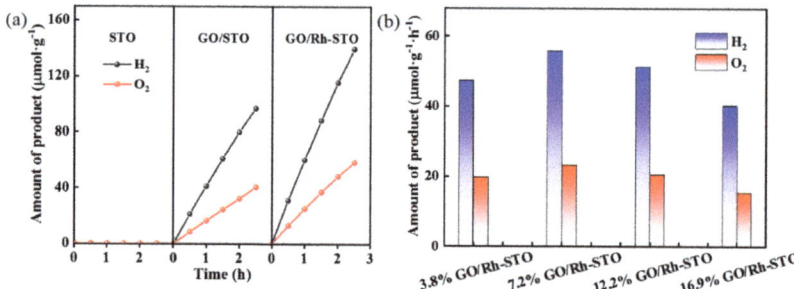

Figure 4. (**a**) Time course of H_2 and O_2 evolution over STO, GO/STO, GO/Rh−STO, and (**b**) photocatalytic activity of xGO/Rh−STO (x = 3.8%, 7.2%, 12.2%, or 16.9%) under visible light ($\lambda \geq 420$ nm) irradiation.

In recent years, there have been many works aimed at enhancing the performance of STO overall water splitting as shown in Table 2. Modulating the morphology of STO is also an effective strategy to improve photocatalytic activity, in addition to the forming of composite materials. Many different morphological STO samples have been reported, such as nanoparticles, nanosheets, nanospheres, nanorods, coral-like and flower-like microspheres, etc. [40–43] The study of STO with different morphologies revealed that highly interconnected porous structures are more favorable for reactant adsorption, light utilization, and carrier migration. For example, flower-like STO with a large number of cavities facilitates a reduction in photogenerated carrier recombination while facilitating the reflection of light to improve light utilization [42,43]. These are the keys to improving the photocatalytic activity of STO. Morphological modulation provides a strategy to further improve the photocatalytic activity of GO/Rh-STO. The modified STO catalysts showed relatively high overall water splitting performance under UV light and low performance under visible light. This work synthesized GO/Rh-STO photocatalyst and exhibited excellent overall water splitting performance compared to other works under visible light. The GO/Rh-STO composites have great potential for large-scale photocatalytic overall water splitting applications, but their performance at present is still insufficient and much work needs to be carried out to improve the overall water splitting performance of GO/Rh-STO under visible light.

Table 2. The comparison of overall water splitting performance of STO-based photocatalysts in other works.

Materials	Catalyst Dosage (mg)	Cocatalysts	Reactant Solution Water (mL)	Light Source	H_2 Rate (μmolh$^{-1}\cdot$g^{-1})	O_2 Rate (μmolh$^{-1}\cdot$g^{-1})	Ref.
GO/Rh-STO	100	/	100	300W Xe Lamp ($\lambda > 420$ nm)	55.83	23.26	This work
SrTiO$_3$-C950	400	/	10	300W Xe Lamp ($\lambda > 420$ nm)	2.3	1.0	[44]
SrTiO$_3$/TiO$_2$	50	Pt (0.3 wt%)	150	300W Xe Lamp ($\lambda > 420$ nm)	10.6	5.1	[45]
SrTiO$_3$:Rh-RGO-BiVO$_4$:Mo	200	Ru (0.7 wt%) Co (0.1 wt%)	120	300W Xe Lamp ($\lambda > 420$ nm)	14	6.1	[46]
Mg-doped SrTiO$_3$	25	Ni (1 wt%)	25	300W Xe Lamp ($\lambda > 300$ nm)	8.8	4.2	[47]
PdCrO$_x$/SrTiO$_3$	100	/	140	300W Xe Lamp ($\lambda > 300$ nm)	15	5	[48]
Ultrafine Pt clusters on SrTiO$_3$	50	/	100	300W Xe Lamp ($\lambda > 300$ nm)	23	12	[49]
SrTiO$_3$ (impregnation methods)	300	Pt (0.3 wt%)	150	300W Xe Lamp ($\lambda > 300$ nm)	38	20	[50]
TiO$_2$/SrTiO$_3$	100	Rh (0.1 wt%) Cr (0.05 wt%) Co (0.05 wt%)	50	300W Xe Lamp ($\lambda < 380$ nm)	38.6	19.2	[51]
oxygen vacancies SrTiO$_3$	100	Pt (0.3 wt%)	60	300W Xe Lamp ($\lambda < 380$ nm)	81	40	[43]

2.4. The Influence Factors of Enhancing Photocatalytic Performance

The valence and conduction band structures are essential for the analysis of the underlying causes of the enhanced photocatalytic activity. As shown in Figure 5a, the UV-vis DRS of pure STO shows an absorption edge at 385 nm, consistent with the published literature [52]. Surprisingly, the GO/STO photocatalyst has a much higher absorption capacity for visible light compared to pure STO. Meanwhile, according to the equation $(\alpha h\nu) = A(h\nu - E_g)^{n/2}$, the plotting of $(\alpha h\nu)^{1/2}$ with respect to the energy $(h\nu)$ is performed to calculate the band gap energy due to the indirect band gap ($n = 1$) property of STO [53], while the plotting of $(\alpha h\nu)^2$ with respect to the energy $(h\nu)$ is performed due to the direct band gap ($n = 4$) property of GO/Rh-STO [33]. In Figure 5b, the band gap energies of STO and GO/Rh-STO are calculated to be 3.20 eV and 2.09 eV, respectively, which indicates that the combination of GO and STO greatly reduces the band gap energy and enhances the absorption and utilization of visible radiation. In addition, Mott–Schottky measurements were performed to investigate the conduction bands of STO and GO/Rh-STO. As shown in Figure 5c,d, the Mott–Schottky slopes of STO and GO/Rh-STO indicate a positive correlation, which indicates that they are all n-type semiconductors. The conduction band potentials of STO and GO/Rh-STO are −0.41 eV and −0.65 eV (vs. NHE), respectively, as calculated by the equation $E_{NHE} = E_{Ag/AgCl} + 0.059 \times PH + 0.197$. The VB potentials of STO and GO/Rh-STO were determined to be 2.79 eV and 1.44 eV from the empirical equation of $E_{VB} = E_{CB} + E_g$ [54,55].

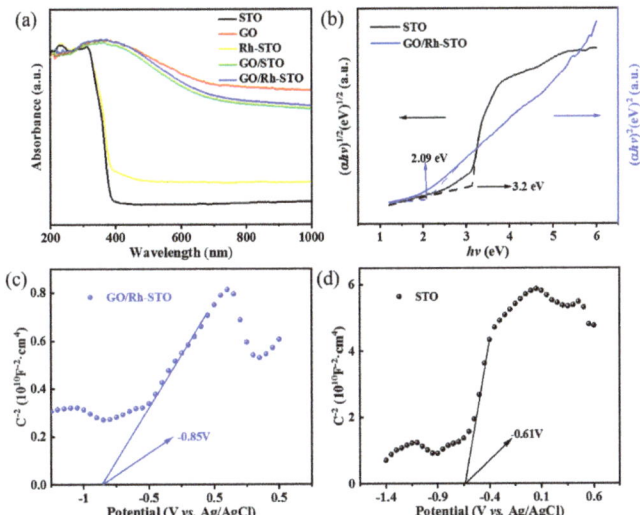

Figure 5. (a) UV−vis diffuse reflectance spectra and (b) curves of $(\alpha h\nu)^{1/2}$ and $(\alpha h\nu)^2$ versus energy $(h\nu)$ of STO and GO/Rh−STO; Mott−Schottky curves of (c) GO/Rh−STO and (d) STO.

The photoelectric properties are closely related to the transfer and separation ability of the charge carriers of the photocatalyst. As shown in Figure 6a, the surface charge transfer efficiency of the samples was studied by EIS and the GO/Rh-STO samples show a much smaller semicircular diameter and much lower charge transfer resistance than pure STO. By calculating the diameter of the semicircle obtained by EIS, the charge transfer resistance (R_{ct}) of GO/STO, GO, and pure STO were 103.2, 41.3, and 279.5 Ω, respectively, and the conductivity (R_{ct}^{-1}) was from the largest to the smallest sample as GO>GO/STO>STO. This is also demonstrated by the linear sweep voltage plot (LSV) shown in Figure 6c, where the overpotential of GO/Rh-STO (1.19V) is smaller than that of pure STO (1.93 V) at −10 mA·cm^{-2}, indicating that the conductivity of GO/Rh-STO samples is better than that of STO. The improved electrical conductivity of GO/Rh-STO composites is due to the introduction of GO with excellent electrical conductivity. Furthermore, Figure 6b shows the photocurrent responses of the prepared STO, GO, and GO/Rh-STO heterostructures, which all exhibit stable photocurrent signals over six cycles. Notably, the photocurrent intensity of the GO/Rh-STO samples is stronger than that of pure STO and GO, indicating that the GO/Rh-STO composites have better separation of charge–hole pairs.

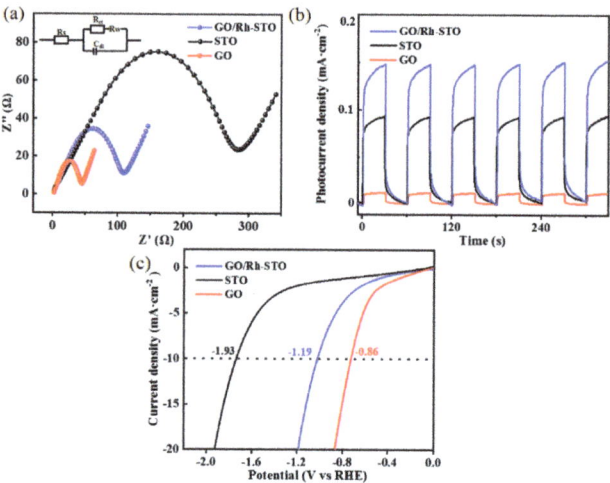

Figure 6. (a) EIS Nyquist plots, the electrical equivalent circuit model of as-prepared samples is shown in the inset of (a) including charge transfer resistance (R_{ct}), resistance of solution (R_s), double layer capacitance (C_{dl}) and Warburg resistance (R_W); (b) the transient photocurrent responses and; (c) LSV curves of STO, GO and GO/Rh–STO.

In this study of the behavior of charge separation of samples by photoluminescence (PL) and transient photoluminescence spectroscopy (TRPL), the PL spectra of photocatalysts at the excitation wavelength of 385 nm are depicted in Figure 7a, in which PL emission at 480 nm on as-prepared photocatalysts is shown. As shown in Figure 7b, under the condition of λ_{ex} = 385 nm and λ_{em} = 480 nm, we tracked the transient photoluminescence (TRPL) emission profile. From Figure 7b, the average decay lifetimes of GO/STO (1.6 ns) and GO/Rh-STO (1.9 ns) composite catalysts are longer compared to pure STO (1.1 ns), which further demonstrates the effective facilitative effect of the separation and transfer of photoinduced charges.

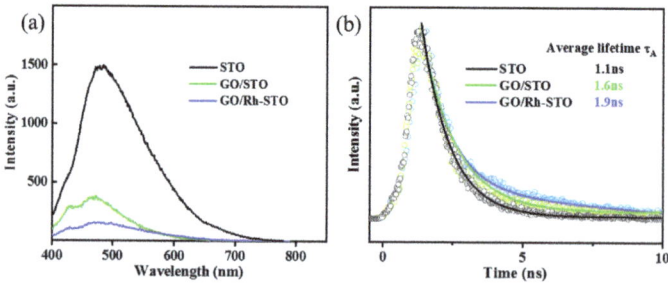

Figure 7. (a) PL spectra and (b) TRPL spectra of STO, GO/STO, and GO/Rh-STO (λ_{ex} = 385 nm and λ_{em} = 480 nm).

Furthermore, the specific surface area was investigated in order to investigate the factors affecting the activity of photocatalytic water splitting. As is shown in Figure 8a–c, the N_2 adsorption–desorption isotherms show that the samples GO, STO, and GO/Rh-STO all exhibit typical type IV isotherms with H3-type hysteresis loops. This indicates the existence of multi-porous structures in GO, STO, and GO/Rh-STO. The BET surface area of the GO/Rh-STO (22.537 m^2/g, Figure 8a) composites was significantly larger compared to that of STO (4.15 m^2/g, Figure 8c), which was mainly attributed to the large specific surface area of GO. In addition, the composition of GO/Rh-STO nanocomposites was analyzed by Raman spectroscopy. As shown in Figure 8d, STO has secondary Raman scattering from 160 cm^{-1} to 451 cm^{-1}, which leads to a continuous broadband. STO shows its characteristic peaks at 171, 245, 307, 609, 674, and 1044 cm^{-1} [56,57] and GO shows its characteristic peaks at 1393 cm^{-1} (D-band) and 1596 cm^{-1} (G-band). Two additional bands of GO/Rh-STO were obtained at 1392 (D-band) and 1595 cm^{-1} (G-band) [58], which confirmed the presence of GO. Therefore, the Raman spectra illustrate the successful preparation of GO/Rh-STO nanocomposites.

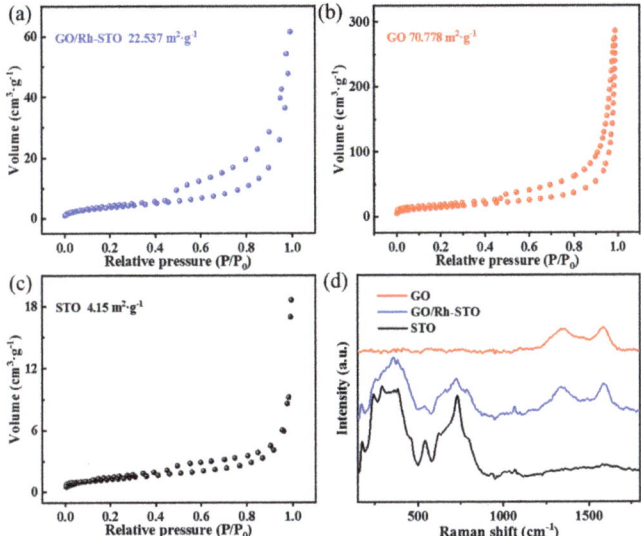

Figure 8. Nitrogen adsorption–desorption curves of (**a**) GO/Rh−STO, (**b**) GO, (**c**) STO, and (**d**) Raman spectra of GO, STO, and GO/Rh−STO.

2.5. Photocatalytic Mechanism Research

Based on the above experimental results, the mechanism of the photocatalytic reaction on GO/Rh-STO heterostructure is shown in Figure 9. According to the above study, the VB and CB energy levels of STO were confirmed to be 2.79 eV and −0.41 eV, respectively. The Fermi energy level of GO is −0.008 V with respect to NHE [59]. The CB energy level of STO is −0.41 eV, which is positive for the Fermi energy level of GO. As a result, the photoexcited electrons are rapidly transferred from the CB of STO to GO, thereby suppressing the photoexcited e$^-$/h$^+$ pairs' recombination and improving their separation efficiency. During the photocatalytic experiments, when the light starts being exposed on the photocatalyst, STO absorbs the light and the e$^-$ production of STO transfers from VB to CB and the h$^+$ were left at VB of STO. Due to the solid-solid tight contact interface, GO facilitates the rapid transfer of e$^-$ to the Rh active sites. The e$^-$ transfers quickly to the Rh active sites and produces H_2 with H$^+$. Meanwhile, h$^+$ on the VB of STO produces O_2 with water, because the valence band maximum for the STO is located at a more positive position than the O_2/H_2O energy level [60].

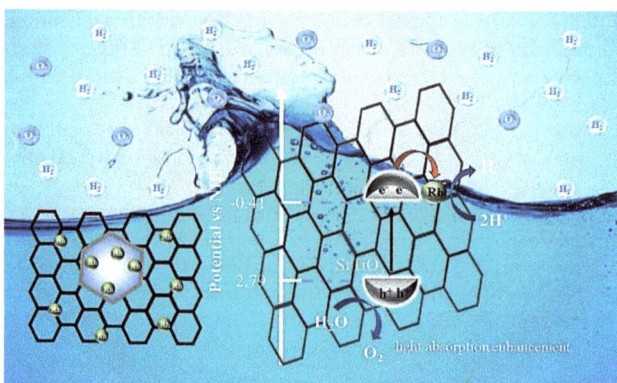

Figure 9. Schematic representation of possible photocatalytic degradation mechanism.

3. Experiment

3.1. Materials

Titanium butoxide ($C_{16}H_{36}O_4Ti$), sodium hydroxide (NaOH), strontium chloride ($SrCl_2$), rhodium(III) chloride trihydrate ($RhCl_3 \cdot 3H_2O$), and graphite powder were obtained from Shanghai Aladdin Technology Co., Ltd. (Shanghai, China). Ethanol (C_2H_5OH), hydrochloric acid (HCl), sulphuric acid (H_2SO_4), hydrogen peroxide solution (H_2O_2), sodium nitrate ($NaNO_3$), and potassium permanganate ($KMnO_4$) were obtained from Shanghai Sinopharm Co., Ltd. (Shanghai, China). All reagents involved in the experiments were of analytical grade (purity $\geq$ 99.0%) and the water used throughout the study was deionized water (DI).

3.2. Synthesis and Preparation

3.2.1. Synthesis of Graphene Oxide (GO)

GO was synthesized using a modified Hummer method which is reported in the other literature [61]. In short, 100 mL H_2SO_4 (98%) was added to a round-bottom flask (500 mL) in an ice bath with stirring at 450 rpm. Graphite powder (5 g), $NaNO_3$ (1.6 g), and $KMnO_4$ (10 g) were added to concentrated sulfuric acid. An amount of 100 mL H_2O was added to the mixture when the color of the reaction mixture turned light gray. After increasing the temperature of the reaction vessel to 102 °C for 2 h, H_2O_2 (40 mL) was added. The reaction mixture was washed with 5% HCl until the filtrate became colorless after stirring the reaction mixture for 1 h. Then, we obtained graphite oxide after washing 3 times with H_2O and dried it in a desiccator at 60 °C. Finally, we obtained GO by ultrasonication of the dispersed graphite oxide in water.

3.2.2. Synthesis of Strontium Titanate (STO)

The STO nanoparticle was synthesized via a hydrothermal route [62]. Firstly, $SrCl_2$ (0.01 mol) was added to an autoclave (50 mL) followed by 30 mL of water. After stirring for 20 min, 3.4 mL titanium butoxide was added, then 0.8 g NaOH. After stirring for 2 h, the autoclave was heated to 195 °C and held for 23 h. After the completion of the reaction, the reaction was washed by centrifugation several times and dried at 60 °C. Finally, we obtained STO white powder.

3.2.3. Synthesis of GO/Rh-STO Composite Catalysts

GO/Rh-STO composites were synthesized by the hydrothermal method [63]. Firstly, $SrCl_2$ (0.01 mol) and $RhCl_3 \cdot 3H_2O$ (50 mg) were added to an autoclave (50 mL) followed by 30 mL of water. After stirring for 20 min, 3.4 mL titanium butoxide was added, then 0.8 g NaOH. After stirring for 30 min, 20 mg GO was added. After stirring for 2 h, the autoclave was heated to 195 °C and held for 23 h. After the completion of the reaction, the reaction was washed by centrifugation several times and dried at 60 °C to obtain GO/Rh-STO precursor powder. Finally, we obtained the GO/Rh-STO composite catalysts by annealing at 200 °C in ambient for 2 h. The xGO/Rh-STO composites were synthesized by changing the amount of GO (10 mg, 20 mg, 40 mg, 60 mg). The percentages of C atomic were 3.8%, 7.2%, 12.2%, and 16.9%, which named 3.8% GO/Rh-STO, 7.2% GO/Rh-STO, 12.2% GO/Rh-STO, and 16.9% GO/Rh-STO, respectively, according to the C atomic percentage analysis of XPS in Table 1.

3.3. Evaluation on Efficiency of Photocatalysts

The evaluation of the overall water splitting performance was performed on a CEL-PAEM-D8 photocatalytic activity evaluation system (CEL-PF300-T9) equipped with a 300W xenon lamp (Beijing Zhongjiao Jinyuan Technology Co., Ltd., Beijing, China) and a cutoff filter allowing $\lambda > 420$ nm. Briefly, 0.1 g of catalyst was dispersed in an light reactor containing 100 mL of water, connected to the photoreaction system, and then evacuated. The reactor was irradiated with a 300w xenon lamp as the light source for performance testing. The produced H_2 and O_2 were evaluated hourly on a gas chromatograph (GC 7920) using high-purity Ar as the carrier gas, and the amounts of H_2 and O_2 were calculated from their standard curves.

3.4. Characterization

The structure and crystalline phase of the samples were analyzed by XPS (Thermo Fisher Scientific K-Alpha, Waltham, MA, USA) and XRD (Rigaku Ultima IV, Tokyo, Japan). X-ray photoelectron spectroscopy (XPS) analysis was conducted on a Thermo Fisher Scientific K-Alpha photoelectron spectroscopy at 5.0×10^{-10} mbar. X-ray diffraction (XRD) patterns of the samples were collected with a scan rate of 0.02°. The morphology was observed on a field emission scanning electron microscope (Hitachi Regulus 8100, Tokyo, Japan) and a field emission transmission electron microscope (FEI Talos F200X G2, Waltham, MA, USA). Transmission electron microscopy (TEM) and high-resolution TEM (HRTEM) images were obtained by an FEI Talos F200X G2 instrument at an accelerating voltage of 200 kV. The diffuse reflectance UV-visible (DRS) was recorded with a PE Lambda 750. The specific surface area was analyzed on an N_2 adsorption–desorption device (TriStar II 20, Waltham, MA, USA) at 77 K, and the samples were degassed at 120 °C for 6 h in a vacuum before the measurements were taken. The photoluminescence (PL) spectra and transient photoluminescence spectra (TRPL) were measured using an Edinburgh FLS-1000 spectrofluorometer.

3.5. Electrochemical Tests

The electrochemical properties of the samples were tested on a CHI600E electrochemical workstation. The platinum electrode, ITO glass (1 cm × 2 cm) coated by sample, and Ag^+/AgCl electrode served as counter electrode, working electrode, and reference electrode to build the three-electrode system in 0.5 M H_2SO_4 electrolyte.

4. Conclusions

In conclusion, we reported a GO/Rh-SrTiO$_3$ composite that was prepared by a simple hydrothermal process for overall water splitting. The champion catalyst realized H_2 production rate of 55.83 $\mu mol \cdot g^{-1} \cdot h^{-1}$ and O_2 production rate of 23.26 $\mu mol \cdot g^{-1} \cdot h^{-1}$ under visible light ($\lambda \geq 420$ nm). The studies of microstructure, physicochemical properties, and photoelectric behavior demonstrated that the GO/Rh-SrTiO$_3$ heterojunction can work under visible light, which greatly improves the utilization of sunlight. Moreover, the Rh metal as the electron acceptor, which greatly increases the active sites and promotes the electron–hole separation and transfer. This work provides a new strategy to transform traditional UV catalysts into visible-light-responsive catalysts with abundant active sites.

Author Contributions: Conceptualization, P.H. and Y.Y. (Yongsheng Yan); methodology, Y.Y. (Yan Yan); software, E.J.; validation, S.Z., J.W. and Z.L.; formal analysis, Y.Y. (Yan Yan); investigation, S.Z.; resources, Y.Y. (Yongsheng Yan); data curation, S.Z.; writing—original draft preparation, S.Z.; writing—review and editing, E.J.; visualization, S.Z.; supervision, J.W.; project administration, Y.Y. (Yan Yan); funding acquisition, Y.Y. (Yongsheng Yan). All authors have read and agreed to the published version of the manuscript.

Funding: This research was funded by the National Natural Science Foundation of China (Grant No. 21776117 and 21806060).

Data Availability Statement: Not applicable.

Conflicts of Interest: The authors declare no competing financial interest.

References

1. Wang, W.; Xu, M.; Xu, X.; Zhou, W.; Shao, Z. Perovskite oxide based electrodes for high-performance photoelectrochemical water splitting. *Angew. Chem. Int. Ed.* **2020**, *59*, 136–152. [CrossRef] [PubMed]
2. Niu, W.; Yang, Y. Graphitic carbon nitride for electrochemical energy conversion and storage. *ACS Energy Lett.* **2018**, *3*, 2796–2815. [CrossRef]
3. Zhou, Z.; Pei, Z.; Wei, L.; Zhao, S.; Jian, X.; Chen, Y. Electrocatalytic hydrogen evolution under neutral pH conditions: Current understandings, recent advances, and future prospects. *Energy Environ. Sci.* **2020**, *13*, 3185–3206. [CrossRef]
4. Holladay, J.D.; Hu, J.; King, D.L.; Wang, Y. An overview of hydrogen production technologies. *Catal. Today* **2009**, *139*, 244–260. [CrossRef]
5. LeValley, T.L.; Richard, A.R.; Fan, M. The progress in water gas shift and steam reforming hydrogen production technologies—A review. *Int. J. Hydrogen Energy* **2014**, *39*, 16983–17000. [CrossRef]
6. Hisatomi, T.; Kubota, J.; Domen, K. Recent advances in semiconductors for photocatalytic and photoelectrochemical water splitting. *Chem. Soc. Rev.* **2014**, *43*, 7520–7535. [CrossRef] [PubMed]
7. Jiang, R.; Lu, G.; Yan, Z.; Wu, D.; Zhou, R.; Bao, X. Insights into a CQD-SnNb$_2$O$_6$/BiOCl Z-scheme system for the degradation of benzocaine: Influence factors, intermediate toxicity and photocatalytic mechanism. *Chem. Eng. J.* **2019**, *374*, 79–90. [CrossRef]
8. Che, H.; Che, G.; Zhou, P.; Liu, C.; Dong, H.; Li, C.; Li, C. Nitrogen doped carbon ribbons modified g-C$_3$N$_4$ for markedly enhanced photocatalytic H$_2$-production in visible to near-infrared region. *Chem. Eng. J.* **2020**, *382*, 122870. [CrossRef]
9. Tang, M.; Ao, Y.; Wang, C.; Wang, P. Rationally constructing of a novel dual Z-scheme composite photocatalyst with significantly enhanced performance for neonicotinoid degradation under visible light irradiation. *Appl. Catal. B Environ.* **2020**, *270*, 118918. [CrossRef]
10. Li, C.; Yu, S.; Zhang, X.; Wang, Y.; Liu, C.; Chen, G.; Dong, H. Insight into photocatalytic activity, universality and mechanism of copper/chlorine surface dual-doped graphitic carbon nitride for degrading various organic pollutants in water. *J. Colloid Interface Sci.* **2019**, *538*, 462–473. [CrossRef]
11. Patial, S.; Hasija, V.; Raizada, P.; Singh, P.; Singh, A.A.P.K.; Asiri, A.M. Tunable photocatalytic activity of SrTiO$_3$ for water splitting: Strategies and future scenario. *J. Environ. Chem. Eng.* **2020**, *8*, 103791. [CrossRef]
12. Szafraniak, B.; Fusnik, Ł.; Xu, J.; Gao, F.; Brudnik, A.; Ry-dosz, A. Semiconducting metal oxides: SrTiO$_3$, BaTiO$_3$ and BaSrTiO$_3$ in gas-sensing applications: A review. *Coatings* **2021**, *11*, 185. [CrossRef]
13. Jayabal, P.; Sasirekha, V.; Mayandi, J.; Jeganathan, K.; Ramakrishnan, V. A facile hydrothermal synthesis of SrTiO$_3$ for dye sensitized solar cell application. *J. Alloys Compd.* **2014**, *586*, 456–461. [CrossRef]
14. Ghosh, D.; Giri, S.; Sahoo, S.; Das, C.K. In situ synthesis of graphene/amine-modified graphene, polypyrrole composites in presence of SrTiO$_3$ for supercapacitor applications. *Polym. Plast. Technol. Eng.* **2013**, *52*, 213–220. [CrossRef]
15. Muhamad, N.F.; Osman, R.A.M.; Idris, M.S.; Yasin, M.N.M. Physical and electrical properties of SrTiO$_3$ and SrZrO$_3$//EPJ Web of Conferences. *EDP Sci.* **2017**, *162*, 01052. [CrossRef]
16. Hu, X.J.; Yang, Y.; Hou, C.; Liang, T.X. Thermodynamic and Electronic Properties of Two-Dimensional SrTiO$_3$. *J. Phys. Chem. C* **2021**, *126*, 517–524. [CrossRef]
17. Liu, G.; Zhao, Y.; Sun, C.; Li, F.; Lu, G.Q.; Cheng, H.M. Synergistic effects of B/N doping on the visible-light photocatalytic activity of mesoporous TiO$_2$. *Angew. Chem. Int. Ed.* **2008**, *47*, 4516–4520. [CrossRef]
18. Zhang, X.; Wang, Y.; Liu, B.; Sang, Y.; Liu, H. Heterostructures construction on TiO$_2$ nanobelts: A powerful tool for building high-performance photocatalysts. *Appl. Catal. B Environ.* **2017**, *202*, 620–641. [CrossRef]
19. Zhang, Z.; Huang, L.; Zhang, J.; Wang, F.; Xie, Y.; Shang, X.; Wang, X. In situ constructing interfacial contact MoS$_2$/ZnIn$_2$S$_4$ heterostructure for enhancing solar photocatalytic hydrogen evolution. *Appl. Catal. B Environ.* **2018**, *233*, 112–119. [CrossRef]
20. Maarisetty, D.; Baral, S.S. Defect engineering in photocatalysis: Formation, chemistry, optoelectronics, and interface studies. *J. Mater. Chem. A* **2020**, *8*, 18560–18604. [CrossRef]
21. Bai, S.; Zhang, N.; Gao, C.; Xiong, Y. Defect engineering in photocatalytic materials. *Nano Energy* **2018**, *53*, 296–336. [CrossRef]
22. Putri, L.K.; Ong, W.J.; Chang, W.S.; Chai, S.P. Heteroatom doped graphene in photocatalysis: A review. *Appl. Surf. Sci.* **2015**, *358*, 2–14. [CrossRef]

23. Cui, D.; Hao, W.; Chen, J. The synergistic effect of heteroatom doping and vacancy on the reduction of CO_2 by photocatalysts. *ChemNanoMat* **2021**, *7*, 894–901. [CrossRef]
24. Zhang, L.H.; Shi, Y.; Wang, Y.; Shiju, N.R. Nanocarbon catalysts: Recent understanding regarding the active sites. *Adv. Sci.* **2020**, *7*, 1902126. [CrossRef] [PubMed]
25. Li, Y.; Li, X.; Zhang, H.; Fan, J.; Xiang, Q. Design and application of active sites in g-C_3N_4-based photocatalysts. *J. Mater. Sci. Technol.* **2020**, *56*, 69–88. [CrossRef]
26. Ha, M.N.; Zhu, F.; Liu, Z.; Wang, L.; Liu, L.; Lu, G.; Zhao, Z. Morphology-controlled synthesis of $SrTiO_3/TiO_2$ heterostructures and their photocatalytic performance for water splitting. *RSC Adv.* **2016**, *6*, 21111–21118. [CrossRef]
27. Deng, Y.; Shu, S.; Fang, N.; Wang, R.; Chu, Y.; Liu, Z.; Cen, W. One-pot synthesis of $SrTiO_3$-$SrCO_3$ heterojunction with strong interfacial electronic interaction as a novel photocatalyst for water splitting to generate H_2. *Chin. Chem. Lett.* **2023**, *34*, 107323. [CrossRef]
28. Pan, J.H.; Shen, C.; Ivanova, I.; Zhou, N.; Wang, X.; Tan, W.C.; Wang, Q. Self-template synthesis of porous perovskite titanate solid and hollow submicrospheres for photocatalytic oxygen evolution and mesoscopic solar cells. *ACS Appl. Mater. Interfaces* **2015**, *7*, 14859–14869. [CrossRef]
29. Xiang, Q.; Yu, J.; Jaroniec, M. Graphene-based semiconductor photocatalysts. *Chem. Soc. Rev.* **2012**, *41*, 782–796. [CrossRef]
30. Zhang, N.; Zhang, Y.; Xu, Y.J. Recent progress on graphene-based photocatalysts: Current status and future perspectives. *Nanoscale* **2012**, *4*, 5792–5813. [CrossRef]
31. Johra, F.T.; Jung, W.G. RGO–TiO_2–ZnO composites: Synthesis, characterization, and application to photocatalysis. *Appl. Catal. A Gen.* **2015**, *491*, 52–57. [CrossRef]
32. Gao, P.; Liu, J.; Sun, D.D.; Ng, W. Graphene oxide–CdS composite with high photocatalytic degradation and disinfection activities under visible light irradiation. *J. Hazard. Mater.* **2013**, *250*, 412–420. [CrossRef] [PubMed]
33. Hunge, Y.M.; Yadav, A.A.; Dhodamani, A.G.; Suzuki, N.; Terashima, C.; Fujishima, A.; Mathe, V.L. Enhanced photocatalytic performance of ultrasound treated GO/TiO_2 composite for photocatalytic degradation of salicylic acid under sunlight illumination. *Ultrason. Sonochem.* **2020**, *61*, 104849. [CrossRef] [PubMed]
34. Wang, Y.; Du, Y.; Fu, Z.; Ren, J.; Fu, Y.; Wang, L. Construction of Ru/FeCoP heterointerface to drive dual active site mechanism for efficient overall water splitting. *J. Mater. Chem. A* **2022**, *10*, 16071–16079. [CrossRef]
35. Shahabuddin, S.; Muhamad Sarih, N.; Mohamad, S.; Joon Ching, J. $SrTiO_3$ nanocube-doped polyaniline nanocomposites with enhanced photocatalytic degradation of methylene blue under visible light. *Polymers* **2016**, *8*, 27. [CrossRef]
36. Zhang, K.; Zhang, L.L.; Zhao, X.S.; Wu, J. Graphene/polyaniline nanofiber composites as supercapacitor electrodes. *Chem. Mater.* **2010**, *22*, 1392–1401. [CrossRef]
37. Kumar, N.A.; Choi, H.J.; Shin, Y.R.; Chang, D.W.; Dai, L.; Baek, J.B. Polyaniline-grafted reduced graphene oxide for efficient electrochemical supercapacitors. *ACS Nano* **2012**, *6*, 1715–1723. [CrossRef]
38. Kiss, B.; Manning, T.D.; Hesp, D.; Didier, C.; Taylor, A.; Pickup, D.M.; Rosseinsky, M.J. Nano-structured rhodium doped $SrTiO_3$–Visible light activated photocatalyst for water decontamination. *Appl. Catal. B Environ.* **2017**, *206*, 547–555. [CrossRef]
39. Kawasaki, S.; Nakatsuji, K.; Yoshinobu, J.; Komori, F.; Takahashi, R.; Lippmaa, M.; Kudo, A. Epitaxial Rh-doped $SrTiO_3$ thin film photocathode for water splitting under visible light irradiation. *Appl. Phys. Lett.* **2012**, *101*, 033910. [CrossRef]
40. Kiran, K.S.; Ashwath Narayana, B.S.; Lokesh, S.V. Enhanced photocatalytic activity of perovskite $SrTiO_3$ nanorods. *Solid State Technol.* **2020**, *63*, 1913–1920.
41. Zhao, W.; Wang, H.; Liu, N.; Rong, J.; Zhang, Q.; Li, M.; Yang, X. Hydrothermal synthesis of Litchi-like $SrTiO_3$ with the help of ethylene glycol. *J. Am. Ceram. Soc.* **2019**, *102*, 981–987. [CrossRef]
42. Yang, D.; Sun, Y.; Tong, Z.; Nan, Y.; Jiang, Z. Fabrication of bimodal-pore $SrTiO_3$ microspheres with excellent photocatalytic performance for Cr (VI) reduction under simulated sunlight. *J. Hazard. Mater.* **2016**, *312*, 45–54. [CrossRef] [PubMed]
43. Kong, C.; Su, X.; Qing, D.; Zhao, Y.; Wang, J.; Zeng, X. Controlled synthesis of various $SrTiO_3$ morphologies and their effects on photoelectrochemical cathodic protection performance. *Ceram. Int.* **2022**, *48*, 20228–20236. [CrossRef]
44. Fan, Y.; Liu, Y.; Cui, H.; Wang, W.; Shang, Q.; Shi, X.; Tang, B. Photocatalytic overall water splitting by $SrTiO_3$ with surface oxygen vacancies. *Nanomaterials* **2020**, *10*, 2572. [CrossRef] [PubMed]
45. Wei, Y.; Wang, J.; Yu, R.; Wan, J.; Wang, D. Constructing $SrTiO_3$–TiO_2 heterogeneous hollow multi-shelled structures for enhanced solar water splitting. *Angew. Chem. Int. Ed.* **2019**, *58*, 1422–1426. [CrossRef]
46. Iwase, A.; Udagawa, Y.; Yoshino, S.; Ng, Y.H.; Amal, R.; Kudo, A. Solar Water Splitting under Neutral Conditions Using Z-Scheme Systems with Mo-Doped $BiVO_4$ as an O_2-Evolving Photocatalyst. *Energy Technol.* **2019**, *7*, 1900358. [CrossRef]
47. Han, K.; Lin, Y.C.; Yang, C.M.; Jong, R.; Mul, G.; Mei, B. Promoting photocatalytic overall water splitting by controlled magnesium incorporation in $SrTiO_3$ photocatalysts. *ChemSusChem* **2017**, *10*, 4510–4516. [CrossRef]
48. Kanazawa, T.; Nozawa, S.; Lu, D.; Maeda, K. Structure and photocatalytic activity of $PdCrO_x$ cocatalyst on $SrTiO_3$ for overall water splitting. *Catalysts* **2019**, *9*, 59. [CrossRef]
49. Qureshi, M.; Garcia-Esparza, A.T.; Jeantelot, G.; Ould-Chikh, S.; Aguilar-Tapia, A.; Hazemann, J.L.; Takanabe, K. Catalytic consequences of ultrafine Pt clusters supported on $SrTiO_3$ for photocatalytic overall water splitting. *J. Catal.* **2019**, *376*, 180–190. [CrossRef]
50. Zhang, X.; Li, Z.; Liu, T.; Li, M.; Zeng, C.; Matsumoto, H.; Han, H. Water oxidation sites located at the interface of $Pt/SrTiO_3$ for photocatalytic overall water splitting. *Chin. J. Catal.* **2022**, *43*, 2223–2230. [CrossRef]

51. Zhuo, Z.; Wang, X.; Shen, C.; Cai, M.; Jiang, Y.; Xue, Z.; Sun, S. Construction of $TiO_2/SrTiO_3$ Heterojunction Derived from Monolayer Ti_3C_2 MXene for Efficient Photocatalytic Overall Water Splitting. *Chem. A Eur. J.* **2023**, *29*, e202203450. [CrossRef] [PubMed]
52. Liu, J.W.; Chen, G.; Li, Z.H.; Zhang, Z.G. Electronic structure and visible light photocatalysis water splitting property of chromium-doped $SrTiO_3$. *J. Solid State Chem.* **2006**, *179*, 3704–3708. [CrossRef]
53. Ouyang, S.; Tong, H.; Umezawa, N.; Cao, J.; Li, P.; Bi, Y.; Ye, J. Surface-alkalinization-induced enhancement of photocatalytic H_2 evolution over $SrTiO_3$-based photocatalysts. *J. Am. Chem. Soc.* **2012**, *134*, 1974–1977. [CrossRef] [PubMed]
54. Zhao, Y.; Wang, Y.; Liang, X.; Shi, H.; Wang, C.; Fan, J.; Liu, E. Enhanced photocatalytic activity of $Ag-CsPbBr_3/CN$ composite for broad spectrum photocatalytic degradation of cephalosporin antibiotics 7-ACA. *Appl. Catal. B Environ.* **2019**, *247*, 57–69. [CrossRef]
55. Xu, B.; Li, Y.; Gao, Y.; Liu, S.; Lv, D.; Zhao, S.; Ge, L. $Ag-AgI/Bi_3O_4Cl$ for efficient visible light photocatalytic degradation of methyl orange: The surface plasmon resonance effect of Ag and mechanism insight. *Appl. Catal. B Environ.* **2019**, *246*, 140–148. [CrossRef]
56. Yu-Lei, D.; Guang, C.; Ming-Sheng, Z.; Sen-Zu, Y. Phonon characteristics of polycrystalline cubic $SrTiO_3$ thin films. *Chin. Phys. Lett.* **2003**, *20*, 1561. [CrossRef]
57. Rahman, J.U.; Du, N.V.; Nam, W.H.; Shin, W.H.; Lee, K.H.; Seo, W.S.; Lee, S. Grain boundary interfaces controlled by reduced graphene oxide in nonstoichiometric $SrTiO_3$-δ thermoelectrics. *Sci. Rep.* **2019**, *9*, 8624. [CrossRef]
58. Kogularasu, S.; Govindasamy, M.; Chen, S.M.; Akilarasan, M.; Mani, V. 3D graphene oxide-cobalt oxide polyhedrons for highly sensitive non-enzymatic electrochemical determination of hydrogen peroxide. *Sens. Actuators B Chem.* **2017**, *253*, 773–783. [CrossRef]
59. Huang, C.; Li, C.; Shi, G. Graphene based catalysts. *Energy Environ. Sci.* **2012**, *5*, 8848–8868. [CrossRef]
60. Chen, S.; Takata, T.; Domen, K. Particulate photocatalysts for overall water splitting. *Nat. Rev. Mater.* **2017**, *2*, 17050. [CrossRef]
61. Wang, X.; Dou, W. Preparation of graphite oxide (GO) and the thermal stability of silicone rubber/GO nanocomposites. *Thermochim. Acta* **2012**, *529*, 25–28. [CrossRef]
62. Wei, X.; Xu, G.; Ren, Z.; Xu, C.; Weng, W.; Shen, G.; Han, G. Single-Crystal-like Mesoporous $SrTiO_3$ Spheres with Enhanced Photocatalytic Performance. *J. Am. Ceram. Soc.* **2010**, *93*, 1297–1305.
63. Wei, X.; Xu, G.; Ren, Z.; Xu, C.; Shen, G.; Han, G. PVA-Assisted Hydrothermal Synthesis of $SrTiO_3$ Nanoparticles with Enhanced Photocatalytic Activity for Degradation of RhB. *J. Am. Ceram. Soc.* **2008**, *91*, 3795–3799. [CrossRef]

Disclaimer/Publisher's Note: The statements, opinions and data contained in all publications are solely those of the individual author(s) and contributor(s) and not of MDPI and/or the editor(s). MDPI and/or the editor(s) disclaim responsibility for any injury to people or property resulting from any ideas, methods, instructions or products referred to in the content.

Article

Fabrication of FeTCPP@CNNS for Efficient Photocatalytic Performance of p-Nitrophenol under Visible Light

Shiyun Li [1,†], Yuqiong Guo [1,†], Lina Liu [1], Jiangang Wang [1], Luxi Zhang [1], Weilong Shi [1,*], Malgorzata Aleksandrzak [2], Xuecheng Chen [2,*] and Jie Liu [3]

1. School of Materials Science and Engineering, Jiangsu University of Science and Technology, Zhenjiang 212003, China
2. Faculty of Chemical Technology and Engineering, West Pomeranian University of Technology, Piastów Ave. 42, 71-065 Szczecin, Poland
3. State Key Laboratory of Polymer Physics and Chemistry, Changchun Institute of Applied Chemistry, Chinese Academy of Sciences, Changchun 130022, China; liujie@ciac.ac.cn
* Correspondence: shiwl@just.edu.cn (W.S.); xchen@zut.edu.pl (X.C.)
† These authors contributed equally to this work.

Abstract: A photocatalyst of iron–porphyrin tetra-carboxylate (FeTCPP)-sensitized g-C_3N_4 nanosheet composites (FeTCPP@CNNS) based on g-C_3N_4 nanosheet (CNNS) and FeTCPP have been fabricated by in situ hydrothermal self-assembly. FeTCPP is uniformly introduced to the surface of CNNS. Only a small amount of FeTCPP is introduced, and the stacked lamellar structure is displayed in the composite. As compared with pure CNNS, the FeTCPP@CNNS composites exhibit significantly improved photocatalytic performance by the photodegradation of p-nitrophenol (4-NP). At the optimum content of FeTCPP to CNNS (3 wt%), the photodegradation activity of the FeTCPP@CNNS photocatalyst can reach 92.4% within 1 h. The degradation rate constant for the 3% FeTCPP@CNNS composite is 0.037 min^{-1} (4-NP), which is five times that of CNNS (0.0064 min^{-1}). The results of recycling experiments show that 3% FeTCPP@CNNS photocatalyst has excellent photocatalytic stability. A possible photocatalytic reaction mechanism of FeTCPP@CNNS composite for photocatalytic degradation of 4-NP has been proposed. It is shown that superoxide radical anions played the major part in the degradation of 4-NP. The appropriate content of FeTCPP can enhance the charge transfer efficiency. The FeTCPP@CNNS composites can provide more active sites and accelerate the transport and separation efficiency of photogenerated carriers, thus further enhancing the photocatalytic performance.

Keywords: FeTCPP@CNNS; g-C_3N_4 nanosheets; photocatalytic; visible light

1. Introduction

With the fast development of economy and industry, environmental deterioration—especially water pollution by organic dyes, which result in a serious threat to human health—has garnered wide attention from the government and society. At present, many effective methods have been used to resolve water pollutant problems, including the membrane oxidation method, adsorption method, separation method, and photocatalysis method [1–4]. Photocatalysis are proved to be a safe, economical, and renewable method to solve the aforementioned pollution problems by photocatalysts at ambient pressure and room temperature under solar light, which is considered one of the most promising wastewater treatment methods [5]. Although the application of photocatalytic technology has been wildly used, there is still a puzzle to develop photocatalysts with physicochemical stability, photocatalytic activity, and enhancing visible light utilization efficiency for practice application [6]. Photocatalytic reaction can be divided into three basic processes, including light capture process, carrier separation and migration process, and photocatalytic redox reaction [7–12].

Citation: Li, S.; Guo, Y.; Liu, L.; Wang, J.; Zhang, L.; Shi, W.; Aleksandrzak, M.; Chen, X.; Liu, J. Fabrication of FeTCPP@CNNS for Efficient Photocatalytic Performance of p-Nitrophenol under Visible Light. Catalysts 2023, 13, 732. https://doi.org/10.3390/catal13040732

Academic Editor: Weilin Dai

Received: 9 March 2023
Revised: 6 April 2023
Accepted: 10 April 2023
Published: 12 April 2023

Copyright: © 2023 by the authors. Licensee MDPI, Basel, Switzerland. This article is an open access article distributed under the terms and conditions of the Creative Commons Attribution (CC BY) license (https://creativecommons.org/licenses/by/4.0/).

Metal-free polymeric semiconductor graphite-like carbon nitride (g-C_3N_4), with a large amount of of pendant amine and unique two-dimensional structure, is a promising candidate for solar energy conversion and organic pollutions degradation by solar irradiation due to its easy fabrication, good chemical and physical stability, nontoxicity, low cost, and visible light activity. The g-C_3N_4 possesses a bandgap of approx. 2.7 eV, which has presented good chemical stability in the removal of organic dyes and high visible light absorption ability. However, the photocatalytic activity and the practical application of g-C_3N_4 are restricted by the fast recombination of the photogenerated electron–hole carriers [13]. To break these limitations, many studies have been adopted to depress the rapid recombination of carriers, such as the morphology control of g-C_3N_4, doping, combination with other semiconductors, surface sensitization, and dye sensitization [14,15].

As one of the light-harvesting materials, porphyrins play an important role in photocatalysis. Porphyrin compounds act as excellent photosensitizers for photocatalysts due to their wide absorption band, large conjugate structure, and good electron-donating properties [14,16,17]. In general, under UV and visible light irradiation, metalloporphyrins can catalyze a great many oxidative transformations. The metalloporphyrins can be combined with photocatalysts in the outer of the porphyrin ring through covalent interaction among the different functional groups (such as COOH and OH) [18,19]. The covalent bands can be used as the electron transfer channels between metalloporphyrins and photocatalysts and can further give rise to better selectivity and/or efficiency in catalytic processes [20–23]. Indeed, due to its two-dimensional flexible structure, g-C_3N_4 may be easily modified with organic small molecules as a promising photocatalyst [24–33]. Accordingly, on the basis of the latent characteristics of metalloporphyrins, it can be anticipated that the combination between g-C_3N_4 and metalloporphyrins could be supported to provide a synergistic effect of enhancing the photocatalytic activity with considerable visible light utilization efficiency.

In this work, tetra(4-carboxyphenyl)porphyrin (TCPP), FeTCPP, and CNNS were successfully prepared firstly. Different mass contents of FeTCPP were introduced on the surface of g-C_3N_4 nanosheets, forming FeTCPP@CNNS photocatalysts by π–π stacking interactions and hydrogen bonding. Illustrated in Figure 1, the photocatalysts were fabricated by integrating FeTCPP with g-C_3N_4 sheets via a mechanical mix method. The FeTCPP acts as the light-harvesting part, and CNNS as the catalytic center, which can accelerate the separation rate of the photogenerated electron and hole carriers. Under visible light irradiation, the sensitized photocatalysts 3% FeTCPP@CNNS shows a high photocatalytic activity for 4-NP degradation due to the efficient transfer to CNNS of the photogenerated electrons of the excited FeTCPP. On the basis of the results of the active radical identification experiments, the possible photocatalytic mechanism for the TCPP/CNNS composites was also elucidated. This work shows that CNNS sensitized by FeTCPP could enhance the photocatalytic degradation of 4-NP for more efficiently utilizing solar radiation.

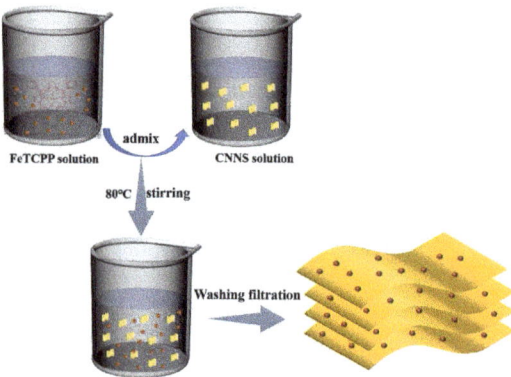

Figure 1. Schematic illustration of preparation process for FeTCPP@CNNS.

2. Results and Discussion

Figure 2a displays the X-ray diffraction pattern of the prepared materials. It can be found from Figure 2a that the bulk g-C_3N_4 presents peaks at 12.8° and 27.7°, corresponding to the (100) crystal plane and (002) crystal plane of g-C_3N_4, respectively. The peak at 12.7° is weak, which reflects the regular arrangement of triazine rings in g-C_3N_4, and the peak strength is strong at 27.5°, which reflects the typical graphite interlayer stacking structure [34,35]. Compared with bulk g-C_3N_4, the peak (002) of CNNS became wider and slightly more weakened, indicating that the crystallinity of CNNS was not as good as bulk g-C_3N_4. Meanwhile, the peak (100) of CNNS almost disappeared, indicating that the nanosheet was successfully exfoliated [8]. FeTCPP has a very wide diffraction peak at approximately 21.4°, indicating that TCPP has an amorphous structure [16]. Figure 2b shows the X-ray diffraction of FeTCPP@CNNS materials and CNNS. It can be found that when CNNS is sensitized by a small amount of FeTCPP, the peak (002) generated by interlamellar deposition of graphite is slightly larger than that of pure CNNS. It is caused by the interaction between CNNS and FeTCPP through the triazine unit of porphyrin [36]. No obvious characteristic peak of FeTCPP was observed in FeTCPP@CNNS composites, on the one hand, because of the low content of FeTCPP in the composites, and on the other hand, because of the weak peak width of the diffraction characteristics of FeTCPP [37]. By further comparison, it can be observed that the XRD spectra of composites are very similar to those of the CNNS monomer, which also indicates that the addition of FeTCPP will not damage the crystal structure of CNNS [38].

Figure 2c,d show FTIR spectra of TCPP, FeTCPP, CNNS, and FeTCPP@CNNS. FTIR spectra of TCPP and FeTCPP are reflected in Figure 2c, where the characteristic peak at 963 cm^{-1} represents the N-H telescopic vibration pattern on the pyrrole ring of TCPP [39]. This feature peak disappears in the FTIR spectrogram of FeTCPP. In addition, a new characteristic peak at 1001 cm^{-1} can be observed in the FTIR spectrum of FeTCPP, which indicates that after the metal ions enter the porphyrin ring, the deformation vibration of the ring is enhanced. In addition, the telescopic vibration characteristic peak of Fe-N is generated, which demonstrates that the porphyrin ligand can form with the metal ion form the complex [38]. For FeTCPP materials, the characteristic peaks at 1276, 1405, and 1604 cm^{-1} belong to the -OH tensile vibration in -COOH, the C-N in-plane vibration of pyrrole. The tensile vibration of C-C at 1710 cm^{-1} indicates the telescopic vibration absorption of the -COOH and -NH_2 functional groups in their molecular structures [39,40]. In Figure 2d, for pure CNNS, the spike absorption peak at 809 cm^{-1} is attributed to the typical vibration pattern of the graphite phase carbon nitride triazine ring, and the presence of four more obvious characteristic absorption peaks in the range of 1700–1200 cm^{-1} is due to the telescopic vibration of the surface C-N heterocyclic ring [41,42]. In the FTIR spectra of the FeTCPP@CNNS, it can be observed that the characteristic peaks are almost consistent with those of pure CNNS, which demonstrate that the structure of CNNS has not changed during mechanical stirring. The -NH_2 deformation vibration band at 1573 cm^{-1} in the CNNS monomer disappeared in the FeTCPP@CNNS composite because of the covalent formation of -N-O by the COOH of porphyrin and NH_2 of g-C_3N_4 [39]. The characteristic tensile bands of amide groups formed between metalloporphyrin and g-C_3N_4 at 1640 and 1260 cm^{-1} were not clearly observed, possibly because that the peaks were too small and similar to the peaks of pure CNNS. The results confirm that the hybrid effect between FeTCPP and CNNS molecules may come mainly from non-covalent interactions [14].

The UV-visible diffuse reflection spectra of CNNS, FeTCPP, and FeTCPP@CNNS composites are exhibited in Figure 3a; it can be seen that pure CNNS has an absorption edge at 450 nm, and the DRS spectra of FeTCPP@CNNS composites also exhibit the absorption characteristic peaks of CNNS. As illustrated in the DRS spectrum of FeTCPP, there is an absorption peak at 403 nm, related to the Soret band (B band) of the porphyrin compound, and absorption peaks at 520 nm, 578 nm, and 693 nm, corresponding to the Q band of the porphyrin compound [43,44]. With the increasing content of FeTCPP, the absorption characteristic peak of the FeTCPP@CNNS composites is slightly enhanced. As shown in

Figure 3a, there is a slight redshift phenomenon indicating that there was a π-π interaction between CNNS and FeTCPP [38] and further illustrating the successful formation of FeTCPP@CNNS composites. Figure 3b shows the band gap according to the Kubelka Munk transform [45]: $(\alpha h v)^2 = A(hv - Eg)$. The corresponding band gap of CNNS and FeTCPP are, respectively, calculated to be approximately 2.74 eV and 2.01 eV. The band gap of 3% FeTCPP@CNNS composite material is the smallest one at approximately 2.36 eV, illustrating that the 3% FeTCPP@CNNS composite can mostly improve the utilization rate of visible light.

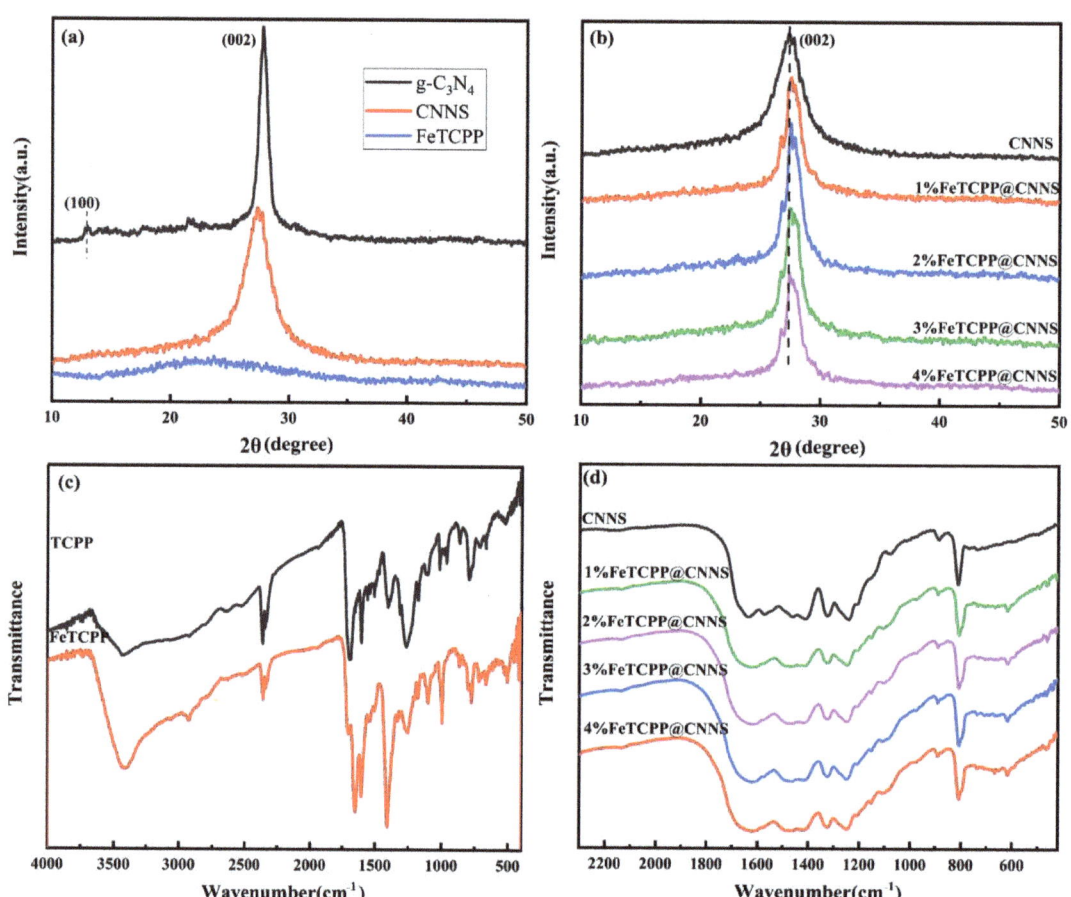

Figure 2. XRD spectra of (**a**) g-C_3N_4, CNNS, and FeTCPP and (**b**) FeTCPP@CNNS composites; FTIR spectra of (**c**) TCPP and FeTCPP and (**d**) CNNS and FeTCPP@CNNS composites.

The typical SEM images and EDS patterns of as-prepared samples are shown in Figure 4. In Figure 4a, it can be seen that the FeTCPP presents an irregular small particle shape. Figure 4b,c are SEM images of CNNS and 3% FeTCPP@CNNS. The CNNS sample presents a thin sheet-shaped morphology with a wrinkle, facilitating the transport of electrons. With only a small amount of FeTCPP introduced, as shown in Figure 4c, the stacked lamellar structure is displayed. FeTCPP is deposited on the surface of CNNS. In Figure 4d, the C:N atomic ratio is approximately 1:1, higher than that of g-C_3N_4 (3:4), which indicates that FeTCPP has been successfully loaded onto CNNS [16]. The elemental mapping of the 3% FeTCPP@CNNS composite shown in Figure 4e–h reveals a uniform distribution of C, N, O, and Fe elements in the 3% FeTCPP@CNNS framework, highlighting the C, N, O, and Fe co-doped nature of the 3% FeTCPP@CNNS composite.

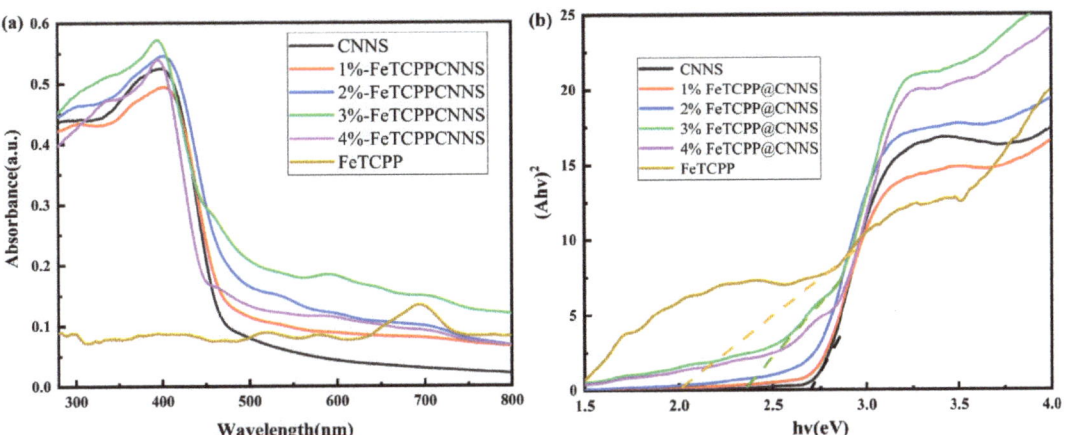

Figure 3. (**a**) UV-vis diffuse reflection spectra and (**b**) the band gap determined from Kubelka Munk transformation of CNNS, FeTCPP, and FeTCPP@CNNS composites.

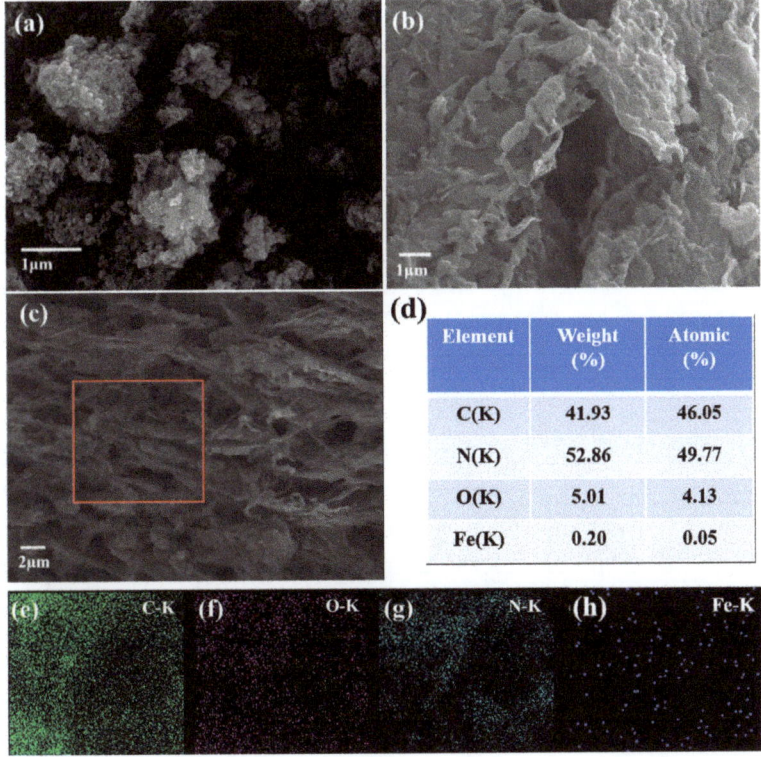

Figure 4. SEM images of (**a**) FeTCPP, (**b**) CNNS, and (**c**) 3% FeTCPP@CNNS; (**d**) EDS at 3% FeTCPP@CNNS (red area in figure (**c**)); (**e**–**h**) The corresponding EDS elemental mapping of C, O, N, and Fe elements in 3% FeTCPP@CNNS composites.

In order to further investigate the chemical composition and surface element valence states of FeTCPP@CNNS composite, X-ray photoelectron spectroscopy (XPS) analysis of 3% FeTCPP@CNNS composite is carried out. For comparison, the composition of CNNS and FeTCPP are also determined by XPS measurement. As shown in Figure 5a, C, N, and O signals are detected in the CNNS sample, while C, N, O, and Fe signals are detected in the FeTCPP sample. In the 3% FeTCPP@CNNS composite, the corresponding C, N, O, and Fe signals are found. The results show that the FeTCPP sample is successfully introduced in the composite material. Figure 5b–e show the spectra of C 1s, N 1s, O 1s, and Fe 2p, respectively. Figure 5b shows the C 1s spectra of CNNS and 3% FeTCPP@CNNS. There are two primary peaks at 284.6 eV and 288.2 eV, corresponding to the C-C bond of graphite and the Sp^2 hybrid carbon in the N=C-N aromatic ring [46,47]. Figure 5c shows the N 1s spectra of CNNS and 3% FeTCPP@CNNS materials. In the N 1s spectrum of CNNS, the peaks at 398.2 eV, 399.3 eV, and 401 eV correspond to C-N=C bond, (N-(C)3) bond, and secondary amino (C-N-H) bond, respectively [48,49]. The three peaks can also be observed in 3% FeTCPP@CNNS; however, the peak of C-N=C bond shifted to higher binding energy by 0.6 eV, and the peak of CN-H shifted to lower binding energy by 0.4 eV, which should be caused by the formation of type-II heterojunction photocatalyst [50] FeTCPP@CNNS. The O 1s spectra are shown in Figure 5d. The peak at 531.9 eV can be attributed to the -OH group, which means that only -OH forms on the surface due to the combustion of g-C_3N_4 in air [16,41]. The Fe 2p spectra of FeTCPP materials are shown in Figure 5e. The two peaks at 710.7 and 724.1 eV correspond to Fe 2p1/2 of Fe 2p3/2 and Fe^{3+} at octahedral positions, respectively [51–53]. However, no obvious Fe 2p peak was detected in 3% FeTCPP@CNNS, which may be because the low content of Fe doped in the composite and wrapped in porphyrin molecules.

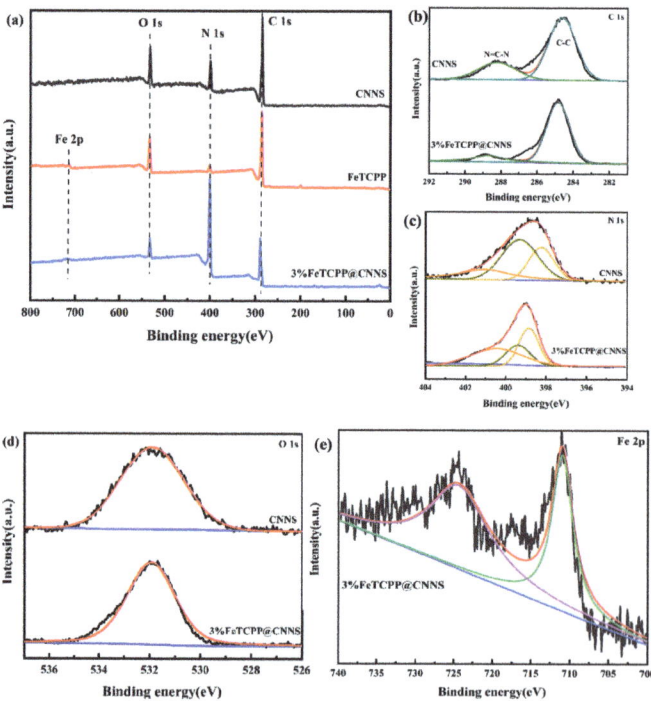

Figure 5. (a) XPS survey of CNNS, FeTCPP, and 3% FeTCPP@CNNS, (b) C 1s spectra, (c) N 1s spectra, (d) O 1s spectra of CNNS and 3% FeTCPP@CNNS, and (e) Fe 2p spectra of 3% FeTCPP@CNNS.

Figure 6 shows that the N_2 adsorption–desorption isotherms of CNNS and 3% FeTCPP@CNNS composite materials are similar and belong to the type IV isotherm, which manifests that both materials display mesoporous structures [54]. Additionally, the specific surface area (S_{BET}), average pore size, and pore volume of all samples are presented in Table 1. The BET surface area of CNNS and 3% FeTCPP@CNNS composite are approximately 11.8 and 24.9 $m^2 g^{-1}$, respectively. Compared with CNNS, the 3% FeTCPP@CNNS composite has a larger specific surface area, indicating that it can provide more active sites and promote the transport and separation efficiency of photogenerated carriers [55], thus further improving the utilization of light.

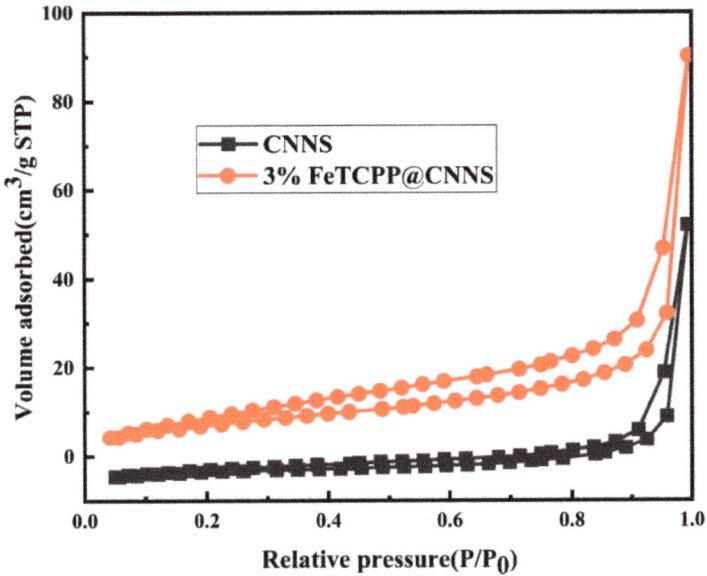

Figure 6. N_2 adsorption–desorption isotherms curves of CNNS and 3% FeTCPP@CNNS.

Table 1. Specific surface, pore characteristics, and crystallite sizes of the as-prepared samples.

Samples	SBET (m^2g^{-1})	Pore Volume (cm^3g^{-1})	Pore Size (nm)
CNNS	11.8	0.064	21.8
3% FeTCPP@CNNS	24.9	0.100	16.1

The photoluminescence spectra of CNNS, FeTCPP, and FeTCPP@CNNS composites are depicted in Figure 7a at an excitation wavelength of 320 nm. As reported [56], the weak PL emission peak indicates that the separation efficiency of the photoexcitation electron-hole pair is higher, resulting in higher photocatalytic performance. It can be found that CNNS and FeTCPP@CNNS composites show emission peaks in the region from 420 nm to 440 nm, while pure FeTCPP has no obvious emission peaks [36]. CNNS shows the strongest emission peak in the region from 440 nm to 550 nm, demonstrating the highest recombination rate of photogenerated photoelectrons and holes on the materials' surface. With a certain content of FeTCPP, it can inhibit the recombination of photogenerated carries in CNNS. With the increasing content of FeTCPP, the PL intensity of FeTCPP@CNNS composites decreases gradually. The 3% FeTCPP@CNNS exhibits the weakest PL intensity. However, when the content of FeTCPP is further increased, the PL intensity becomes slightly stronger, indicating that the appropriate content of FeTCPP can enhance the charge transfer efficiency.

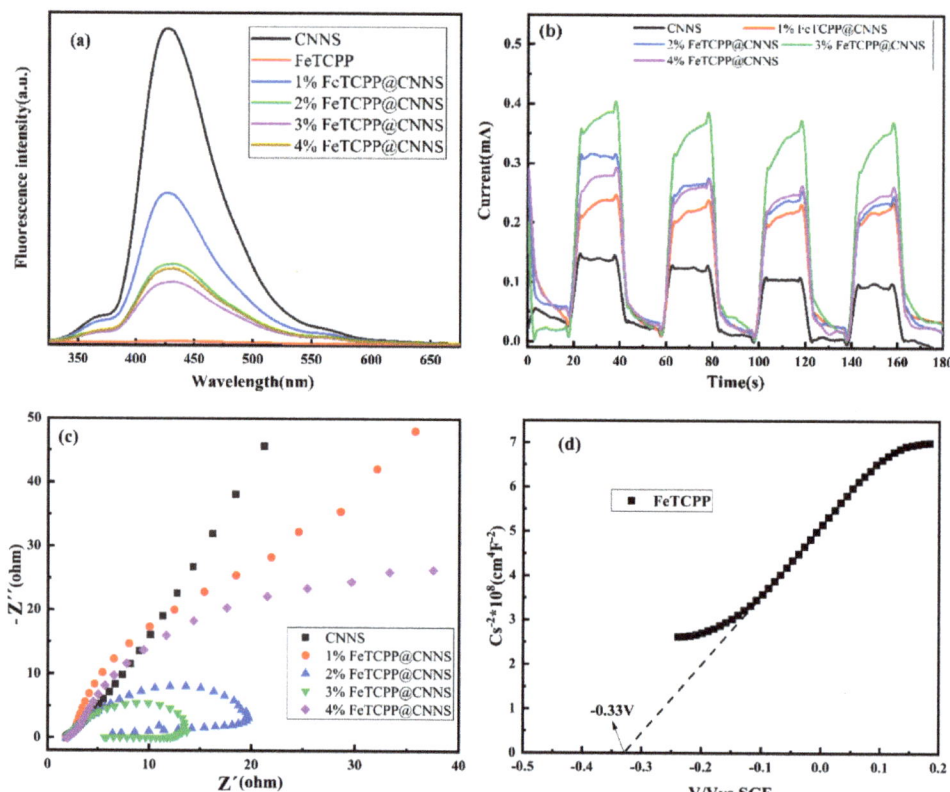

Figure 7. (a) PL spectra of CNNS, FeTCPP, and FeTCPP@CNNS composites, (b) photocurrent response diagram, (c) EIS spectra of CNNS and FeTCPP@CNNS composites, and (d) Mott-Schottky plot of FeTCPP.

To further study the electron transfer process, the photocurrent response of CNNS and FeTCPP@CNNS composites are measured, as shown in Figure 7b. Obviously, the photocurrent intensity of all samples increases sharply after the light irradiation is turned on, and when the irradiation is interrupted, it drops sharply to zero. The results indicate that they all have photocatalytic capability. The higher photocurrent intensity indicates more efficient separation of photogenerated carries. Among them, the photocurrent intensity of 3% FeTCPP@CNNS composite is the highest. In addition, the photocurrent intensity of 3% FeTCPP@CNNS composite exhibits approximately 0.4 μA under simulated sunlight, which is twice more than that of the pure CNNS photocatalyst. It indicates that the 3% FeTCPP@CNNS composite could effectively improve the separation and transfer of photogenerated carries under visible light. The stability of the photocurrent response of all the prepared samples is determined by intermittent illumination for 20 s multiple cycles. It shows that there is only a slight reduction in the photocurrent intensity after four cycle operations, resulting in prepared samples that have good stability. Figure 7c presents electrochemical impedance spectroscopy. The minimum arc radius of the 3% FeTCPP@CNNS composite indicates the lowest interfacial resistance. Meanwhile, spectral line of CNNS has the largest slope in the low frequency region, indicating the largest diffusion resistance of CNNS [9,57]. The 3% FeTCPP@CNNS composite exhibits excellent charge separation, which is consistent with the above photocurrent results. In Figure 7d, the Mott–Schottky (MS) plot of FeTCPP is recorded by an electrochemical analyzer. The flat band potential (E_{fb}) of FeTCPP is −0.39 eV (vs. SCE). The E_{CB} of FeTCPP is −0.35 eVB based on the formula [58]: $E_{CB(NHE, pH=7)} = E_{fb(SCE, pH=7)} + 0.24 − 0.2$.

In Figure 8a, the degradation of 4-NP was negligible without photocatalysts, showing that 4-NP has almost no degradation by only direct visible light irradiation. A 50 mg catalyst is added to the 4-NP solution (20 mg L^{-1}) in the dark for 30 min to achieve adsorption–desorption equilibrium. After 60 min of illumination, it can be observed that the corresponding photodegradation efficiency of CNNS, 1% FeTCPP@CNNS, 2% FeTCPP@CNNS, 3% FeTCPP@CNNS, and 4% FeTCPP@CNNS are 37.2%, 63.7%, 77.9%, 92.4%, and 76.3%, respectively, as shown in Figure 8c. Compared with other photocatalysts, this shows that when the content of FeTCPP reaches 3%, the photodegradation efficiency reaches a maximum of 92.4% in 60 min. After the first-order kinetic equation fitting, it shows that the largest constant value of 3% FeTCPP@CNNS (k) is 0.037 min^{-1}, which is approximately 5 times that of CNNS. The results show that after being sensitized by FeTCPP, the FeTCPP@CNNS can quickly capture the visible light source and more easily produce photogenerated electrons under the illumination by visible light. Moreover, the separation rate of photogenerated electrons and holes is improved, resulting in effectively improved photocatalytic efficiency. As shown in Figure 8d, the absorbance of 4-NP significantly decreased with the increased illumination time. For comparison, recent photocatalysis performances of g-C_3N_4-based and TCPP-based materials under visible light irradiation are shown in Table 2. This information reveals that the 3% FeTCPP@CNNS exhibits higher photodegradation efficiency and higher degradation rate constants.

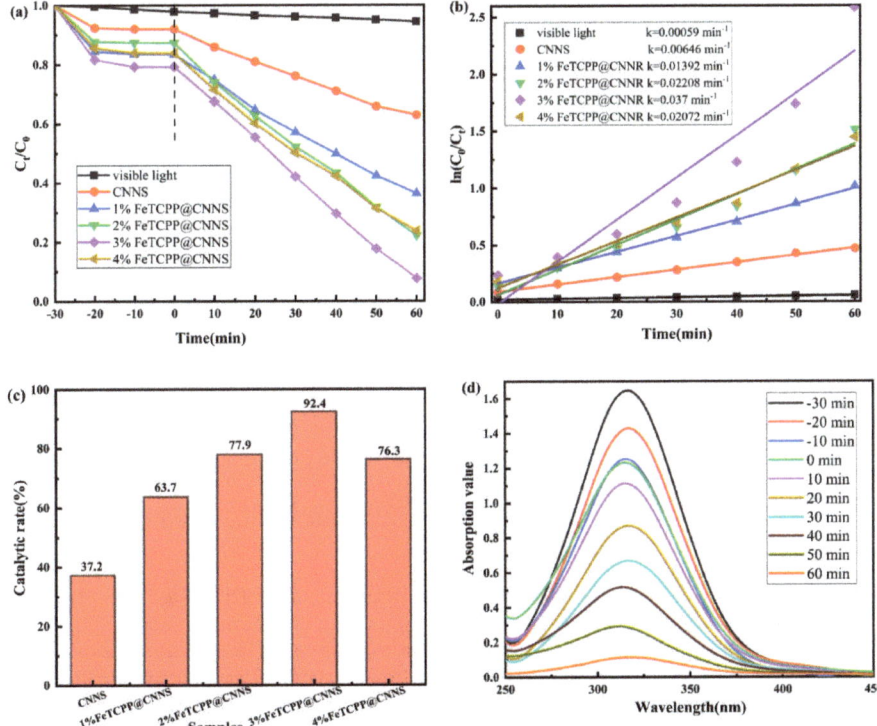

Figure 8. (**a**) Photocatalytic degradation for 4-NP aqueous solution over different photocatalysts under simulated solar light irradiation, (**b**) kinetics curves, (**c**) photodegradation efficiency of the as-prepared samples, and (**d**) temporal UV-vis absorption spectral changes of 4-NP in aqueous solution with presence of 3% FeTCPP@CNNS composite during the photocatalytic degradation.

Table 2. Some reported materials based on g-C_3N_4 or TCPP studied for photocatalysis under visible light irradiation in recent years.

Composite	Catalyst Dose	Concentration	Light Source	Degradation and Time (%)	Degradation Rate Constant (k)
TCPP/$ZnFeO_4$@ZnO [59]	50 mg	10 mg/L, 50 mL (4-NP)	5 W LED lamp	67% in 3 h	-
g-C_3N_4@MoS_2/TiO_2(CMT10) [60]	50 mg	1 × 10^{-5} mol/L (4-NP)	500 W tungsten halogen lamp	78% in 1 h	-
g-C_3N_4/$CoFe_2O_4$ [61]	25 mg	20 mg/L (4-NP)	Visible-light	-	0.0156 min^{-1}
g-C_3N_4-30%@Ti-MIL125 [62]	-	-	Visible-light	75% in 4 h	-
1 $ZnFe_2O_4$/g-C_3N_4 [63]	50 mg	20 mg/L, 100 mL (4-NP)	Sunlight	-	0.02876 min^{-1}
0.4 S/Cl-g-C_3N_4 [64]	50 mg	5 mg/L, 100 mL (4-NP)	Xenon lamp	-	0.0095 min^{-1}
30% ZrO_2/g-C_3N_4 [65]	360 mg	30 mg/L, 100 mL (4-NP)	300 W Xe	-	0.0167 min^{-1}
0.75% CuTCPP/g-C_3N_4 [14]	25 mg	5 ppm, 50 mL (phenol)	500 W Xe	-	0.024 h^{-1}
3% FeTCPP@CNNS in this work	20 mg	20 mg/L, 100 mL (4-NP)	150 W Xe	94.2% in 1 h	0.037 min^{-1}

In order to evaluate the photocatalytic stability of the 3% FeTCPP@CNNS composite, the cycle test of it is performed five times, as shown in Figure 9a. After five cycling experiments, the photocatalytic activity remains the same. In addition, there is no obvious difference in XRD patterns of the prepared 3% FeTCPP@CNNS and the used one after five cycles. Therefore, the 3% FeTCPP@CNNS sample exhibits excellent degradation performance and possesses high stability under visible light.

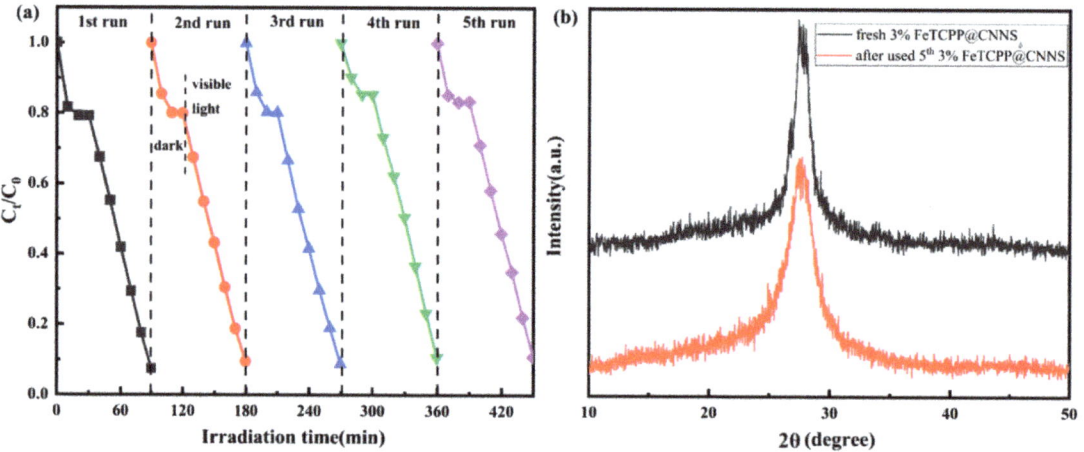

Figure 9. The 3% FeTCPP@CNNS photocatalytic degradation of 4-NP (**a**) cycle test curve and (**b**) XRD contrast after 5 cycles.

To further explore the reaction mechanism of the 3% FeTCPP@CNNS composite, the reactive species are determined in 1 mmol p-benzoquinone (BQ), isopropanol (IPA), and disodium edetate (EDTA-2Na), which were treated as superoxide radicals (·O^{2-}), hydroxyl radicals (·OH), and an inhibitor of the photo-excited hole (h^+), respectively [66]. From Figure 10a, the addition of IPA and EDTA-2Na inhibitors had little effect on the photocatalytic activity of the 3% FeTCPP@CNNS photocatalyst, and the addition of BQ significantly inhibited its photocatalytic activity. These results mean that ·O^{2-} is the main substance for photocatalytic degradation of 4-NP.

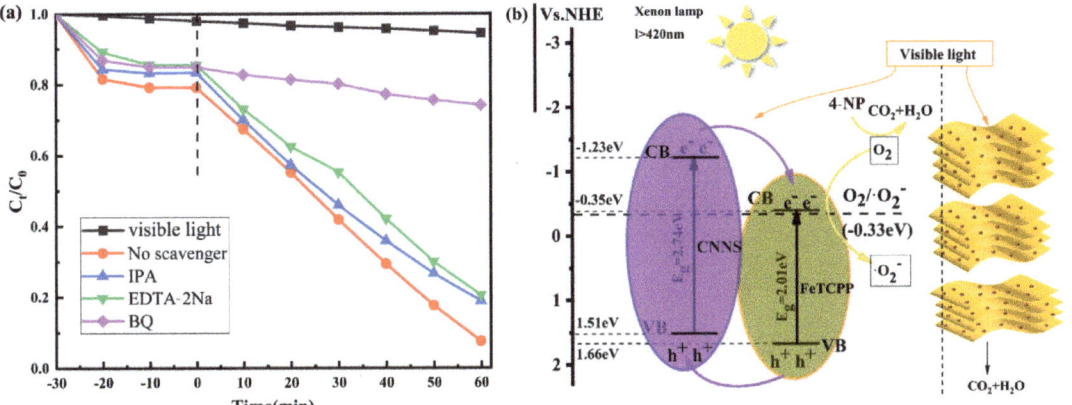

Figure 10. (a) Active substance capture experiment of 3% FeTCPP@CNNS photocatalyst under visible light irradiation and (b) photocatalytic mechanism of FeTCPP@CNNS photocatalyst under visible light irradiation.

The photocatalytic degradation capacity of pure CNNS and FeTCPP@CNNS composites are evaluated by using 4-NP as a contaminant according to the formula:

$$\eta = (1 - C_t/C_0) \times 100\%$$

The photodegradation rate (η) is known. At the same time, the adsorption experimental data conform to the pseudo-first-order model:

$$\ln(C_t/C_0) = kt$$

where C_0 and C_t signify initial concentration and instantaneous concentration within the reaction time t, respectively. k is the first-order reaction rate constants [67].

To further account for the photocatalytic mechanism, the band gap potential of the sample needs to be tested, and the conduction band (CB), the valence band (VB) position of CNNS, and the valence band (VB) position of FeTCPP are estimated using the following formulas [68]:

$$E_{VB} = \chi - E^e + 1/2 E_g$$

$$E_{CB} = E_{VB} - E_g$$

where χ is electronegativity, E_{VB} and E_{CB} denote VB and CB marginal potentials, respectively, E_g denotes the energy band gap, E_e represents the energy of free electrons at the hydrogen scale (approximately 4.5 eV vs. NHE) [69], χ is the geometric mean of the constituent atoms, and g-C_3N_4 is 4.64 eV [70]. According to the above analysis, the E_{VB} and E_{CB} of CNNS are calculated, respectively, to be 1.51 eV and −1.23 eV. Through DRS and MS tests, the FeTCPP E_g is 2.01 eV and the E_{CB} is −0.35 eV, so that the E_{VB} of FeTCPP is 1.66 eV. Relying on the above experimental results, a possible photocatalytic mechanism is proposed and demonstrated in Figure 10b. Under visible light illumination, CNNS can form a photogenerated electron–hole pair under the excitation of visible light. Because the E_{CB} edge and E_{VB} edge charges of FeTCPP are smaller than those of CNNS, parts of the photosensitive electrons (e^-) at the CB position of CNNS can migrate to the CB of FeTCPP, and parts of the photosensitive electrons (e^-) at the CB position of CNNS can also be captured and generated by nearby O_2. At the same time, $\cdot O_2^-$ reacts with 4-NP to decompose into CO_2 and H_2O [36], and photogenerated h^+ can be transferred from the VB of FeTCPP to the VB of CNNS, reducing the chance of recombination of photogenerated e^--h^+ pairs, which can effectively improve photocatalytic activity. Furthermore, the improvement on the photocatalytic performance is also related to the photo-Fenton effect, as reported in previous reports [71,72].

3. Experimental Section

3.1. Materials

Melamine ($C_3H_6N_6$), ethanol (C_2H_6O), ethylene glycol ($C_2H_6O_2$), barium sulfate ($BaSO_4$), anhydrous sodium sulfate (Na_2SO_4), Carboxy benzaldehyde ($C_8H_6O_3$), propionic acid (CH_3CH_2COOH), pyrrole (Py), methanol (CH_3OH), N-N dimethylformamide (DMF)), ferric chloride hexahydrate ($FeCl_3 \cdot 6H_2O$), isopropyl alcohol (IPA), and EDTA-2NA were purchased from China, Shanghai Sinopharm Chemical Reagents Co., Ltd.; Nitric acid (HNO_3) was ordered from China, Shanghai Suyi Chemical Reagent Co., Ltd.; and Para-benzoquinone ($C_6H_4O_2$) was ordered from China, Shanghai Maclin Biochemical Technology Co., Ltd. All chemicals used in this experiment are reagent grade and were used as received.

3.2. Synthesis of FeTCPP

Tetracarboxylic phenyl porphyrin was synthesized by the Adler method [73]. Firstly, Carboxy benzaldehyde (2 g) and propionic acid (150 mL) were mixed, being stirred in a three-neck flask equipped with reflux condenser at 135 °C. Pyrrole (1 mL) was dissolved in 20 mL propionic acid, which was added drop by drop into the upper reaction solution over 1 h. The reaction solution was refluxed for 2 h. The propionic acid was removed under vacuum and then the residue was dispersed in $CHCl_3$, filtered, and washed. The remaining solid powder was dissolved in the mixture solvent (propionic acid/$CHCl_3$ with 3:2 volume), and the insoluble portion was removed by filtration. Using propionic acid/$CHCl_3$ (3:2, v/v) as eluent, the mixture was separated by chromatography on silica gel column. The first colored band was collected concentrated and dried. The resulting product was tetracarboxylic phenyl porphyrin, labelled as TCPP.

TCPP (0.5 g) and $FeCl_3 \cdot 6H_2O$ (1 g) in 100 mL DMF were dissolved in DMF, and then heated to 150 °C for 3 h with stirring. DMF was removed under vacuum after cooling to room temperature. The remaining solid was dissolved in ethanol, and insoluble impurities were removed via filtration. By rotary evaporation, the residue was dried. The obtained composites were labelled as FeTCPP.

3.3. Synthesis of CNNS

Bulk g-C_3N_4 was prepared first. Then, 5 g melamine powder was mixed with ethylene glycol (120 mL) by ultrasonication for 2 h, and then dilute nitric acid (120 mL, 0.36 mol L^{-1}) was added to the above mixture solution under stirring. The sediment was washed and dried after stirring for 12 h. The obtained powder was heated to 550 °C for 4 h at a heating rate of 5 °C/min in a muffle furnace. Then, the bulk g-C_3N_4 was obtained. As previously reported, the g-C_3N_4 nanosheets were produced by thermal oxidation etching of bulk g-C_3N_4 directly at 550 °C for 2 h [36]. The obtained light yellow powder was dried at 60 °C for 12 h in a vacuum oven and was named CNNS.

3.4. Preparation of FeTCPP@CNNS

Typically, CNNS powder (1 g) was dispersed in 50 mL ethanol by ultrasonication. Then, a certain amount of FeTCPP was added to ethanol (10 mL), and the mixture was mixed into the above solution with magnetic stirring at 80 °C. Then, FeTCPP@CNNS composite material was obtained. The preparation process of FeTCPP@CNNS materials are shown in Figure 1. By this method, FeTCPP@CNNS composites with different FeTCPP contents (10 mg, 20 mg, 30 mg, and 40 mg) were prepared, which were denoted as X% FeTCPP@CNNS (X% = 1%, 2%, 3%, and 4%).

3.5. Photocatalytic Assessment

The photocatalytic degradation of 4-NP was studied using xenon lamp (150 W) with filter as visible light source. Here, 50 mg photocatalyst was dispersed by magnetic stirring into 100 mL 4-NP aqueous solution (20 mg L^{-1}). Firstly, the adsorption–desorption equilibrium was obtained by ultrasound for 0.5 h in the dark, then under visible light, and the reaction mixture was irradiated for 1 h. Then, 5 mL of the mixture was removed from the reactor, and at the same time interval, the concentration of 4-NP was determined by UV-Vis spectrophotometer.

During the photocatalytic degradation of 4-NP, the effect of reactive oxygen species on the best-performing photocatalyst was tested for a scavengers, namely P-benzoquinone (BQ), isopropanol (IPA), and ethylenediamine tetraacetic acid disodium salt (EDTA-2Na). For this test, 1 mM scavenger was, respectively, added into 100 mL 4-NP solution (20 mg L^{-1}), then was added 20 mg of catalyst. An additional procedure was carried out, which was the same as the process but without scavengers. Thus, all experiments were conducted under the identical conditions.

3.6. Description of the Characteristics

XRD-600 (Rigaku, Japan) was used to determine the crystal phase of X-ray samples. FTS2000 (Thermos, Waltham, MA, USA) Fourier infrared spectrometer was used for qualitative analysis of the chain structure of the samples. Using barium sulfate as blank samples, the samples were detected by diffuse reflectivity spectroscopy (DRS) spectrophotometer on UV-2550 (Shimadzu, Tokyo, Japan). Merlin Compact (Merlin, Forchtenberg, Germany) was used for obtaining the field emission scanning electron microscope (FESEM) images. The surface chemical composition of the samples was analyzed by X-ray photoelectron spectroscopy (XPS, Thermos Fish Scientific, USA). Nitrogen adsorption and desorption tests (BET) were measured at 77 k using the TA Instruments SDT Q600 analyzer (Quadrasorb, WI, USA). Quantachrome Instrument nitrogen adsorption device was used to record the adsorption and desorption isotherms. The photoluminescence spectrum was performed via Spectro fluorometer FS5 (Picoquant, Berlin, Germany) with a slit of 10 nm and an excitation wavelength of 320 nm. The prepared sample was considered to be the working electrode, a platinum wire to be a counter electrode, and a saturated Ag/AgCl electrode to be a reference electrode. Additionally, 0.5 mol L^{-1} aqueous solution of Na_2SO_4 was the electrolyte. Under the disturbance signal of 8 mV, electrochemical impedance spectroscopy (EIS) was measured in the frequency range of 1 MHz to 1000 MHz, and photocurrent test and electrochemical impedance–potential test were measured on the material. In addition, the MPC-3100 UV-NEAR infrared spectrophotometer (USA) was used for analyzing the degradation concentration of the sample.

4. Conclusions

In summary, the FeTCPP@CNNS photocatalysts with stacked lamellar structure have been successfully fabricated by an in situ hydrothermal self-assembly approach. The FeTCPP@CNNS composites exhibit higher photocatalytic efficiency and stability than CNNS by the photodegradation of 4-NP dyes. The photocatalytic degradation rate reached the maximum value of approximately 92.4% of 3% FeTCPP@CNNS.

The degradation rate constant of the 3% FeTCPP@CNNS photocatalyst is 0.037 min^{-1} (4-NP), which is 5 times that of CNNS, indicating that proper FeTCPP introduced into CNNS can effectively improve transformation of photoexcitation electrons and holes. In addition, the results of the active species trapping experiments for the photodegradation of 4-NP show that $\cdot O^{2-}$ plays a major role in photocatalytic reactions. A possible photocatalytic reaction mechanism of FeTCPP@CNNS composite for photocatalytic degradation of 4-NP has been proposed. This work enables the application of CNNS-based photocatalysis under sunlight irradiation in wastewater treatment.

Author Contributions: Conceptualization, J.L.; methodology, Y.G.; software, L.L.; investigation, M.A.; data curation, J.W.; writing—original draft preparation, L.Z.; writing—review and editing, S.L.; supervision, W.S.; funding acquisition, X.C. All authors have read and agreed to the published version of the manuscript.

Funding: This research was funded by Postgraduate Research and Practice Innovation Program of Jiangsu Province, China (SJCX21_1765 and SJCX22_1932); NCN, Poland (UMO-2020/39/B/ST8/02937); and NAWA, (2020 PPN/BEK/2020/1/00129/ZAS/00001). This research was also supported by the Open Research Fund of the State Key Laboratory of Polymer Physics and Chemistry, Changchun Institute of Applied Chemistry, Chinese Academy of Sciences (2022-06). We also appreciate the support of the funding project by National Natural Science Foundation of China: 22006057.

Data Availability Statement: All data generated or analyzed during this study are included in this published article.

Conflicts of Interest: The authors declare no conflict of interest.

References

1. Jin, X.; Zhou, X.; Sun, P.; Lin, S.; Cao, W.; Li, Z.; Liu, W. Photocatalytic degradation of norfloxacin using N-doped TiO_2: Optimization, mechanism, identification of intermediates and toxicity evaluation. *Chemosphere* **2019**, *237*, 124433. [CrossRef]
2. Wang, S.; Wang, F.; Su, Z.; Wang, X.; Han, Y.; Zhang, L.; Xiang, J.; Du, W.; Tang, N. Controllable Fabrication of Heterogeneous p-TiO_2 QDs@g-C_3N_4 p-n Junction for Efficient Photocatalysis. *Catalysts* **2019**, *9*, 439. [CrossRef]
3. Li, S.; Zhang, Q.; Liu, L.; Wang, J.; Zhang, L.; Shi, M.; Chen, X. Ultra-stable sandwich shaped flexible MXene/CNT@Ni films for high performance supercapacitor. *J. Alloys Compd.* **2023**, *941*, 168963. [CrossRef]
4. Li, S.; Zhang, L.; Zhang, L.; Guo, Y.; Chen, X.; Holze, R.; Tang, T. Preparation of Fe_3O_4@polypyrrole composite materials for asymmetric supercapacitor applications. *New J. Chem.* **2021**, *45*, 16011. [CrossRef]
5. Lu, S.; Weng, B.; Chen, A.; Li, X.; Huang, H.; Sun, X.; Feng, W.; Lei, Y.; Qian, Q.; Yang, M.Q. Facet Engineering of Pd Nanocrystals for Enhancing Photocatalytic Hydrogenation: Modulation of the Schottky Barrier Height and Enrichment of Surface Reactants. *ACS Appl. Mater. Interfaces* **2021**, *13*, 13044. [CrossRef] [PubMed]
6. Gao, M.; Zhou, W.-Y.; Mo, Y.-X.; Sheng, T.; Deng, Y.; Chen, L.; Wang, K.; Tan, Y.; Zhou, H. Outstanding long-cycling lithium−sulfur batteries by core-shell structure of S@Pt composite with ultrahigh sulfur content. *Adv. Powder Mater.* **2022**, *1*, 100006. [CrossRef]
7. Li, M.; Li, Z.; Wang, X.; Meng, J.; Liu, X.; Wu, B.; Han, C.; Mai, L. Comprehensive understanding of the roles of water molecules in aqueous Zn-ion batteries: From electrolytes to electrode materials. *Energy Environ. Sci.* **2021**, *14*, 3796. [CrossRef]
8. Asadzadeh-Khaneghah, S.; Habibi-Yangjeh, A.; Vadivel, S. Fabrication of novel g-C_3N_4 nanosheet/carbon dots/$Ag_6Si_2O_7$ nanocomposites with high stability and enhanced visible-light photocatalytic activity. *J. Taiwan Inst. Chem. Eng.* **2019**, *103*, 94. [CrossRef]
9. Gu, W.; Lu, F.; Wang, C.; Kuga, S.; Wu, L.; Huang, Y.; Wu, M. Face-to-Face Interfacial Assembly of Ultrathin g-C_3N_4 and Anatase TiO_2 Nanosheets for Enhanced Solar Photocatalytic Activity. *ACS Appl. Mater. Interfaces* **2017**, *9*, 28674. [CrossRef]
10. Pan, T.; Chen, D.; Xu, W.; Fang, J.; Wu, S.; Liu, Z.; Wu, K.; Fang, Z. Anionic polyacrylamide-assisted construction of thin 2D-2D WO_3/g-C_3N_4 Step-scheme heterojunction for enhanced tetracycline degradation under visible light irradiation. *J. Hazard. Mater.* **2020**, *393*, 122366. [CrossRef]
11. He, S.; Mo, Z.; Shuai, C.; Liu, W.; Yue, R.; Liu, G.; Pei, H.; Chen, Y.; Liu, N.; Guo, R. Pre-intercalation δ-MnO_2 Zinc-ion hybrid supercapacitor with high energy storage and Ultra-long cycle life. *Appl. Surf. Sci.* **2022**, *577*, 151904. [CrossRef]
12. Weng, B.; Liu, S.; Zhang, N.; Tang, Z.-R.; Xu, Y.-J. A simple yet efficient visible-light-driven CdS nanowires-carbon nanotube 1D–1D nanocomposite photocatalyst. *J. Catal.* **2014**, *309*, 146. [CrossRef]
13. Sun, M.W.; Shao, Y.; He, Y.; Zeng, Q.; Liang, H.; Yan, T.; Du, B. Fabrication of a novel Z-scheme g-C_3N_4/Bi_4O_7 heterojunction photocatalyst with enhanced visible light-driven activity toward organic pollutants. *J. Colloid Interface Sci.* **2017**, *501*, 123. [CrossRef]
14. Chen, D.; Wang, K.; Hong, W.; Zong, R.; Yao, W.; Zhu, Y. Visible light photoactivity enhancement via CuTCPP hybridized g-C_3N_4 nanocomposite. *Appl. Catal. B Environ.* **2015**, *166*, 366. [CrossRef]
15. Ma, D.; Wu, J.; Gao, M.; Xin, Y.; Chai, C. Enhanced debromination and degradation of 2,4-dibromophenol by an Z-scheme Bi_2MoO_6/CNTs/g-C_3N_4 visible light photocatalyst. *Chem. Eng. J.* **2017**, *316*, 461. [CrossRef]
16. Mei, S.; Gao, J.; Zhang, Y.; Yang, J.; Wu, Y.; Wang, X.; Zhao, R.; Zhai, X.; Hao, C.; Li, R.; et al. Enhanced visible light photocatalytic hydrogen evolution over porphyrin hybridized graphitic carbon nitride. *J. Colloid Interface Sci.* **2017**, *506*, 58. [CrossRef] [PubMed]
17. Zhu, M.; Li, Z.; Xiao, B.; Lu, Y.; Du, Y.; Yang, P.; Wang, X. Surfactant assistance in improvement of photocatalytic hydrogen production with the porphyrin noncovalently functionalized graphene nanocomposite. *ACS Appl. Mater. Interfaces* **2013**, *5*, 1732. [CrossRef] [PubMed]
18. Khusnutdinova, D.; Beiler, A.M.; Wadsworth, B.L.; Jacob, S.I.; Moore, G.F. Metalloporphyrin-modified semiconductors for solar fuel production. *Chem. Sci.* **2017**, *8*, 253. [CrossRef]

19. Li, X.; Liu, L.; Kang, S.-Z.; Mu, J.; Li, G. Differences between Zn-porphyrin-coupled titanate nanotubes with various anchoring modes: Thermostability, spectroscopic, photocatalytic and photoelectronic properties. *Appl. Surf. Sci.* **2011**, *257*, 5950. [CrossRef]
20. Gao, W.Y.; Chrzanowski, M.; Ma, S. Metal-metalloporphyrin frameworks: A resurging class of functional materials. *Chem. Soc. Rev.* **2014**, *43*, 5841. [CrossRef]
21. Kilian, K.; Pegier, M.; Pyrzynska, K. The fast method of Cu-porphyrin complex synthesis for potential use in positron emission tomography imaging. *Spectrochim. Acta A Mol. Biomol. Spectrosc.* **2016**, *159*, 123. [CrossRef] [PubMed]
22. Liu, S.; Zhou, S.; Hu, C.; Duan, M.; Song, M.; Huang, F.; Cai, J. Coupling graphitic carbon nitrides with tetracarboxyphenyl porphyrin molecules through π–π stacking for efficient photocatalysis. *J. Mater. Sci. Mater. Electron.* **2020**, *31*, 10677. [CrossRef]
23. Wang, J.; Zheng, Y.; Peng, T.; Zhang, J.; Li, R. Asymmetric Zinc Porphyrin Derivative-Sensitized Graphitic Carbon Nitride for Efficient Visible-Light-Driven H_2 Production. *ACS Sustain. Chem. Eng.* **2017**, *5*, 7549. [CrossRef]
24. Guo, F.; Li, L.; Shi, Y.; Shi, W.; Yang, X. Synthesis of N-deficient g-C_3N_4/epoxy composite coating for enhanced photocatalytic corrosion resistance and water purification. *J. Mater. Sci.* **2023**, *58*, 4223. [CrossRef]
25. Li, L.; Zhang, Y.; Shi, Y.; Guo, F.; Yang, X.; Shi, W. A hydrophobic high-crystalline g-C_3N_4/epoxy resin composite coating with excellent durability and stability for long-term corrosion resistance. *Mater. Today Commun.* **2023**, *35*, 105692. [CrossRef]
26. Lu, J.; Shi, Y.; Chen, Z.; Sun, X.; Yuan, H.; Guo, F.; Shi, W. Photothermal effect of carbon dots for boosted photothermal-assisted photocatalytic water/seawater splitting into hydrogen. *Chem. Eng. J.* **2023**, *453*, 139784. [CrossRef]
27. Shi, W.; Cao, L.; Shi, Y.; Chen, Z.; Cao, F.; Du, X. Environmentally friendly supermolecule self-assembly preparation of S-doped hollow porous tubular g-C_3N_4 for boosted photocatalytic H_2 production. *Ceram. Int.* **2023**, *49*, 11989. [CrossRef]
28. Shi, W.; Cao, L.; Shi, Y.; Zhong, W.; Chen, Z.; Wei, Y.; Guo, F.; Chen, L.; Du, X. Boosted built-in electric field and active sites based on Ni-doped heptazine/triazine crystalline carbon nitride for achieving high-efficient photocatalytic H_2 evolution. *J. Mol. Struct.* **2023**, *1280*, 135076. [CrossRef]
29. Shi, W.; Sun, W.; Liu, Y.; Zhang, K.; Sun, H.; Lin, X.; Hong, Y.; Guo, F. A self-sufficient photo-Fenton system with coupling in-situ production H_2O_2 of ultrathin porous g-C_3N_4 nanosheets and amorphous FeOOH quantum dots. *J. Hazard. Mater.* **2022**, *436*, 129141. [CrossRef]
30. Shi, Y.; Li, L.; Sun, H.; Xu, Z.; Cai, Y.; Shi, W.; Guo, F.; Du, X. Engineering ultrathin oxygen-doped g-C_3N_4 nanosheet for boosted photoredox catalytic activity based on a facile thermal gas-shocking exfoliation effect. *Sep. Purif. Technol.* **2022**, *292*, 121038. [CrossRef]
31. Shi, Y.; Li, L.; Xu, Z.; Guo, F.; Li, Y.; Shi, W. Synergistic coupling of piezoelectric and plasmonic effects regulates the Schottky barrier in Ag nanoparticles/ultrathin g-C_3N_4 nanosheets heterostructure to enhance the photocatalytic activity. *Appl. Surf. Sci.* **2023**, *616*, 156466. [CrossRef]
32. Sun, H.; Shi, Y.; Shi, W.; Guo, F. High-crystalline/amorphous g-C_3N_4 S-scheme homojunction for boosted photocatalytic H_2 production in water/simulated seawater: Interfacial charge transfer and mechanism insight. *Appl. Surf. Sci.* **2022**, *593*, 153281. [CrossRef]
33. Yuan, H.; Shi, W.; Lu, J.; Wang, J.; Shi, Y.; Guo, F.; Kang, Z. Dual-channels separated mechanism of photo-generated charges over semiconductor photocatalyst for hydrogen evolution: Interfacial charge transfer and transport dynamics insight. *Chem. Eng. J.* **2023**, *454*, 140442. [CrossRef]
34. Chen, T.; Zhong, L.; Yang, Z.; Mou, Z.; Liu, L.; Wang, Y.; Sun, J.; Lei, W. Enhanced Visible-light Photocatalytic Activity of g-C_3N_4/Nitrogen-doped Graphene Quantum Dots/TiO_2 Ternary Heterojunctions for Ciprofloxacin Degradation with Narrow Band Gap and High Charge Carrier Mobility. *Chem. Res. Chin. Univ.* **2020**, *36*, 1083. [CrossRef]
35. Pourhashem, S.; Duan, J.; Guan, F.; Wang, N.; Gao, Y.; Hou, B. New effects of TiO_2 nanotube/g-C_3N_4 hybrids on the corrosion protection performance of epoxy coatings. *J. Mol. Liq.* **2020**, *317*, 114214. [CrossRef]
36. Lin, L.; Hou, C.; Zhang, X.; Wang, Y.; Chen, Y.; He, T. Highly efficient visible-light driven photocatalytic reduction of CO_2 over g-C_3N_4 nanosheets/tetra(4-carboxyphenyl)porphyrin iron(III) chloride heterogeneous catalysts. *Appl. Catal. B Environ.* **2018**, *221*, 312. [CrossRef]
37. Zhu, K.; Zhang, M.; Feng, X.; Qin, L.; Kang, S.-Z.; Li, X. A novel copper-bridged graphitic carbon nitride/porphyrin nanocomposite with dramatically enhanced photocatalytic hydrogen generation. *Appl. Catal. B Environ.* **2020**, *268*, 118434. [CrossRef]
38. Wang, D.H.; Pan, J.N.; Li, H.H.; Liu, J.J.; Wang, Y.B.; Kang, L.T.; Yao, J.N. A pure organic heterostructure of µ-oxo dimeric iron(iii) porphyrin and graphitic-C_3N_4 for solar H_2 roduction from water. *J. Mater. Chem. A* **2016**, *4*, 290. [CrossRef]
39. Li, W.; He, X.; Ge, R.; Zhu, M.; Feng, L.; Li, Y. Cobalt porphyrin (CoTCPP) advanced visible light response of g-C_3N_4 nanosheets. *Sustain. Mater. Technol.* **2019**, *22*, e00114. [CrossRef]
40. Machado, G.S.; Wypych, F.; Nakagaki, S. Immobilization of anionic iron(III) porphyrins onto in situ obtained zinc oxide. *J. Colloid Interface Sci.* **2012**, *377*, 379. [CrossRef]
41. Jiang, G.; Yang, X.; Wu, Y.; Li, Z.; Han, Y.; Shen, X. A study of spherical TiO_2/g-C_3N_4 photocatalyst: Morphology, chemical composition and photocatalytic performance in visible light. *Mol. Catal.* **2017**, *432*, 232. [CrossRef]
42. Liu, C.; Raziq, F.; Li, Z.; Qu, Y.; Zada, A.; Jing, L. Synthesis of TiO_2/g-C_3N_4 nanocomposites with phosphate–oxygen functional bridges for improved photocatalytic activity. *Chin. J. Catal.* **2017**, *38*, 1072. [CrossRef]
43. Huang, C.; Lv, Y.; Zhou, Q.; Kang, S.; Li, X.; Mu, J. Visible photocatalytic activity and photoelectrochemical behavior of TiO_2 nanoparticles modified with metal porphyrins containing hydroxyl group. *Ceram. Int.* **2014**, *40*, 7093. [CrossRef]

44. Rezaeifard, A.; Jafarpour, M. The catalytic efficiency of Fe-porphyrins supported on multi-walled carbon nanotubes in the heterogeneous oxidation of hydrocarbons and sulfides in water. *Catal. Sci. Technol.* **2014**, *4*, 1960. [CrossRef]
45. Hao, R.; Wang, G.; Tang, H.; Sun, L.; Xu, C.; Han, D. Template-free preparation of macro/mesoporous g-C_3N_4/TiO_2 heterojunction photocatalysts with enhanced visible light photocatalytic activity. *Appl. Catal. B Environ.* **2016**, *187*, 47. [CrossRef]
46. Lv, B.; Lu, L.; Feng, X.; Wu, X.; Wang, X.; Zou, X.; Zhang, F. Efficient photocatalytic hydrogen production using an NH_4TiOF_3/TiO_2/g-C_3N_4 composite with a 3D camellia-like Z-scheme heterojunction structure. *Ceram. Int.* **2020**, *46*, 26689. [CrossRef]
47. Mousavi, M.; Soleimani, M.; Hamzehloo, M.; Badiei, A.; Ghasemi, J.B. Photocatalytic degradation of different pollutants by the novel gCN-NS/Black-TiO_2 heterojunction photocatalyst under visible light: Introducing a photodegradation model and optimization by response surface methodology (RSM). *Mater. Chem. Phys.* **2021**, *258*, 123912. [CrossRef]
48. Kong, L.; Ji, Y.; Dang, Z.; Yan, J.; Li, P.; Li, Y.; Liu, S.F. g-C_3N_4 Loading Black Phosphorus Quantum Dot for Efficient and Stable Photocatalytic H_2 Generation under Visible Light. *Adv. Funct. Mater.* **2018**, *28*, 1800668. [CrossRef]
49. Yang, S.; Gong, Y.; Zhang, J.; Zhan, L.; Ma, L.; Fang, Z.; Vajtai, R.; Wang, X.; Ajayan, P.M. Exfoliated graphitic carbon nitride nanosheets as efficient catalysts for hydrogen evolution under visible light. *Adv. Mater.* **2013**, *25*, 2452. [CrossRef]
50. Barrocas, B.T.; Ambrozova, N.; Koci, K. Photocatalytic Reduction of Carbon Dioxide on TiO_2 Heterojunction Photocatalysts—A Review. *Materials* **2022**, *15*, 967. [CrossRef]
51. Kumar, S.; Kumar, B.; Baruah, A.; Shanker, V. Synthesis of Magnetically Separable and Recyclable g-C_3N_4–Fe_3O_4 Hybrid Nanocomposites with Enhanced Photocatalytic Performance under Visible-Light Irradiation. *J. Phys. Chem. C* **2013**, *117*, 26135. [CrossRef]
52. Wei, M.; Wan, J.; Hu, Z.; Peng, Z.; Wang, B.; Wang, H. Preparation, characterization and visible-light-driven photocatalytic activity of a novel Fe(III) porphyrin-sensitized TiO_2 nanotube photocatalyst. *Appl. Surf. Sci.* **2017**, *391*, 267. [CrossRef]
53. Zhu, X.; Zhang, F.; Wang, M.; Ding, J.; Sun, S.; Bao, J.; Gao, C. Facile synthesis, structure and visible light photocatalytic activity of recyclable $ZnFe_2O_4$/TiO_2. *Appl. Surf. Sci.* **2014**, *319*, 83. [CrossRef]
54. Li, C.; Sun, Z.; Xue, Y.; Yao, G.; Zheng, S. A facile synthesis of g-C_3N_4/TiO_2 hybrid photocatalysts by sol–gel method and its enhanced photodegradation towards methylene blue under visible light. *Adv. Powder Technol.* **2016**, *27*, 330. [CrossRef]
55. Li, Y.; Lv, K.; Ho, W.; Dong, F.; Wu, X.; Xia, Y. Hybridization of rutile TiO_2 (rTiO_2) with g-C_3N_4 quantum dots (CN QDs): An efficient visible-light-driven Z-scheme hybridized photocatalyst. *Appl. Catal. B Environ.* **2017**, *202*, 611. [CrossRef]
56. Liu, C.; Dong, S.; Chen, Y. Enhancement of visible-light-driven photocatalytic activity of carbon plane/g-C_3N_4/TiO_2 nanocomposite by improving heterojunction contact. *Chem. Eng. J.* **2019**, *371*, 706. [CrossRef]
57. Hu, L.; Yan, J.; Wang, C.; Chai, B.; Li, J. Direct electrospinning method for the construction of Z-scheme TiO_2/g-C_3N_4/RGO ternary heterojunction photocatalysts with remarkably ameliorated photocatalytic performance. *Chin. J. Catal.* **2019**, *40*, 458. [CrossRef]
58. Wang, J.; Shi, W.; Liu, D.; Zhang, Z.; Zhu, Y.; Wang, D. Supramolecular organic nanofibers with highly efficient and stable visible light photooxidation performance. *Appl. Catal. B Environ.* **2017**, *202*, 289. [CrossRef]
59. Rabbani, M.; Heidari-Golafzani, M.; Rahimi, R. Synthesis of TCPP/$ZnFe_2O_4$@ZnO nanohollow sphere composite for degradation of methylene blue and 4-nitrophenol under visible light. *Mater. Chem. Phys.* **2016**, *179*, 35. [CrossRef]
60. Mahalakshmi, G.; Rajeswari, M.; Ponnarasi, P. Synthesis of few-layer g-C_3N_4 nanosheets-coated MoS_2/TiO_2 heterojunction photocatalysts for photo-degradation of methyl orange (MO) and 4-nitrophenol (4-NP) pollutants. *Inorg. Chem. Commun.* **2020**, *120*, 108146. [CrossRef]
61. Palanivel, B.; Lallimathi, M.; Arjunkumar, B.; Shkir, M.; Alshahrani, T.; Al-Namshah, K.S.; Hamdy, M.S.; Shanavas, S.; Venkatachalam, M.; Ramalingam, G. rGO supported g-C_3N_4/$CoFe_2O_4$ heterojunction: Visible-light-active photocatalyst for effective utilization of H_2O_2 to organic pollutant degradation and OH radicals production. *J. Environ. Chem. Eng.* **2021**, *9*, 104698. [CrossRef]
62. Abazari, R.; Mahjoub, A.R.; Salehi, G. Preparation of amine functionalized g-C_3N_4@(H/S)MOF NCs with visible light photocatalytic characteristic for 4-nitrophenol degradation from aqueous solution. *J. Hazard. Mater.* **2019**, *365*, 921. [CrossRef] [PubMed]
63. Palanivel, B.; Jayaraman, V.; Ayyappan, C.; Alagiri, M. Magnetic binary metal oxide intercalated g-C_3N_4: Energy band tuned p-n heterojunction towards Z-scheme photo-Fenton phenol reduction and mixed dye degradation. *J. Water Process Eng.* **2019**, *32*, 100968. [CrossRef]
64. Yi, F.; Gan, H.; Jin, H.; Zhao, W.; Zhang, K.; Jin, H.; Zhang, H.; Qian, Y.; Ma, J. Sulfur- and chlorine-co-doped g-C_3N_4 nanosheets with enhanced active species generation for boosting visible-light photodegradation activity. *Sep. Purif. Technol.* **2020**, *233*, 115997. [CrossRef]
65. Zarei, M.; Bahrami, J.; Zarei, M. Zirconia nanoparticle-modified graphitic carbon nitride nanosheets for effective photocatalytic degradation of 4-nitrophenol in water. *Appl. Water Sci.* **2019**, *9*, 175. [CrossRef]
66. Ma, L.; Wang, G.; Jiang, C.; Bao, H.; Xu, Q. Synthesis of core-shell TiO_2@g-C_3N_4 hollow microspheres for efficient photocatalytic degradation of rhodamine B under visible light. *Appl. Surf. Sci.* **2018**, *430*, 263. [CrossRef]
67. Huang, K.; Li, C.; Zhang, X.; Meng, X.; Wang, L.; Wang, W.; Li, Z. TiO_2 nanorod arrays decorated by nitrogen-doped carbon and g-C_3N_4 with enhanced photoelectrocatalytic activity. *Appl. Surf. Sci.* **2020**, *518*, 146219. [CrossRef]

68. Zhang, X.; Li, L.; Zeng, Y.; Liu, F.; Yuan, J.; Li, X.; Yu, Y.; Zhu, X.; Xiong, Z.; Yu, H.; et al. TiO_2/Graphitic Carbon Nitride Nanosheets for the Photocatalytic Degradation of Rhodamine B under Simulated Sunlight. *ACS Appl. Nano Mater.* **2019**, *2*, 7255. [CrossRef]
69. Huang, Y.; Wang, W.; Zhang, Q.; Cao, J.J.; Huang, R.J.; Ho, W.; Lee, S.C. In situ Fabrication of alpha-Bi_2O_3/$(BiO)_2CO_3$ Nanoplate Heterojunctions with Tunable Optical Property and Photocatalytic Activity. *Sci. Rep.* **2016**, *6*, 23435. [CrossRef] [PubMed]
70. Chen, Y.; Huang, W.; He, D.; Situ, Y.; Huang, H. Construction of heterostructured g-C_3N_4/Ag/TiO_2 microspheres with enhanced photocatalysis performance under visible-light irradiation. *ACS Appl. Mater. Interfaces* **2014**, *6*, 14405. [CrossRef]
71. Barrocas, B.; Chiavassa, L.D.; Conceicao Oliveira, M.; Monteiro, O.C. Impact of Fe, Mn co-doping in titanate nanowires photocatalytic performance for emergent organic pollutants removal. *Chemosphere* **2020**, *250*, 126240. [CrossRef] [PubMed]
72. Barrocas, B.T.; Osawa, R.; Oliveira, M.C.; Monteiro, O.C. Enhancing Removal of Pollutants by Combining Photocatalysis and Photo-Fenton Using Co, Fe-Doped Titanate Nanowires. *Materials* **2023**, *16*, 2051. [CrossRef] [PubMed]
73. Adler, A.D.; Longo, F.R.; Finarelli, J.D.; Goldmacher, J.; Assour, J.; Korsakoff, L. A simplified synthesis for meso-tetraphenylporphine. *J. Org. Chem.* **1967**, *32*, 476. [CrossRef]

Disclaimer/Publisher's Note: The statements, opinions and data contained in all publications are solely those of the individual author(s) and contributor(s) and not of MDPI and/or the editor(s). MDPI and/or the editor(s) disclaim responsibility for any injury to people or property resulting from any ideas, methods, instructions or products referred to in the content.

Article

Hollow g-C_3N_4@$Cu_{0.5}In_{0.5}S$ Core-Shell S-Scheme Heterojunction Photothermal Nanoreactors with Broad-Spectrum Response and Enhanced Photocatalytic Performance

Yawei Xiao [1], Zhezhe Wang [1], Bo Yao [1], Yunhua Chen [2], Ting Chen [3,*] and Yude Wang [4,*]

[1] National Center for International Research on Photoelectric and Energy Materials, School of Materials and Energy, Yunnan University, Kunming 650504, China; zzwang@mail.ynu.edu.cn (Z.W.); yaobo@mail.ynu.edu.cn (B.Y.)
[2] Department of Physics, Yunnan University, Kunming 650504, China; yhchen@ynu.edu.cn
[3] Institute of Materials Science & Devices, School of Materials Science and Engineering, Suzhou University of Science and Technology, Suzhou 215009, China
[4] Yunnan Key Laboratory of Carbon Neutrality and Green Low-Carbon Technologies, Yunnan University, Kunming 650504, China
* Correspondence: chenting@mail.usts.edu.cn (T.C.); ydwang@ynu.edu.cn (Y.W.)

Abstract: Improving spectral utilization and carrier separation efficiency is a key point in photocatalysis research. Herein, we prepare hollow g-C_3N_4 nanospheres by the template method and synthesize a g-C_3N_4@$Cu_{0.5}In_{0.5}S$ core-shell S-scheme photothermal nanoreactor by a simple chemical deposition method. The unique hollow core-shell structure of g-C_3N_4@$Cu_{0.5}In_{0.5}S$ is beneficial to expand the spectral absorption range and improving photon utilization. At the same time, the photogenerated carriers can be separated, driven by the internal electric field. In addition, g-C_3N_4@$Cu_{0.5}In_{0.5}S$ also has a significantly enhanced photothermal effect, which promotes the photocatalytic reaction by increasing the temperature of the reactor. The benefit from the synergistic effect of light and heat, the H_2 evolution rate of g-C_3N_4@$Cu_{0.5}In_{0.5}S$ is as high as 2325.68 $\mu mol\ h^{-1}\ g^{-1}$, and the degradation efficiency of oxytetracycline under visible light is 95.7%. The strategy of combining S-scheme heterojunction with photothermal effects provides a promising insight for the development of an efficient photocatalytic reaction.

Keywords: g-C_3N_4@$Cu_{0.5}In_{0.5}S$; S-scheme heterojunction; hollow nanostructure; photothermal effect; photocatalysis

1. Introduction

With the development of industry, the problem of energy shortage and environmental pollution is becoming more and more serious [1–3]. It is extremely urgent to develop green, renewable energy and control environmental pollution. The development of photocatalysis technology provides an effective way for the conversion of solar energy into clean fuel and the degradation of organic pollutants [4–7]. The key to the effective use of photocatalysis technology is to develop efficient semiconductor photocatalysts. Graphite phase carbon nitride (g-C_3N_4) is a new type of conjugated polymeric semiconductor photocatalyst. It has become a new research hotspot in the field of energy conversion and environmental remediation because of its simple synthesis, attractive band structure, high stability, and rich reserves of constituent elements [8–11]. Although g-C_3N_4 has many excellent properties as a photocatalyst, there are still some shortcomings in the bulk g-C_3N_4 prepared by direct calcination of precursors: (i) low utilization of the solar spectrum; (ii) serious recombination of the photogenerated carrier; and (iii) a lack of enough active sites on the homogenized surface. Therefore, it is necessary to modify g-C_3N_4 to improve its photocatalytic activity.

Recombination with narrow band gap semiconductors is one of the effective methods to improve the spectral response of g-C_3N_4. Different from doping or manufacturing defects, the introduction of narrow band gap semiconductors does not affect the band structure of g-C_3N_4, which means it has the ability of a broad-spectrum light response while maintaining a strong enough redox ability [12,13]. Huang et al. loaded g-C_3N_4 nanowires on the surface of $ZnIn_2S_4$ nanotubes, which effectively expanded the range of the g-C_3N_4 light response with stronger light absorption capacity and can effectively carry out photocatalytic CO_2 reduction reaction [14]. In addition, some narrow-band gap metal sulfides with strong photothermal effects generate heat by using near-infrared or even infrared sunlight [15,16]. The photothermal effect is rarely considered or even directly ignored in traditional photocatalytic systems, but this part of the energy is very important for the construction of efficient photocatalytic systems. In the heat-assisted photocatalytic reaction based on photochemistry, the catalytic reaction is not directly driven by heat, but this part of the heat can increase the local temperature of the catalyst and help to excite the carrier [17]. Thermal energy can help to further reduce the apparent activation energy of photocatalysis, promote the mobility and mass transfer of photogenerated carriers, and accelerate the transport of materials on the material surface, which is beneficial to increase the rate of the catalytic reaction [18–20].

The morphology of the catalyst affects the exposure of the active site and thus the photocatalytic activity. The low specific surface area of the bulk g-C_3N_4 leads to the lack of a reactive center. The photocatalyst with a hollow structure has an inner surface and an outer surface, which can provide more active sites [21,22]. Compared with bulk g-C_3N_4, g-C_3N_4 with hollow structure has a higher specific surface area and more surface-active centers. In addition, the hollow structure made the distance of charge conduction shorter, reduced the recombination of carriers, enhanced the effects of light scattering and multiple refractions, and improved the absorption and utilization of light [23–26]. More importantly, the well-designed hollow g-C_3N_4 provides different loading surfaces for both oxidizing and reducing co-catalysts, the electrons, and holes migrate to the outer and inner surfaces, respectively. The wall of the hollow structure acted as a cell membrane, which enabled the spatial separation of redox-active centers on the nanometer scale [27].

The construction of heterojunctions can inhibit the rapid recombination of photogenerated carriers by the charge transfer between semiconductors. In recent years, a new S-scheme heterojunction has been extensively investigated, which drives the spatial separation of carriers by creating an internal electric field [28–30]. S-scheme heterojunctions consist of reduction semiconductors (RS) and oxidation semiconductors (OS), where differences in Fermi energy levels lead to the transfer of electrons from RS to OS, creating internal electric fields and energy band bending at the interface. Under light, the free electrons of photocatalysts are transferred from the conduction band (CB) of OS to the valence band (VB) of RS driven by the internal electric field; at the same time, the band bending prevents the transfer of electrons from RS to OS. This special charge transfer pathway not only suppresses the rapid recombination of electrons and holes but also maintains the strong redox ability of the photocatalyst [31–33]. The g-C_3N_4 is an ideal reduction semiconductor because of its rather negative CB position. The S-scheme heterojunction formed by oxidation semiconductors with suitable band structures and g-C_3N_4 not only facilitates charge separation but also enhances the oxidation ability of photocatalysts [34–36]. Fan et al. successfully constructed S-scheme heterostructures with g-C_3N_4 nanowires and TiO_2 nanoparticles. The low reductive electrons on TiO_2 recombine with holes in g-C_3N_4 and show effective charge separation and excellent photocatalytic hydrogen production [37].

In this work, we have constructed a hollow g-C_3N_4@$Cu_{0.5}In_{0.5}S$ (HCN@CIS) core-shell S-scheme photothermal nanoreactor by combining S-scheme heterojunctions with photothermal effects. Hollow g-C_3N_4 nanospheres (HCN) were first prepared by the template method and then narrow band gap $Cu_{0.5}In_{0.5}S$ (CIS) nanosheets were grown on its surface. The hollow core-shell structure enhances light absorption, promotes charge separation, and provides more active centers. The photothermal effect converts some of the light energy into heat, which raises the near-field temperature of the catalyst by radiation and conduction. Benefiting from the effective charge separation and photothermal effect, the HCN@CIS photothermal catalyst offers improved photocatalytic hydrogen production efficiency and excellent oxytetracycline (OTC) degradation performance. In addition, the formation of S-scheme heterojunction is demonstrated by a series of characterization and experiments, and the performance enhancement mechanism of the HCN@CIS photothermal catalyst is analyzed. We believe that this photothermal nanoreactor with fast charge transfer and high efficiency will provide some valuable insights for the development of g-C_3N_4-based photocatalysts for energy conversion and environmental remediation.

2. Results and Discussion

2.1. Morphological and Structural Analysis

The preparation process of the HCN@CIS hollow core-shell nanoreactor is observed by a scanning electron microscope (SEM). Figure S1 shows that the diameter of the SiO_2 nanospheres is about 200 nm, which determines the size of the HCN cavity. Due to the uniform size of the template SiO_2 nanospheres, the prepared HCN morphology is also uniform, and the diameter is about 300 nm (Figure 1a). Figure 1b is a magnified SEM diagram of a single HCN. The notch on the surface shows the typical hollow spherical structure of HCN, and the cross-section of the hollow spherical wall is shown near the notch, which shows the relatively loose particle stacking structure of HCN. The cyanamide is polymerized in SiO_2 mesoporous shell to form g-C_3N_4, and the SiO_2 template forms this stacking shape after being etched. From Figure 1c, the surface of the HCN sphere is wrapped by a CIS nanosheet, which is in the shape of flower ball, and the multi-stage structure of HCN@CIS is more favorable for the exposure of active parts [38,39]. A transmission electron microscope (TEM) can observe this hollow multi-stage core-shell structure more clearly. Figure 1d shows that SiO_2@g-C_3N_4 is a core-shell structure with a dense SiO_2 core and a loose mesoporous SiO_2/g-C_3N_4 shell before the SiO_2 template is etched. The HCN presents a unique hollow spherical structure after the SiO_2 is etched. The cavity size is the same as the dense SiO_2 core, and the thickness of the spherical wall is about 50 nm (Figure 1e). The surface of HCN@CIS still has a hollow structure after the formation of CIS, and the CIS nanosheets are arranged vertically on the surface of the HCN, and the overall size of the nanosphere increases significantly (Figure 1f). Figure 1f,g shows that the thickness of the CIS nanosheets on the surface of the HCN is between 50–100 nm. The lattice fringes of 0.195 nm (Figure 1h,i) can be observed in the high-resolution transmission electron microscope images of HCN@ZIS, which can be assigned to the 024 crystal plane of CIS. Figure 1j shows the element distribution map of HCN@CIS, which clearly shows that the distribution of C (red) and N (blue) elements shows the outline of a hollow sphere, and the element groups of Cn (green), In (yellow) and S (purple) are mainly distributed outside the hollow sphere. These results further prove that the HCN@CIS hollow core-shell structure was prepared successfully.

2.2. Phase Structure, Elemental Composition, and Band Structure Analysis

The crystal structure and composition of photocatalysts are studied by X-ray diffraction (XRD). Figure 2a shows that HCN has two diffraction peaks at 13.1° and 27.3°. The two peaks refer to the interlaminar accumulation of aromatic compounds and the in-plane periodic structure of melon compounds, which correspond to the graphite phase carbon nitride (100) and (002) crystal planes, respectively [40,41]. There are three obvious diffraction peaks of CIS at 29.9°, 46.3°, and 55.1°, which correspond to the (112), (024), and (132) crystal planes of $Cu_{0.5}In_{0.5}S$, respectively. The result is consistent with the characteristic peak in the standard sample $Cu_{0.5}In_{0.5}S$ (JCPDS No.47-1372). The diffraction peak of HCN@CIS contained all the characteristic peaks of HCN and CIS, indicating that CIS is successfully loaded on HCN. Figure 2b is the survey X-ray photoelectron spectroscopy (XPS) of HCN, CIS, and HCN@CIS. The C and N elements are visible in the survey XPS spectrum of HCN, the survey XPS spectrum of HCN contains the signals of S,

In, and Cu elements, and the survey XPS spectrum of HCN@CIS contains the characteristic signals of HCN and CIS, which further indicates that the HCN@CIS composite has been prepared successfully. UV-visible diffuse reflectance spectroscopy (UV-vis DRS) is used to analyze the spectral absorption properties and determine the absorption edge of the powder photocatalysts. Figure 2c shows that the light absorption threshold of HCN is about 470 nm, which can absorb ultraviolet light and part of visible light. The CIS has a broad-spectrum response capability that covers almost the whole spectrum, which is related to its narrow band gap and dark characteristics [42]. Compared with the original HCN, the absorption range of HCN@CIS is significantly expanded and still has a strong spectral response in the visible-near infrared region. Figure 2d shows the optical band gap (E_g) of HCN and CIS are 2.61 eV and 2.17 eV, respectively. The valence band energies of HCN and CIS measured by XPS valence band spectroscopy (VB-XPS) are 1.31 eV and 1.52 eV, respectively. The energy of the samples relative to the standard hydrogen electrode (E_{NHE}) is calculated by the following formula: $E_{NHE} = \varphi + E_{VB\text{-}XPS} - 4.44$, where φ is the work function of the instrument (4.5 eV); therefore, the valence band energy of HCN and CIS relative to the standard hydrogen electrode is about 1.37 and 1.58 eV [43]. According to the relationship between the E_{CB} potential and E_g ($E_{CB} = E_{VB} - E_g$), the CB potentials of HCN and CIS are determined to be −1.19 eV and −0.24 eV, respectively. Therefore, the band structure arrangement of HCN and CIS can be determined as shown in Figure 2f. The CB position of CIS is between CB and VB of HCN, and the VB position of HCN is between CB and VB of CIS, which accords with the staggered energy level arrangement and meets the formation basis of S-scheme heterojunction [44].

2.3. Charge Transfer Pathways of Catalysts

The work function of a semiconductor represents the minimum energy required for an electron in a semiconductor whose energy is equal to the Fermi level to escape into a vacuum. A large work function means that it is difficult for the electrons to leave the semiconductor and a smaller one means that the semiconductor is prone to losing electrons. Therefore, the possible electron transfer pathways of the contact interface of the two kinds of semiconductors can be studied by measuring the work function [45]. The relative surface potential of HCN and CIS are measured by a scanning kelvin probe (SKP), and the real work function is calculated by the difference between the semiconductor surface and the highly sensitive probe [46]. Figure 3a shows that the surface potential of Au, HCN, and CIS are 127.87 eV, −103.02 eV, and −562.13 eV, respectively. The real work function of the Au is 5.1 eV; therefore, the real work functions of HCN and CIS are 4.41 eV and 4.83 eV, respectively. The Fermi level of CIS is relatively lower than that of HCN, and the electrons on HCN will transfer to CIS when they are contacted. As a highly sensitive analytical method, XPS can identify the changes in element binding energy and chemical environment in materials. The change in binding energy is closely related to the changes in electron density and element valence state. Normally, if a material loses electrons, its binding energy moves to a higher energy, and when it gains electrons, the binding energy moves to a lower energy [47]. Figure 3b shows the high-resolution XPS spectrum of C1s. The peak at 284.8 eV is an indeterminate carbon. The peaks at 286.5 and 288.34 eV in HCN correspond to C-N and N-C=N, respectively. The peaks of binding energies 398.24 eV, 400.08 eV, and 401.02 eV in the N1s spectra are attributed to sp^2 coordination nitrogen (N-C=N), tertiary nitrogen (N-(C)$_3$), and surface amino group (N-H$_x$) (Figure 3c) [48]. The peaks of Cu 2p in CIS at 931.79 eV and 951.68 eV correspond to Cu $2p_{3/2}$ and Cu $2p_{1/2}$, respectively (Figure 3d) [49]. The two peaks at 444.72 eV and 452.35 eV are attributed to the In $3d_{3/2}$ and $3d_{1/2}$, respectively (Figure 3e). The S 2p spectrum in Figure 3f is fitted into two peaks, S $2p_{3/2}$ (161.76 eV) and S $2p_{1/2}$ (163.05 eV). Figure 3b–f shows that the binding energy of C and N elements in the HCN@CIS samples increases when the heterojunction is formed, which indicates the loss of electrons in HCN. The binding energies of C and N elements in HCN@CIS samples shift to low energy under light, indicating that HCN obtains electrons again. Correspondingly, the binding energies of Cu, In, and S in HCN@CIS increase after contact, indicating that the electrons lost by HCN are transferred to CIS. The binding energy of Cu, In, and S elements in HCN@CIS samples shifts to a high energy under light, indicating that the electrons on the CIS flow back to the HCN. The results of in situ irradiation of XPS further verified the charge transfer mechanism in HCN@CIS [50]. When the HCN and CIS come into contact, the free electrons on the HCN will be transferred to the CIS through the contact interface until the Fermi level is balanced. At this time, electrons accumulate on the CIS and a large number of holes remain on the HCN. Therefore, an internal electric field orienting from the HCN to the CIS is formed. Due to charge interactions, the band of the HCN at the interface bends upward, while the band of the ZIS bends downward. The HCN and CIS are excited by light to produce carriers. Under the action of the internal electric field and band bending, the electrons in the CB of the CIS will recombine with the holes in the VB of the HCN, thus completing the charge transfer of the S-scheme pathway.

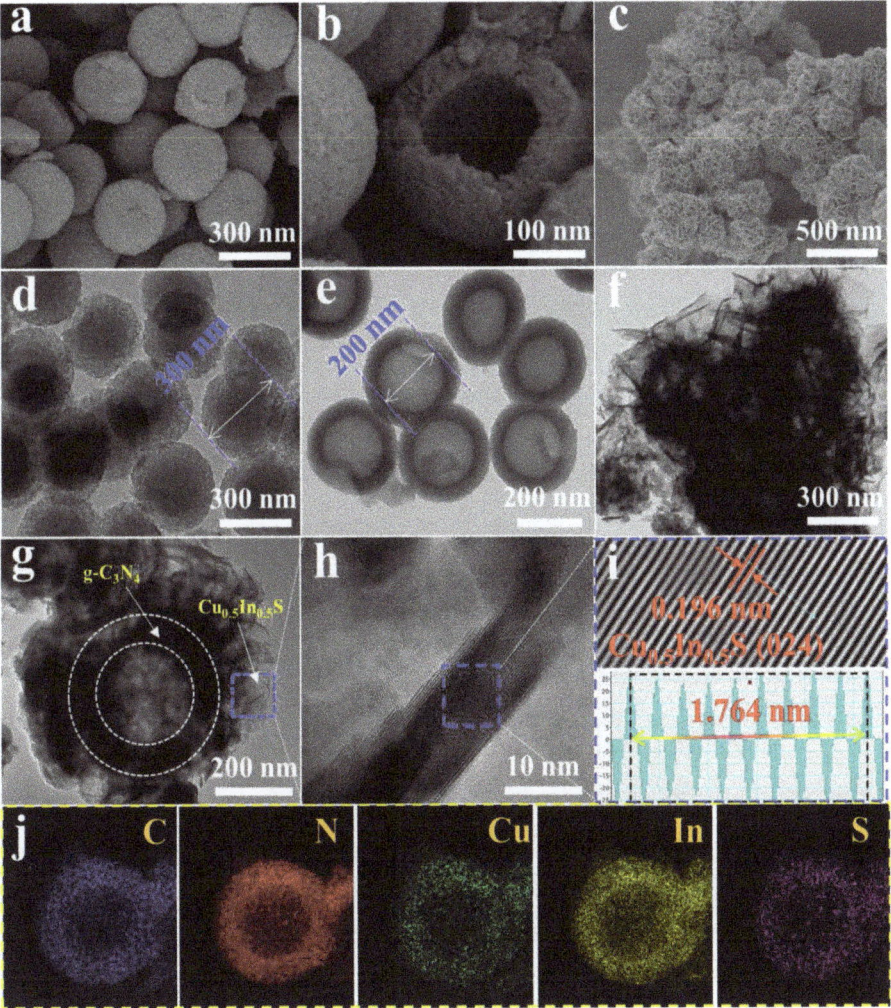

Figure 1. SEM images of HCN (**a**,**b**) and HCN@CIS (**c**). TEM images of SiO_2@g-C_3N_4 (**d**), HCN (**e**), and HCN@CIS (**f**,**g**). HRTEM images of HCN@ZIS (**h**,**i**). Elemental mapping images of HCN@CIS (**j**).

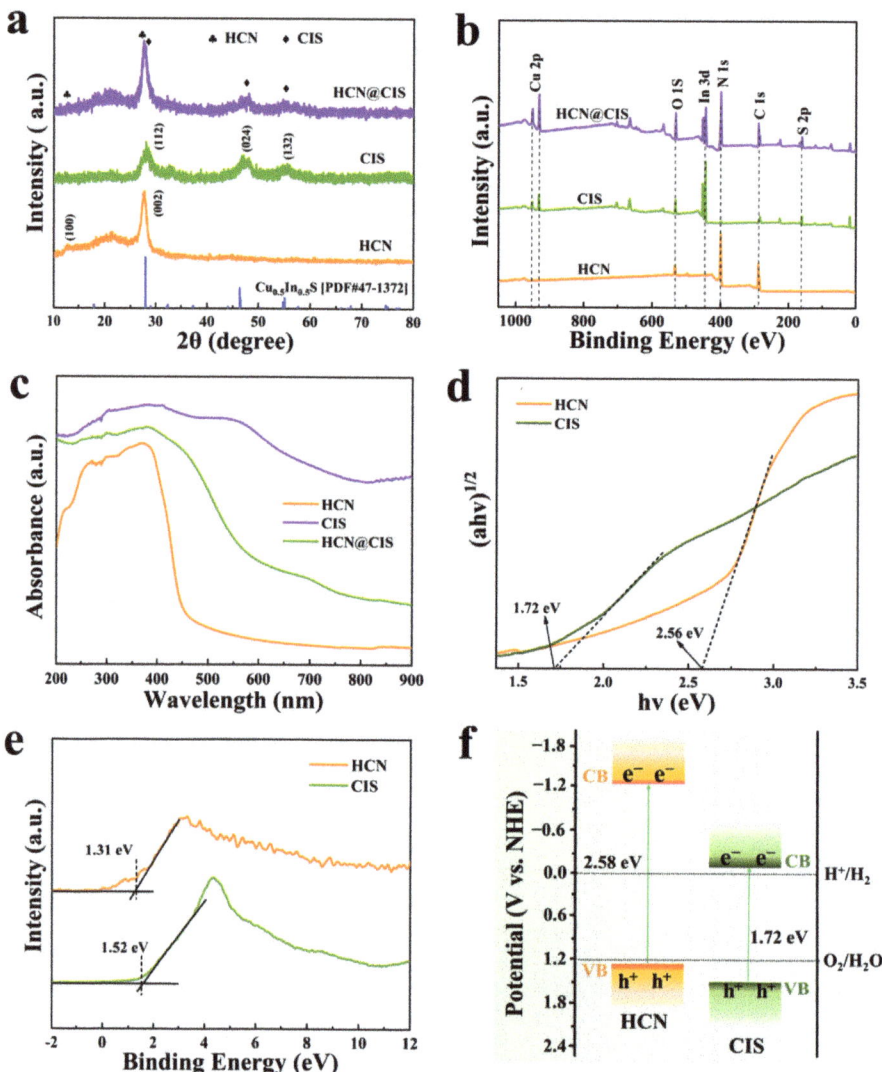

Figure 2. XRD pattern (**a**), XPS spectra (**b**), and UV-vis spectra (**c**) of HCN, CIS, and HCN@CIS. Tauc plots (**d**) and VB-XPS spectra (**e**) of HCN and CIS. Band structure alignments of HCN and CIS (**f**).

2.4. Photothermal Properties of Catalysts

The Arrhenius formula shows that the chemical reaction rate is positively related to the reaction temperature, so the photothermal effect of a photocatalyst can promote the catalytic reaction [51]. The change in temperature of the powder photocatalyst with illumination time was recorded by an infrared camera. A 300 W xenon lamp illuminates the powder photocatalyst vertically at a distance of 30 cm. The initial temperature of the HCN powder is 21.7 °C. After 90 s of irradiation, the surface temperature rises to 62.7 °C (Figure 4a), and the surface temperature increased by 41 °C. Under the same conditions, the heating range of the CIS powder is 116.2 °C, which is 2.8 times that of the HCN. The strong photothermal effect of CIS is related to its narrow band gap and dark color, which is easy to absorb light (Figure S3). It is worth noting that after 90 s of irradiation, the surface temperature of HCN@CIS powder increases from 22.8 °C to 153 °C, which is 3.2 times higher than that of pure HCN and 14 °C higher than that of pure CIS. This may be due to the layered structure formed by the CIS arranged vertically on the HCN surface, facilitating enhanced light absorption. The multiple reflections and absorptions in the hollow structure are also reasons for the improvement of the photothermal performance of HCN@CIS. In addition, the rapid carrier transfer process between heterojunctions can also convert part of the mechanical energy into thermal energy [52]. The unique hollow core-shell structure of HCN@CIS provides a basis for the effective utilization of heat generated by the photothermal effect. We assume that the heat generated by the photothermal effect of HCN@CIS spreads to the external aqueous solution and the internal space at the same time. It is difficult to increase the overall temperature of the solution by the photothermal effect alone, but there is only a small amount of water inside the hollow sphere that needs to be heated. Therefore, under the action of double shell heat preservation and heat collection, the internal temperature of the HCN@CIS reactor should be much higher than that of the external solution [53]. The increase in HCN@CIS temperature will accelerate the movement of surrounding water molecules, which is beneficial to increase the collision probability of carriers and active radicals with reactant molecules and improve the photocatalytic reaction rate by promoting local mass transfer kinetics.

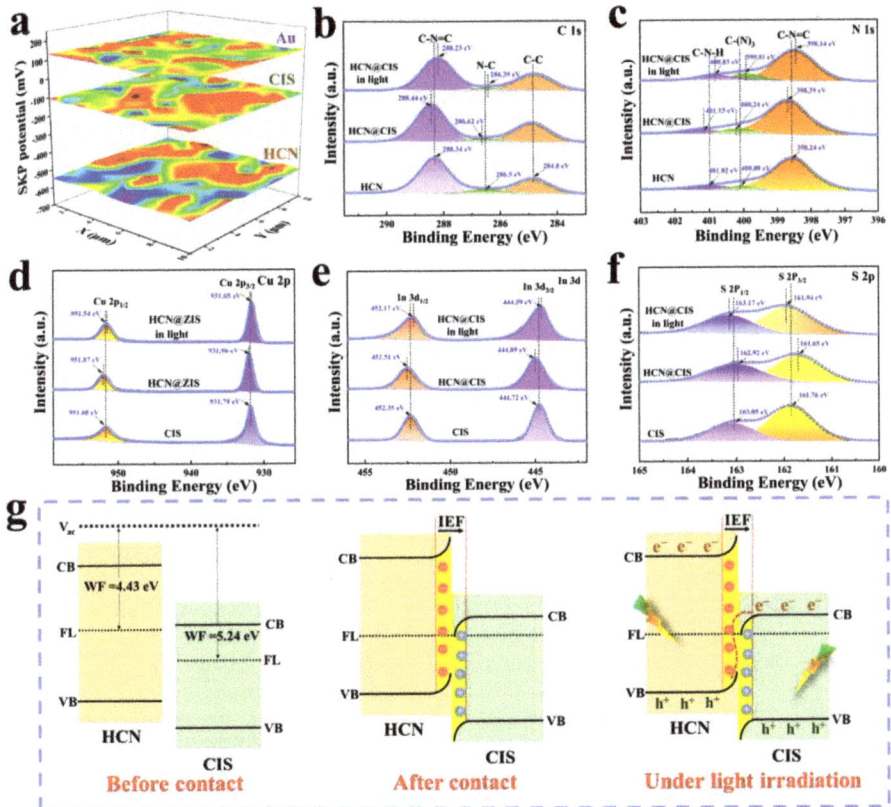

Figure 3. (**a**) Work functions of the HCN and the CIS. XPS spectra of C 1 s (**b**) and N 1 s (**c**) of the HCN and the HCN@CIS. XPS spectra of Cu 1 s (**d**), In 2p (**e**), and S 1 s (**f**) of the CIS and the HCN@CIS. Schematic representation of the formation process and charge transfer mechanism of the HCN@CIS S-scheme heterojunction (**g**).

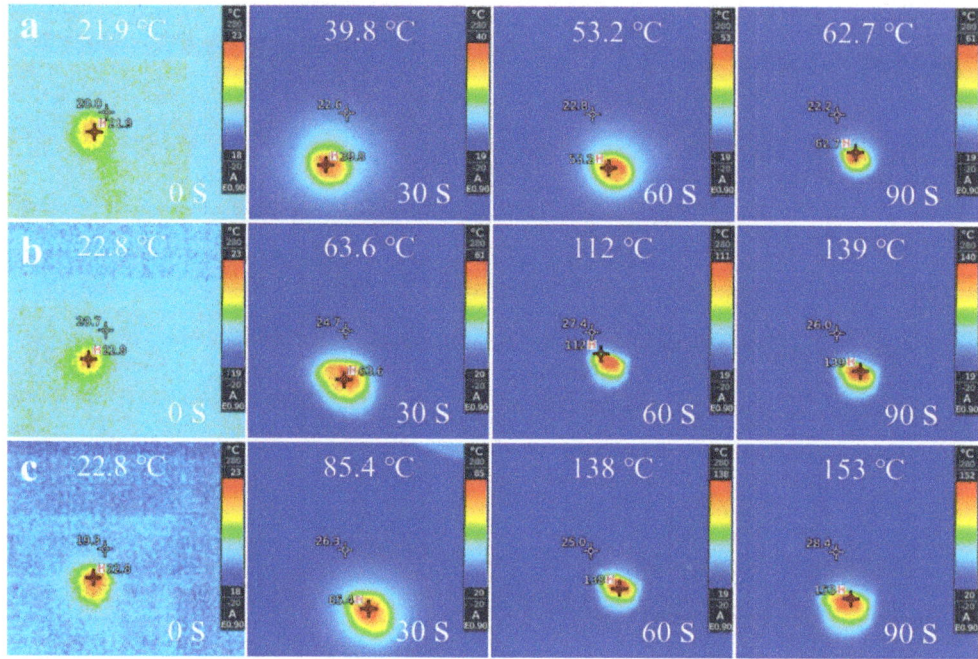

Figure 4. The IR images of temperature variation with light time for HCN (**a**), CIS (**b**), and HCN@CIS (**c**) (The light source is a 300 W Xe lamp with a 420 nm filter).

2.5. Photocatalytic H_2 Production and Pollutant Removal

Figure 5a shows the photocatalytic H_2 production performance of the different samples. HCN@CIS with 20% CIS load has the best photocatalytic H_2 production performance (unless otherwise specified, the HCN@ZIS in this paper refers to the optimum sample), and the hydrogen production rate is as high as 2325.68 µmol h^{-1} g^{-1}, which is 62.8 and 12.2 times higher than that of pure HCN and CIS, respectively. Compared with other heterojunction photocatalysts reported recently, the HCN@CIS has excellent performance for H_2 evolution (Table S1). The HCN@CIS shows excellent stability in the 20 h cycle test (Figure 5b). There are two reasons why the HCN@ZIS sample has good hydrogen production performance. On the one hand, the S-scheme heterojunction provides fast carrier separation efficiency, and on the other hand, the photothermal effect activates the water molecules in the cavity, which promotes the evolution of active H* species and accelerates the formation of H_2 [9]. Figure 5c shows the photocatalytic degradation curves of different samples for OTC. The solution reached dynamic adsorption equilibrium after the reaction was conducted in the dark for 30 min, and then the

Photocatalytic degradation experiment was carried out. It can be seen that the HCN@CIS with a 20% ZIS load has the best degradation effect of OTC, and the degradation rate in 120 min is as high as 95.7%, indicating that the reaction active site can be fully exposed with an appropriate load. Figure 5d shows that the apparent reaction rate of the HCN@CIS is 9.4 and 5.9 times higher that of pure HCN and ZIS, respectively, indicating that the efficient charge separation of the S-scheme heterojunction can improve the photocatalytic reaction rate. In addition, the HCN@CIS composite photocatalyst has excellent stability, with only a slight decrease in degradation efficiency after 600 min of cycling (Figure 5e). The active species in photocatalytic reactions are analyzed by a radical quenching experiment. Benzoquinone (BQ), silver nitrate (AgNO$_3$), triethanolamine (TEOA), and isopropanol (IPA) were used as quenching agents of ·O$_2^-$, e$^-$, h$^+$ and ·OH, respectively [54]. It can be seen from Figure 5f that the contribution rate of free radicals in the process of photocatalysis is $·O_2^- > e^- > h^+ > ·OH$. Among them, $·O_2^-$, e^-, and h^+ play an important role, and electron trapping may decrease the yield of $·O_2^-$ and affect the photocatalytic activity. The VB potential of CIS in HCN@CIS is not enough to oxidize water to produce ·OH, so the addition of IPA has almost no effect on the efficiency of photocatalytic degradation.

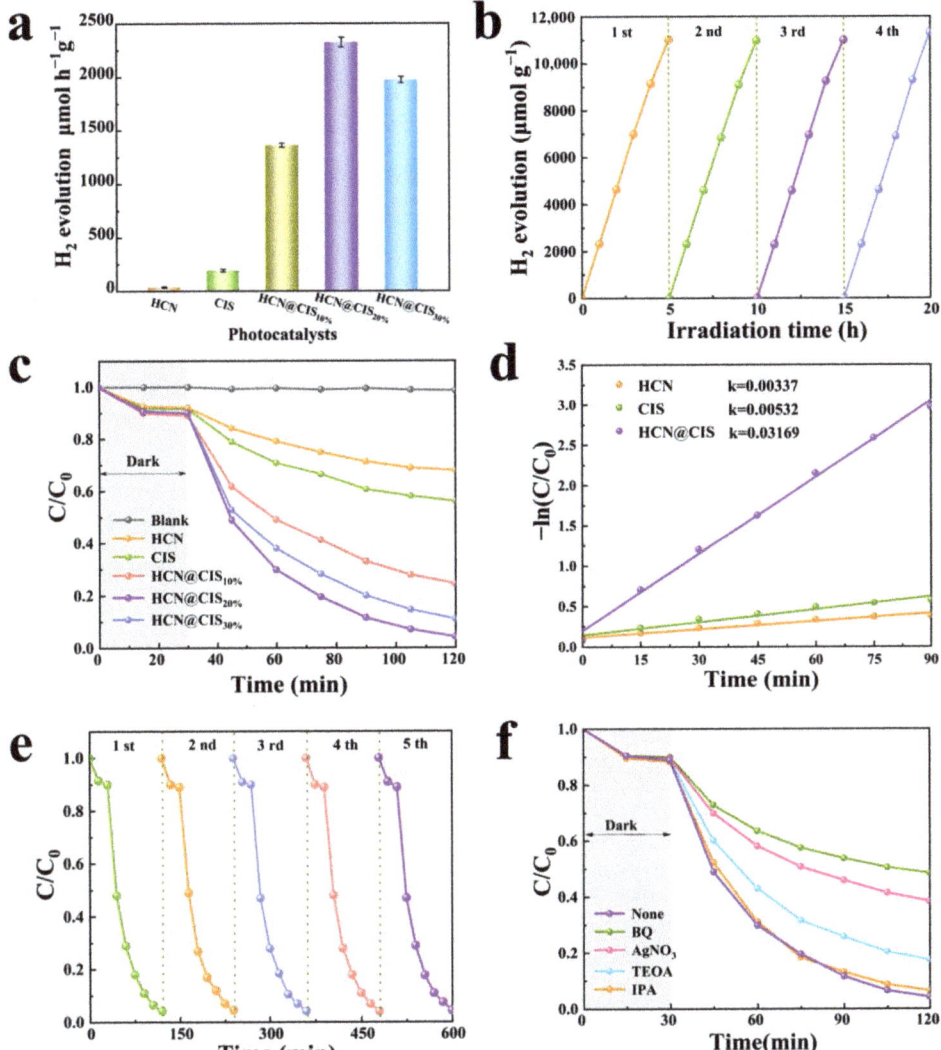

Figure 5. Photocatalytic H_2 evolution for all samples (**a**) and H_2 production cycle test of HCN@CIS (**b**). Photocatalytic degradation of OTC for all samples (**c**) and the corresponding apparent reaction rate constants k (**d**). Cycling tests (**e**) and free radical quenching experiments (**f**) on HCN@CIS degraded OTC.

2.6. Analysis of Charge-Transfer Dynamics

The photoelectrochemical test is used to analyze the charge separation characteristics in the photocatalyst. Figure 6a shows the photocurrent response curve obtained by a cyclic switching light source, and the HCN@CIS shows the highest current density, indicating that the S-scheme heterojunction has better charge separation ability [55,56]. Electrochemical impedance spectroscopy (EIS) shows that the radius of the Nyquist curve of HCN@CIS is the smallest (Figure 6b), which means that the resistance of surface charge transfer is lower, which is beneficial to rapid carrier separation and transfer [57]. The kinetics of charge transfer of the photocatalysts is closely related to the catalytic performance. Figure S4 shows the steady-state photoluminescence (PL) emission spectra of HCN, CIS, and HCN@CIS. It can be seen that the fluorescence quenching intensity of HCN@CIS is much lower than that of pure HCN and CIS, which indicates that S-scheme heterojunctions effectively reduce the recombination of photogenerated carriers [58,59]. The charge transfer kinetics at the interface are further analyzed by time-resolved photoluminescence (TRPL) spectra. The TRPL kinetic spectra of HCN, CIS, and HCN@CIS are shown in Figure 6c. The normalized results show that the average PL lifetime of HCN@CIS (3.57 ns) is longer than that of pure HCN (1.97 ns) and CIS (2.23 ns). The longer carrier lifetime in S-scheme heterojunctions is beneficial to the full contact of electrons and holes with surface adsorption molecules and improves the effective utilization of photogenerated carriers [60]. The analysis of the active species and content of photocatalytic production is considered an important means to determine the mechanism of photocatalysis. The VB of HCN@CIS is not enough to oxidize H_2O to form $\cdot OH$, so the $\cdot O_2^-$ spin signal produced in the process of the photocatalysis is detected by the electron spin resonance (EPR) technique and DMPO (5-dimethyl-1-proline-N-oxide) as spin trap. Figure 6d shows that HCN, CIS, and HCN@CIS can all produce $\cdot O_2^-$, but the HCN@CIS $\cdot O_2^-$ spin signal is the strongest. The reduction of O_2 adsorbed on the surface by conduction band electrons of the photocatalyst affects the formation of $\cdot O_2^-$. The HCN@ZIS can produce $\cdot O_2^-$ more efficiently, indicating that the electron reduction ability of the CB of the HCN@CIS composite photocatalyst has not decreased, and the possibility of forming Type-II heterojunction between HCN and CIS can be ruled out [61]. The interlaced HCN and CIS follow the S-scheme charge-transfer mechanism. The rapid charge separation and excellent photothermal properties of HCN@CIS promote the mass transfer kinetics of the reaction system and accelerate the formation of $\cdot O_2^-$.

Based on a large number of characterization and photocatalytic experiments, the photocatalytic performance enhancement mechanism of the HCN@CIS heterojunction photothermal nanoreactor (Scheme 1) was proposed. The HCN@CIS with a unique hollow core-shell structure generates carriers excited by light, and the carriers follow S-scheme migration under the action of an internal electric field and interface band bending. The electrons in the CB of the CIS recombine with the holes in the VB of the HCN. The electrons with strong reduction ability in the CB of the HCN and the holes with strong oxidation ability in the VB of the ZIS are successfully retained in the HCN@CIS system. In addition, the hollow structure can improve the utilization of light, and the double shell can inhibit the rapid escape of internal heat, provide a local high-temperature place, and accelerate the reaction kinetics. The free electrons in HCN@CIS are retained on the inner HCN, and the energy of the photothermal effect is gathered in the inner space. Therefore, the HCN@CIS photothermal nanoreactor can efficiently carry out photocatalytic hydrogen production and photocatalytic degradation of oxytetracycline by the synergistic action of S-scheme heterojunction and the photothermal effect.

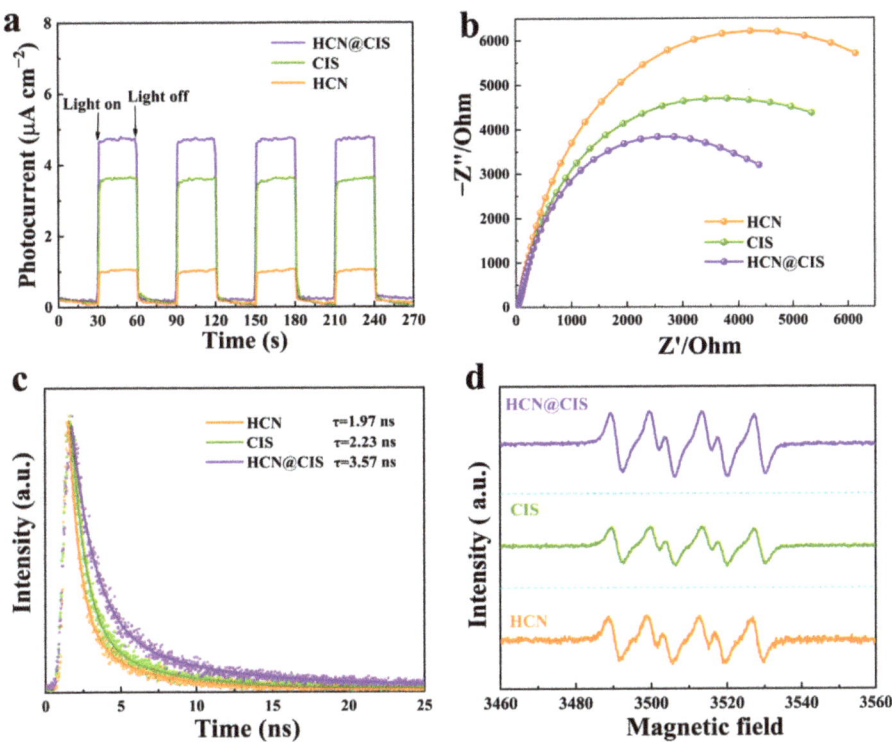

Figure 6. Transient photocurrent density curves (**a**), EIS Nyquist plots (**b**), and TRPL decay spectra (**c**) of HCN, CIS, and HCN@CIS. EPR signals of HCN, CIS, and HCN@CIS in light (**d**).

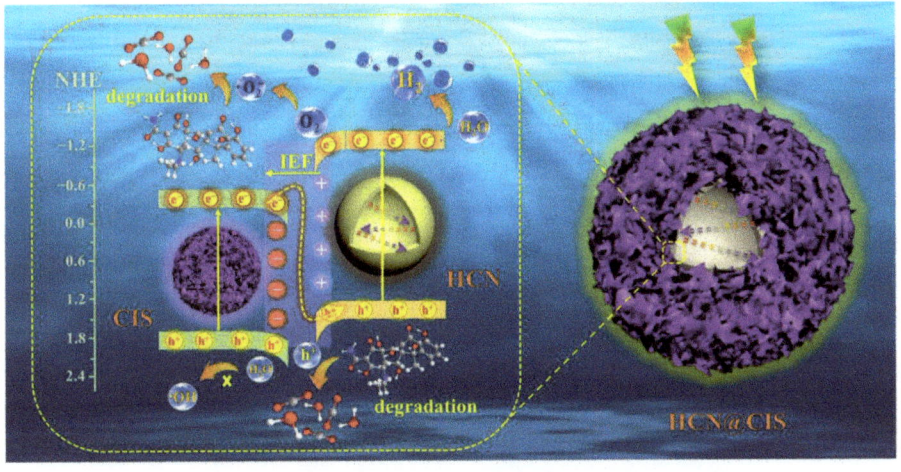

Scheme 1. Schematic diagram of the HCN@ZIS S-scheme hollow core-shell nanoreactor for photocatalytic H_2 production and OTC degradation.

3. Materials and Methods

The chemical reagents and auxiliary materials used in this study are listed in Supporting Information. The chemical reagents are not further treated before use. The steps for the synthesis of the SiO$_2$ template, HCN, and HCN@CIS are shown in Scheme 2. The characterization parameters and experimental details are in Supporting Information, respectively.

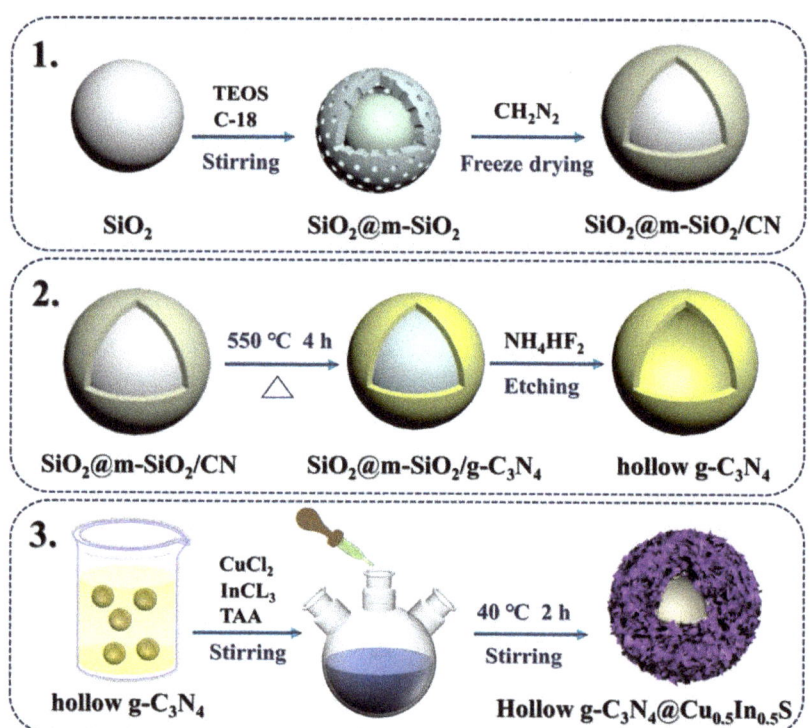

Scheme 2. The schematic diagram for the preparation of the HCN@CIS hollow core-shell nanoreactor. (1) Preparation of mesoporous SiO$_2$/SiO$_2$@CN nanospheres, (2) Synthesis of HCN nanospheres, (3) Preparation of the HCN@CIS core-shell nanoreactor.

3.1. Preparation of Mesoporous SiO$_2$/SiO$_2$ Nanospheres

SiO$_2$ nanosphere templates are prepared by the Stöber method. Typical synthesis methods are as follows: 3.5 mL of TEOS was dropped into the mixed solution of 7.0 mL NH$_3$·H$_2$O, 16.5 mL H$_2$O, and 110 mL C$_2$H$_5$OH under magnetic stirring and stirred at room temperature for 2 h to form uniformly dispersed silica spheres. C18-TMOS and TEOS were mixed at a ratio of 2:3 by volume, then dripped into the mixed solution and aged at room temperature for 3 h. The mixed solution was centrifuged, dried at 60 °C, and calcined at 550 °C for 6 h to obtain mesoporous SiO$_2$/SiO$_2$ nanospheres.

3.2. Synthesis of HCN Nanospheres

The 2 g mesoporous SiO$_2$/SiO$_2$ template was added to the 10 mL cyanamide solution, ultrasonic for 2 h, stirred at 60 °C for 6 h, then centrifuged, washed, and freeze-dried. The dried white powder was placed in a tube furnace, calcined at 550 °C for 4 h under the protection of nitrogen, and the heating rate was 5 °C/min. The gray-white powder was SiO$_2$@g-C$_3$N$_4$, and the SiO$_2$ template was etched with 4 mol/L NH$_4$HF$_2$ solution. Light yellow HCN can be obtained by centrifugation and drying.

3.3. Preparation of the HCN@CIS Core-Shell Nanoreactor

In typical synthesis, the prepared HCN (0.1 g) was added to three beakers containing 40 mL deionized water, respectively, and dispersed by ultrasonic for 30 min, then 0.007 g

$CuCl_2 \cdot 2H_2O$, 0.012 g $InCl_3 \cdot 4H_2O$, and 0.006 g TAA (10 wt%); 0.017 g $CuCl_2 \cdot 2H_2O$, 0.024 g $InCl_3 \cdot 4H_2O$, and 0.012 g TAA (20 wt%); 0.021 g $CuCl_2 \cdot 2H_2O$, 0.036 g $InCl_3 \cdot 4H_2O$, and 0.018 g TAA (30 wt%) were added to 3 beakers under magnetic stirring, respectively (x wt% represents the percentage of $Cu_{0.5}In_{0.5}S$ mass to g-C_3N_4), and the solution pH was adjusted to 2.5 with hydrochloric acid. The mixed solution was transferred to a 150 mL round-bottom flask and refluxed in an oil bath at 40 °C for 2 h. Centrifuge and dry the powder at 60 °C overnight to obtain a black HCN@CIS hollow core-shell nanoreactor.

4. Conclusions

In summary, HCN hollow nanospheres are obtained by the template method, and the hollow HCN@CIS core-shell S-scheme photothermal nanoreactors are obtained by surface continuous growth. The charge-transfer mechanism of the S-scheme in HCN@ZIS was proved by in situ irradiation, XPS, and EPR characterization. In addition, the synergistic effect of the photothermal effect and S-scheme heterojunction on photocatalytic reactions is proposed. The fast charge transfer and redox ability of S-scheme heterojunctions intensely initiate the photocatalytic reaction, and the photothermal effect reduces the activation energy barrier of the chemical reaction and speeds up the reaction rate. Therefore, HCN@CIS has enhanced photothermal catalytic activity, and the photocatalytic H_2 evolution efficiency is 62.8 and 12.2 times higher than that of pure HCN and CIS, respectively, and the degradation efficiency of oxytetracycline is as high as 95.7%. This work fully demonstrates the potential of the synergistic effect of S-scheme heterojunction and photothermal effect and provides a promising strategy for the design of an efficient solar-driven photothermal catalyst platform.

Supplementary Materials: The following supporting information can be downloaded at: https://www.mdpi.com/article/10.3390/catal13040723/s1, Figure S1: SEM image of the SiO_2 spheres; Figure S2: TEM image of $SiO_2@g-C_3N_4$; Figure S3: The sample patterns of HCN, CIS and HCN@CIS; Figure S4: PL emission spectra of HCN, CIS and HCN@CIS; Table S1: Comparison of H_2 generation rates of different photocatalysts. References [62–68] were cited in the Supplementary Materials.

Author Contributions: Y.X.: Data curation, Investigation, Writing—original draft, Methodology. Z.W.: Methodology, Visualization, Writing—review & editing. B.Y.: Writing—review & editing. Y.C.: Supervision. T.C.: Writing—review & editing. Y.W.: Funding acquisition, Writing—review & editing. All authors have read and agreed to the published version of the manuscript.

Funding: This research was funded by Yunnan University's Research Innovation Fund for Graduate Students (KC.22222464).

Data Availability Statement: The data that support the findings of this study are available from the corresponding author upon reasonable request.

Acknowledgments: We gratefully acknowledge the support of this research by the National Natural Science Foundation of China (Nos. 41876055 and 61761047), Yunnan University's Research Innovation Fund for Graduate Students (KC.22222464), and the Program for Innovative Research Team (in Science and Technology) at the University of Yunnan Province.

Conflicts of Interest: The authors declare no conflict of interest.

References

1. Nishiyama, H.; Yamada, T.; Nakabayashi, M.; Maehara, Y.; Yamaguchi, M.; Kuromiya, Y.; Nagatsuma, Y.; Tokudome, H.; Akiyama, S.; Watanabe, T.; et al. Photocatalytic solar hydrogen production from water on a 100-m^2 scale. *Nature* **2021**, *598*, 304–307. [CrossRef] [PubMed]
2. Zhou, P.; Navid, I.A.; Ma, Y.; Xiao, Y.; Wang, P.; Ye, Z.; Zhou, B.; Sun, K.; Mi, Z. Solar-to-hydrogen efficiency of more than 9% in photocatalytic water splitting. *Nature* **2023**, *613*, 66–70. [CrossRef] [PubMed]
3. Nikoloudakis, E.; López-Duarte, I.; Charalambidis, G.; Ladomenou, K.; Ince, M.; Coutsolelos, A.G. Porphyrins and phthalocyanines as biomimetic tools for photocatalytic H_2 production and CO_2 reduction. *Chem. Soc. Rev.* **2022**, *51*, 6965–7045. [CrossRef] [PubMed]
4. Fang, Y.; Hou, Y.; Fu, X.; Wang, X. Semiconducting polymers for oxygen evolution reaction under light illumination. *Chem. Rev.* **2022**, *122*, 4204–4256. [CrossRef] [PubMed]

5. Kosco, J.; Gonzalez-Carrero, S.; Howells, C.T.; Fei, T.; Dong, Y.; Sougrat, R.; Harrison, G.T.; Firdaus, Y.; Sheelamanthula, R.; Purushothaman, B.; et al. Generation of long-lived charges in organic semiconductor heterojunction nanoparticles for efficient photocatalytic hydrogen evolution. *Nat. Energy* **2022**, *7*, 340–351. [CrossRef]
6. Pavliuk, M.V.; Wrede, S.; Liu, A.; Brnovic, A.; Sicong Wang, S.; Axelssona, M.; Tian, H. Preparation, characterization, evaluation and mechanistic study of organic polymer nano-photocatalysts for solar fuel production. *Chem. Soc. Rev.* **2022**, *51*, 6909–6935. [CrossRef]
7. Tao, X.; Zhao, Y.; Wang, S.; Li, C.; Li, R. Recent advances and perspectives for solar-driven water splitting using particulate photocatalysts. *Chem. Soc. Rev.* **2022**, *51*, 3561–3608. [CrossRef]
8. Teng, Z.; Zhang, Q.; Yang, H.; Kato, K.; Yang, W.; Lu, Y.-R.; Liu, S.; Wang, C.; Yamakata, A.; Su, C.; et al. Atomically dispersed antimony on carbon nitride for the artificial photosynthesis of hydrogen peroxide. *Nat. Catal.* **2021**, *4*, 374–384. [CrossRef]
9. Liu, P.; Huang, Z.; Gao, X.; Hong, X.; Zhu, J.; Wang, G.; Wu, Y.; Zeng, J.; Zheng, X. Synergy between palladium single atoms and nanoparticles via hydrogen spillover for enhancing CO_2 photoreduction to CH_4. *Adv. Mater.* **2022**, *34*, 2200057. [CrossRef]
10. Xie, W.; Li, K.; Liu, X.H.; Zhang, X.; Huang, H. P-Mediated Cu–N_4 Sites in Carbon Nitride Realizing CO_2 Photoreduction to C_2H_4 with Selectivity Modulation. *Adv. Mater.* **2023**, *35*, 2208132. [CrossRef]
11. Qian, W.; Hu, W.; Jiang, Z.; Wu, Y.; Li, Z.; Diao, Z.; Li, M. Degradation of Tetracycline Hydrochloride by a Novel CDs/g-C_3N_4/$BiPO_4$ under Visible-Light Irradiation: Reactivity and Mechanism. *Catalysts* **2022**, *12*, 774. [CrossRef]
12. Xie, P.; Ding, J.; Yao, Z.; Pu, T.; Zhang, P.; Huang, Z.; Wang, C.; Zhang, J.; Zhong, Q.; Huang, H.; et al. Oxo dicopper anchored on carbon nitride for selective oxidation of methane. *Nat. Commun.* **2022**, *13*, 1375. [CrossRef] [PubMed]
13. Gao, R.H.; Ge, Q.; Jiang, N.; Cong, H.; Liu, M.; Zhang, Y.Q. Graphitic carbon nitride (g-C_3N_4)-based photocatalytic materials for hydrogen evolution. *Front. Chem.* **2022**, *10*, 1048504. [CrossRef] [PubMed]
14. Li, L.; Ma, D.; Xu, Q.; Huang, S. Constructing hierarchical $ZnIn_2S_4$/g-C_3N_4 S-scheme heterojunction for boosted CO_2 photoreduction performance. *Chem. Eng. J.* **2022**, *437*, 135153. [CrossRef]
15. Yang, T.; Shao, Y.; Hu, J.; Qu, J.; Yang, X.; Yang, F.; Li, C.M. Ultrathin layered 2D/2D heterojunction of ReS_2/high-crystalline g-C_3N_4 for significantly improved photocatalytic hydrogen evolution. *Chem. Eng. J.* **2022**, *448*, 137613. [CrossRef]
16. Chen, Y.; Su, F.; Xie, H.; Wang, R.; Ding, C.; Huang, J.; Xu, Y.; Ye, L. One-step construction of S-scheme heterojunctions of N-doped MoS_2 and S-doped g-C_3N_4 for enhanced photocatalytic hydrogen evolution. *Chem. Eng. J.* **2021**, *404*, 126498. [CrossRef]
17. Wang, Z.; Yang, Y.; Fang, R.; Yan, Y.;Et al, J.; Zhang, L. A State-of-the-art review on action mechanism of photothermal catalytic reduction of CO_2 in full solar spectrum. *Chem. Eng. J.* **2022**, *429*, 132322. [CrossRef]
18. Song, C.; Wang, Z.; Yin, Z.; Xiao, D.; Ma, D. Principles and applications of photothermal catalysis. *Chem. Catal.* **2021**, *2*, 52–83. [CrossRef]
19. Zhu, J.; Shao, W.; Li, X.; Jiao, X.; Zhu, J.; Sun, Y.; Xie, Y. Asymmetric triple-atom sites confined in ternary oxide enabling selective CO_2 photothermal reduction to acetate. *J. Am. Chem. Soc.* **2021**, *143*, 18233–18241. [CrossRef]
20. Xiao, Y.; Yao, B.; Wang, Z.; Chen, T.; Xiao, X.; Wang, Y. Plasma Ag-Modified α-Fe_2O_3/g-C_3N_4 Self-Assembled S-Scheme Heterojunctions with Enhanced Photothermal-Photocatalytic-Fenton Performances. *Nanomaterials* **2022**, *12*, 4212. [CrossRef]
21. Zhang, L.; Wang, Z.Q.; Liao, J.; Zhang, X.; Feng, D.; Deng, H.; Ge, C. Infrared-to-visible energy transfer photocatalysis over black phosphorus quantum dots/carbon nitride. *Chem. Eng. J.* **2022**, *431*, 133453. [CrossRef]
22. Zheng, D.; Pang, C.; Liu, Y.; Wang, X. Shell-engineering of hollow gC_3N_4 nanospheres via copolymerization for photocatalytic hydrogen evolution. *Chem. Commun.* **2015**, *51*, 9706–9709. [CrossRef] [PubMed]
23. Wang, Y.; Liu, M.; Wu, C.; Gao, P.; Li, M.; Xing, Z.; Li, Z.; Zhou, W. Hollow Nanoboxes Cu_{2-x}S@$ZnIn_2S_4$ Core-Shell S-Scheme Heterojunction with Broad-Spectrum Response and Enhanced Photothermal-Photocatalytic Performance. *Small* **2022**, *18*, 2202544. [CrossRef] [PubMed]
24. Tahir, B.; Tahir, M.; Nawawai, M.G.M.; Khoja, A.H.; Haq, B.U.; Farooq, W. Ru-embedded 3D g-C_3N_4 hollow nanosheets (3D CNHNS) with proficient charge transfer for stimulating photocatalytic H_2 production. *Int. J. Hydrog. Energy* **2021**, *46*, 27997–28010. [CrossRef]
25. Zhang, Y.; Wu, Y.; Wan, L.; Ding, H.; Li, H.; Wang, X.; Zhang, W. Hollow core–shell Co_9S_8@$ZnIn_2S_4$/CdS nanoreactor for efficient photothermal effect and CO_2 photoreduction. *Appl. Catal. B Environ.* **2022**, *311*, 121255. [CrossRef]
26. Qiu, Y.; Xing, Z.; Guo, M.; Li, Z.; Wang, N.; Zhou, W. Hollow cubic Cu_{2-xS}/Fe-POMs/$AgVO_3$ dual Z-scheme heterojunctions with wide-spectrum response and enhanced photothermal and photocatalytic-fenton performance. *Appl. Catal. B Environ.* **2021**, *298*, 120628. [CrossRef]
27. Zheng, D.; Cao, X.N.; Wang, X. Precise formation of a hollow carbon nitride structure with a janus surface to promote water splitting by photoredox catalysis. *Angew. Chem. Int. Ed.* **2016**, *55*, 11512–11516. [CrossRef]
28. Xu, Q.; Zhang, L.; Cheng, B.; Fan, J.; Yu, J. S-scheme heterojunction photocatalyst. *Chem* **2020**, *6*, 1543–1559. [CrossRef]
29. Xiao, Y.; Yao, B.; Cao, M.; Wang, Y. Super-Photothermal Effect-Mediated Fast Reaction Kinetic in S-Scheme Organic/Inorganic Heterojunction Hollow Spheres Toward Optimized Photocatalytic Performance. *Small* **2023**, *19*, 2207499. [CrossRef]
30. Ruan, X.; Huang, C.; Cheng, H.; Zhang, Z.; Cui, Y.; Li, Z.; Xie, T.; Ba, K.; Zhang, H.; Zhang, L.; et al. A Twin S-Scheme Artificial Photosynthetic System with Self-Assembled Heterojunctions yields Superior Photocatalytic Hydrogen Evolution Rate. *Adv. Mater.* **2023**, *35*, 2209141. [CrossRef]
31. Le, S.; Zhu, C.; Cao, Y.; Wang, P.; Liu, Q.; Zhou, H.; Chen, C.; Wang, S.; Duan, S. V_2O_5 nanodot-decorated laminar C_3N_4 for sustainable photodegradation of amoxicillin under solar light. *Appl. Catal. B Environ.* **2022**, *303*, 120903. [CrossRef]

32. Cheng, C.; He, B.; Fan, J.; Cheng, B.; Cao, S.; Yu, J. An inorganic/organic S-scheme heterojunction H_2-production photocatalyst and its charge transfer mechanism. *Adv. Mater.* **2021**, *33*, 2100317. [CrossRef] [PubMed]
33. Jia, X.; Hu, C.; Sun, H.; Cao, J.; Lin, H.; Li, X.; Chen, S. A dual defect co-modified S-scheme heterojunction for boosting photocatalytic CO_2 reduction coupled with tetracycline oxidation. *Appl. Catal. B Environ.* **2023**, *324*, 122232. [CrossRef]
34. Zhang, X.; Kim, D.; Yan, J.; Lee, L.Y.S. Photocatalytic CO_2 reduction enabled by interfacial S-scheme heterojunction between ultrasmall copper phosphosulfide and g-C_3N_4. *ACS Appl. Mater. Interfaces* **2021**, *13*, 9762–9770. [CrossRef]
35. Dai, Z.; Zhen, Y.; Sun, Y.; Li, L.; Ding, D. $ZnFe_2O_4$/g-C_3N_4 S-scheme photocatalyst with enhanced adsorption and photocatalytic activity for uranium (VI) removal. *Chem. Eng. J.* **2021**, *415*, 129002. [CrossRef]
36. Khan, A.A.; Tahir, M.; Mohamed, A.R. Constructing S-scheme heterojunction of carbon nitride nanorods (g-CNR) assisted trimetallic CoAlLa LDH nanosheets with electron and holes moderation for boosting photocatalytic CO_2 reduction under solar energy. *Chem. Eng. J.* **2022**, *433*, 133693. [CrossRef]
37. Wang, J.; Wang, G.; Cheng, B.; Yu, J.; Fan, J. Sulfur-doped g-C_3N_4/TiO_2 S-scheme heterojunction photocatalyst for Congo Red photodegradation. *Chin. J. Catal.* **2021**, *42*, 56–68. [CrossRef]
38. Babu, B.; Shim, J.; Kadam, A.N.; Yoo, K. Modification of porous g-C_3N_4 nanosheets for enhanced photocatalytic activity: In-situ synthesis and optimization of NH_4Cl quantity. *Catal. Commun.* **2019**, *124*, 123–127. [CrossRef]
39. Chen, K.; Shi, Y.; Shu, P.; Luo, Z.; Shi, W.; Guo, F. Construction of core–shell FeS_2@$ZnIn_2S_4$ hollow hierarchical structure S-scheme heterojunction for boosted photothermal-assisted photocatalytic H_2 production. *Chem. Eng. J.* **2023**, *454*, 140053. [CrossRef]
40. Suzuki, V.Y.; Amorin, L.H.C.; Fabris, G.S.L.; Dey, S.; Sambrano, J.R.; Cohen, H.; Oron, D.; Porta, F.A.L. Enhanced Photocatalytic and Photoluminescence Properties Resulting from Type-I Band Alignment in the Zn_2GeO_4/g-C_3N_4 Nanocomposites. *Catalysts* **2022**, *12*, 692. [CrossRef]
41. Babu, B.; Shim, J.; Yoo, K. Efficient solar-light-driven photoelectrochemical water oxidation of one-step in-situ synthesized Co-doped g-C_3N_4 nanolayers. *Ceram. Int.* **2020**, *46*, 16422–16430. [CrossRef]
42. Guo, M.; Zhao, T.; Xing, Z.; Qiu, Y.; Pan, K.; Li, Z.; Yang, S.; Zhou, W. Hollow Octahedral $Cu_{2-x}S$/CdS/Bi_2S_3 p–n–p Type Tandem Heterojunctions for Efficient Photothermal Effect and Robust Visible-Light-Driven Photocatalytic Performance. *ACS Appl. Mater. Interfaces* **2020**, *12*, 40328–40338. [CrossRef] [PubMed]
43. Li, X.; Kang, B.; Dong, F.; Zhang, Z.; Luo, X.; Han, L.; Huang, J.; Feng, Z.; Chen, Z.; Xu, J.; et al. Enhanced photocatalytic degradation and H_2/H_2O_2 production performance of S-pCN/$WO_{2.72}$ S-scheme heterojunction with appropriate surface oxygen vacancies. *Nano Energy* **2021**, *81*, 105671. [CrossRef]
44. Zhang, L.; Zhang, J.; Yu, H.; Yu, J. Emerging S-scheme photocatalyst. *Adv. Mater.* **2022**, *34*, 2107668. [CrossRef] [PubMed]
45. Moon, H.S.; Hsiao, K.C.; Wu, M.C.; Yun, Y.; Hsu, Y.J.; Yong, K. Spatial Separation of Cocatalysts on Z-Scheme Organic/Inorganic Heterostructure Hollow Spheres for Enhanced Photocatalytic H_2 Evolution and in-Depth Analysis of the Charge-Transfer Mechanism. *Adv. Mater.* **2023**, *35*, 2200172. [CrossRef] [PubMed]
46. Zhao, D.; Wang, Y.; Dong, C.L.; Huang, Y.C.; Chen, J.; Xue, F.; Shen, S.; Guo, L. Boron-doped nitrogen-deficient carbon nitride-based Z-scheme heterostructures for photocatalytic overall water splitting. *Nat. Energy* **2021**, *6*, 388–397. [CrossRef]
47. Wang, L.; Cheng, B.; Zhang, L.; Yu, J. In situ irradiated XPS investigation on S-scheme TiO_2@$ZnIn_2S_4$ photocatalyst for efficient photocatalytic CO_2 reduction. *Small* **2021**, *17*, 2103447. [CrossRef] [PubMed]
48. Chen, K.; Wang, X.; Li, Q.; Feng, Y.N.; Chen, F.F.; Yu, Y. Spatial distribution of $ZnIn_2S_4$ nanosheets on g-C_3N_4 microtubes promotes photocatalytic CO_2 reduction. *Chem. Eng. J.* **2021**, *418*, 129476. [CrossRef]
49. Li, X.; Tu, D.; Yu, S.; Song, X.; Lian, W.; Wei, J.; Shang, X.; Li, R.; Chen, Y. Highly efficient luminescent I-III-VI semiconductor nanoprobes based on template-synthesized $CuInS_2$ nanocrystals. *Nano Res.* **2019**, *12*, 1804–1809. [CrossRef]
50. Cheng, C.; Zhang, J.; Zhu, B.; Liang, G.; Zhang, L.; Yu, J. Verifying the Charge-Transfer Mechanism in S-Scheme Heterojunctions Using Femtosecond Transient Absorption Spectroscopy. *Angew. Chem. Int. Ed.* **2023**, *135*, e202218688. [CrossRef]
51. Jiang, X.; Huang, J.; Bi, Z.; Ni, W.; Gurzadyan, G.; Zhu, Y.; Zhang, Z. Plasmonic Active "Hot Spots"-Confined Photocatalytic CO_2 Reduction with High Selectivity for CH_4 Production. *Adv. Mater.* **2022**, *34*, 2109330. [CrossRef] [PubMed]
52. Wang, K.; Xing, Z.; Du, M.; Zhang, S.; Li, Z.; Pan, K.; Zhou, W. Hollow $MoSe_2$@Bi_2S_3/CdS core-shell nanostructure as dual Z-scheme heterojunctions with enhanced full spectrum photocatalytic-photothermal performance. *Appl. Catal. B Environ.* **2021**, *281*, 119482. [CrossRef]
53. Zhang, H.C.; Kang, Z.X.; Han, J.J.; Wang, P.; Fan, J.T.; Shen, G.P. Photothermal nanoconfinement reactor: Boosting chemical reactivity with locally high temperature in a confined space. *Angew. Chem. Int. Ed.* **2022**, *134*, e202200093.
54. Xiao, Y.; Wang, K.; Yang, Z.; Xing, Z.; Li, Z.; Pan, K.; Zhou, W. Plasma Cu-decorated TiO_{2-x}/CoP particle-level hierarchical heterojunctions with enhanced photocatalytic-photothermal performance. *J. Hazard. Mater.* **2021**, *414*, 125487. [CrossRef]
55. Babu, B.; Koutavarapu, R.; Shim, J.; Yoo, K. Enhanced visible-light-driven photoelectrochemical and photocatalytic performance of Au-SnO_2 quantum dot-anchored g-C_3N_4 nanosheets. *Sep. Purif. Technol.* **2020**, *240*, 116652. [CrossRef]
56. Jiang, X.H.; Zhang, L.S.; Liu, H.Y.; Wu, D.S.; Wu, F.Y.; Tian, L.; Liu, L.L.; Zou, J.P.; Luo, S.L.; Chen, B.B. Silver single atom in carbon nitride catalyst for highly efficient photocatalytic hydrogen evolution. *Angew. Chem. Int. Ed.* **2020**, *132*, 23312–23316. [CrossRef]
57. Wang, X.; Wang, X.; Huang, J.; Li, S.; Meng, A.; Li, Z. Interfacial chemical bond and internal electric field modulated Z-scheme Sv-$ZnIn_2S_4$/$MoSe_2$ photocatalyst for efficient hydrogen evolution. *Nat. Commun.* **2021**, *12*, 4112. [CrossRef]
58. Amorin, L.H.; Suzuki, V.Y.; de Paula, N.H.; Duarte, J.L.; da Silva, M.A.T.; Taftd, C.A.; Porta, F.A.L. Electronic, structural, optical, and photocatalytic properties of graphitic carbon nitride. *New J. Chem.* **2019**, *43*, 13647–13653. [CrossRef]

59. Zhang, Y.; Zhao, J.; Wang, H.; Xiao, B.; Zhang, W.; Zhao, X.; Lv, T.; Thangamuthu, M.; Zhang, J.; Guo, Y.; et al. Single-atom Cu anchored catalysts for photocatalytic renewable H_2 production with a quantum efficiency of 56%. *Nat. Commun.* **2022**, *13*, 58. [CrossRef]
60. Wang, G.; Huang, R.; Zhang, J.; Mao, J.; Wang, D.; Li, Y. Synergistic Modulation of the Separation of Photo-Generated Carriers via Engineering of Dual Atomic Sites for Promoting Photocatalytic Performance. *Adv. Mater.* **2021**, *33*, 2105904. [CrossRef]
61. Xia, P.; Cao, S.; Zhu, B.; Liu, M.; Shi, M.; Yu, J.; Zhang, Y. Designing a 0D/2D S-scheme heterojunction over polymeric carbon nitride for visible-light photocatalytic inactivation of bacteria. *Angew. Chem. Int. Ed.* **2020**, *59*, 5218–5225. [CrossRef] [PubMed]
62. Zhao, T.; Xing, Z.; Xiu, Z.; Li, Z.; Yang, S.; Zhou, W. Oxygen-doped MoS_2 nanospheres/CdS quantum dots/g-C_3N_4 nanosheets super-architectures for prolonged charge lifetime and enhanced visible-light-driven photocatalytic performance. *ACS Appl. Mater. Interfaces* **2019**, *11*, 7104–7111. [CrossRef] [PubMed]
63. Zhang, G.; Xu, Y.; Yan, D.; He, C.; Li, Y.; Ren, X.; Zhang, P.; Mi, H. Construction of k+ ion gradient in crystalline carbon nitride to accelerate exciton dissociation and charge separation for visible light H_2 production. *ACS Catal.* **2021**, *11*, 6995–7005. [CrossRef]
64. She, P.; Qin, J.; Sheng, J.; Qi, Y.; Rui, H.; Zhang, W.; Ge, X.; Lu, G.; Song, X.; Rao, H. Dual-functional photocatalysis for cooperative hydrogen evolution and benzylamine oxidation coupling over sandwiched-like Pd@TiO_2@$ZnIn_2S_4$ nanobox. *Small* **2022**, *18*, 2015114. [CrossRef]
65. Jiménez-Calvo, P.; Caps, V.; Ghazzal, M.N.; Colbeau-Justin, C.; Keller, V. Au/TiO_2 (P25)-gC_3N_4 composites with low gC_3N_4 content enhance TiO_2 sensitization for remarkable H_2 production from water under visible-light irradiation. *Nano Energy* **2020**, *75*, 104888. [CrossRef]
66. Huang, W.X.; Li, Z.P.; Wu, C.; Zhang, H.J.; Sun, J.; Li, Q. Delaminating Ti_3C_2 MXene by blossom of $ZnIn_2S_4$ microflowers for noble-metal-free photocatalytic hydrogen production. *J. Mater. Sci. Technol.* **2022**, *120*, 89–98. [CrossRef]
67. Sun, M.; Zhou, Y.L.; Yu, T.; Wang, J. Synthesis of g-C_3N_4/WO_3-carbon microsphere composites for photocatalytic hydrogen production. *Int. J. Hydrog. Energy* **2022**, *47*, 10261–10276. [CrossRef]
68. Pan, J.Q.; Dong, Z.J.; Wang, B.B.; Jiang, Z.Y.; Zhao, C.; Wang, J.J.; Song, C.S.; Zheng, Y.Y.; Li, C.R. The enhancement of photocatalytic hydrogen production via Ti_3+ self-doping black TiO_2/g-C_3N_4 hollow core-shell nano-heterojunction. *Appl. Catal. B Environ.* **2019**, *242*, 92–99. [CrossRef]

Disclaimer/Publisher's Note: The statements, opinions and data contained in all publications are solely those of the individual author(s) and contributor(s) and not of MDPI and/or the editor(s). MDPI and/or the editor(s) disclaim responsibility for any injury to people or property resulting from any ideas, methods, instructions or products referred to in the content.

Article

Facile Construction of Intramolecular g-CN-PTCDA Donor-Acceptor System for Efficient CO$_2$ Photoreduction

Jiajia Wei [1,†], Xing Chen [2,†], Xitong Ren [2], Shufang Tian [1,*] and Feng Bai [2,*]

[1] Henan International Joint Laboratory of Medicinal Plants Utilization, College of Chemistry and Chemical Engineering, Henan University, Kaifeng 475004, China

[2] Key Laboratory for Special Functional Materials of Ministry of Education, National and Local Joint Engineering Research Center for High-Efficiency Display and Lighting Technology, Collaborative Innovation Center of Nano Functional Materials and Applications, School of Materials Science and Engineering, Henan University, Kaifeng 475004, China

[*] Correspondence: tianshufang@henu.edu.cn (S.T.); baifengsun@126.com (F.B.)

[†] These authors contributed equally to this work.

Abstract: Due to the different electron affinity, the construction of a donor-acceptor (DA) system in the graphitic carbon nitride (g-CN) matrix is an attractive tactic to accelerate photo-induced electron-holes separation, and then further elevate its photocatalytic performance. In this work, perylene tetracarboxylic dianhydride (PTCDA) with magnificent electron affinity and excellent thermal stability was chosen to copolymerize with urea via facile one-pot thermal copolymerization to fabricate g-CN-PTCDA equipped with DA structures. The specific surface area of g-CN-PTCDA would be enlarged and the visible light absorption range would be broadened simultaneously when adopting this copolymerization strategy. A series of characterizations such as electron paramagnetic resonance (EPR), steady and transient photoluminescence spectra (PL), electrochemical impedance spectroscopy (EIS), and photocurrent tests combined with computational simulation confirmed the charge separation and transfer efficiency dramatically improved due to the DA structures construction. When 0.25% wt PTCDA was introduced, the CO evolution rate was nearly 23 times than that of pristine g-CN. The CO evolution rate could reach up to 87.2 µmol g^{-1} h^{-1} when certain Co^{2+} was added as co-catalytic centers. Meanwhile, g-CN-1 mg PTCDA-Co exhibited excellent long-term stability and recyclability as a heterogeneous photocatalyst. This research may shed light on designing more effective DA structures for solar-to-energy conversion by CO$_2$ reduction.

Keywords: donor-acceptor; graphitic carbon nitride; PTCDA; photocatalysis; CO$_2$ reduction

1. Introduction

CO$_2$ plays a vital role in the global carbon cycle. The earth's environment and ecosystem have been greatly challenged in the past decades due to excessive CO$_2$ being emitted into the atmosphere. Numerous attempts have been made to get around this challenge, as the photocatalytic reduction of CO$_2$ to value-added chemicals under visible light is one of the effective strategies used to fulfill the resource utilization of CO$_2$ [1–5]. However, the high stability of CO$_2$ molecules and the varied reduction used products during the reduction process limit its practical application. To this end, a common strategy to elevate the catalytic performances is to regulate and optimize the structures and properties of the photocatalysts.

g-CN, as a metal-free organic polymeric semiconductor, possesses a suitable bandgap and responses to visible light. It is gifted with excellent thermal and chemical stability and has been extensively explored in photocatalysis fields [6–10]. Nevertheless, the rapid recombination of photo-induced charge carriers, the limited visible light harvesting ability and specific surface area hinder its catalytic performance on CO$_2$ reduction. To promote

the exciton dissociation efficiency, intramolecular DA structures based on g-CN via bottom-up copolymerization have been established [11–16] over the past few years and have exhibited optimized photocatalytic activity. The different electron affinity of the donor and acceptor units would induce electrons to migrate from donor to acceptor parts and further promote charge carriers to disjoin into free electrons and holes. While there are still some aspects that should be noted, the first is that the selected electron-withdraw molecules as co-monomers should maintain extremely high thermal stability. The second is that the picked molecules ought to bear the functional groups that can be covalently imbedded into the skeleton of g-CN during the thermal polymerization. The last is that the properties of g-CN, such as specific surface area or light-harvesting capacity, may also be modulated during copolymerization process.

Based on the above restraints, PTCDA is the optimal choice to fulfill DA structures construction due to its magnificent electron affinity, exceptional thermal stability, and bearing dianhydride functional groups that can react with the amino groups of urea to form diimide covalent bonds. Furthermore, the resultant fragment 3, 4, 9, 10-perylenetetracarboxylic diimide (PTI) as an n-type organic semiconductor owns extraordinary light absorption in the visible region and is widely applied in photocatalytic reactions owing to its more positive valence band [17–21]. A series of Z-scheme PTCDA-C_3N_4 heterostructure photocatalysts have been fabricated previously via imidization reaction and have exhibited remarkable photocatalytic performance in various photocatalytic applications [22–26]. The fact is that the amount of the amine group suspended on the edge of g-CN nanosheets is limited and PTCDA is prone to self-aggregate due to its planar π-conjugated macrocycle structure. Furthermore, the property of a g-CN-like specific surface area can hardly be tuned by traditional post-synthetic modification or other noncovalent composites methods. Nevertheless, it could possibly be accomplished by adopting the copolymerization method because the introduction of a co-monomer may alter the course of the polymerization process so as to regulate the properties of g-CN.

Herein, PTCDA is selected as a co-monomer to bottom-up copolymerize with urea to construct intramolecular DA structures based on g-CN, in which PTCDA served as the acceptor units to capture and store electrons as catalytic active centers due to its distinguished electron affinity, while g-CN composed of heptazine fragments acted as donor units when irradiated under visible light. The obtained g-CN-PTCDA, including DA segments, can not only speed up the photo-induced exciton dissociation but also shorten the migration distances of the electrons or holes to its surface, which results from the expanded specific surface area and the diminished thickness of the g-CN-PTCDA nanosheets. The visible light absorption region was also broadened at the same time due to the photo-sensitiveness of PTCDA. The g-CN-PTCDA with DA structures exhibited a dramatically advanced photocatalytic CO_2 reduction with a maximum CO evolution rate of 5.25 μmol g^{-1} h^{-1}, which was nearly 23-folds that of the unmodified g-CN. This work offers a new concept for the fabrication of all organic photocatalysts for CO_2 reduction, in which PTCDA can function as a catalytic center due to its ability to capture and accumulate electrons.

2. Results and Discussion

All photocatalyst g-CN-x mg PTCDA possess DA structures that can be successfully fabricated via facile one-pot thermal co-polymerization [27,28] by employing a different mass ratio of urea and PTCDA as co-monomers, as shown in Scheme 1.

Firstly, the thermal stability of PTCDA was checked by TG and FT-IR analysis (Figure S1a,b, Supplementary Materials). PTCDA maintained 96.7% of its original mass when the temperature increased to 550 °C. The FT-IR spectrum showed that the typical peaks of PTCDA remained unchanged before and after calcination under the same copolymerization condition, indicating the exceptional high thermal stability of PTCDA. The color of g-CN-x mg PTCDA turned light gray from the yellow powder of the pristine g-CN, it then gradually deepened to dark green along with the increasing amount of PTCDA, and finally presented as a dark purple when the amount of PTCDA increased to 1 g (Figure S2, Supplementary Materials).

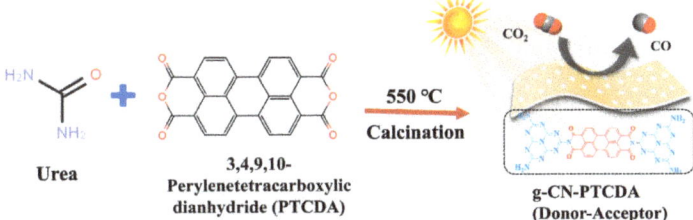

Scheme 1. Schematic illustration of the preparation of the g-CN-PTCDA composite photocatalyst with DA structure using the one pot thermal copolymerization method.

The morphologies of g-CN and the obtained g-CN-x mg PTCDA were investigated by SEM and TEM images. As can be seen, g-CN-1 mg PTCDA retained the fluffy stacking nanosheets (Figure 1b) as g-CN (Figure 1a,d), except that a more abundant mesoporous structure formed, which is in accordance with the semitransparent porous nanosheets with rough edges obtained by TEM characterization (Figure 1e). Some regular nano-rods formed and piled up on the surface of the g-CN nanosheets (Figure 1c,f) when the load of PTCDA increased to 1 g, suggesting that the self-assembly of the excessive unreacted of PTCDA formed via π–π stacking. TEM mapping was also employed to analyze the elemental distribution of the sample (Figure 1g). Carbon, nitrogen, and oxygen (originated from the C=O of PTCDA) elements are evenly distributed in the selected region, which demonstrates the covalent connection of PTCDA and g-CN and that the DA structures were successfully constructed.

Figure 1. SEM images of the g-CN (**a**), g-CN-1 mg PTCDA (**b**), and g-CN-1g PTCDA (**c**). TEM images of the g-CN (**d**), g-CN-1 mg PTCDA (**e**), and g-CN-1g PTCDA (**f**) and the corresponding elemental mappings of the enlarged area of g-CN-1mg PTCDA (**g**).

The specific surface area and mesopores distribution were studied by N_2 adsorption-desorption measurements. As shown in Figure 2a, g-CN-1mg-PTCDA exhibited an obviously enhanced adsorption capacity and the specific surface area increased to 97 $m^2\ g^{-1}$ compared with 36 $m^2\ g^{-1}$ of g-CN. The isotherm for g-CN-1mg PTCDA presented an obvious H1 type hysteresis loop at relative pressure $p/p_0 > 0.8$, indicating the slit-type holes formed by nanosheets stacking [29,30]. The pore volume of g-CN-1 mg PTCDA was also apparently promoted and the pore size distribution shows that the pore size was concentrated at 2.6 and 29 nm (Figure 2b). This enlarged specific surface area can hardly be fulfilled through traditional post-synthetic modification.

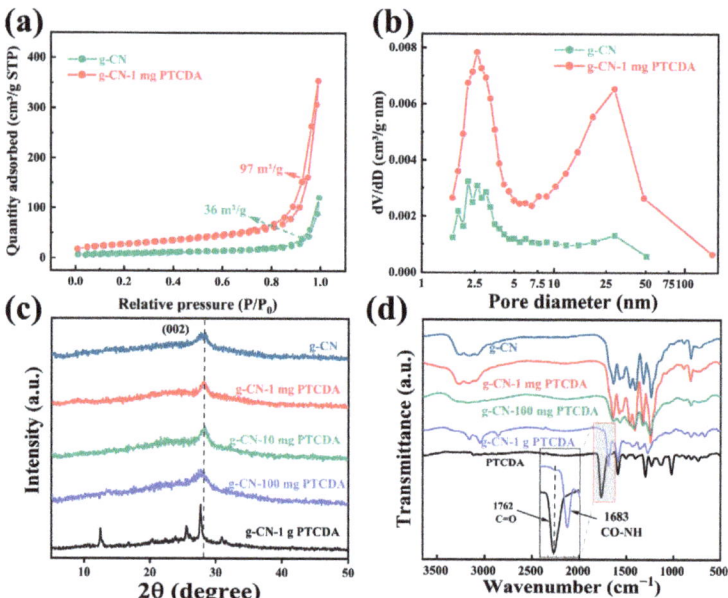

Figure 2. N$_2$ adsorption-desorption isotherms (**a**) and the corresponding pore size distribution curves (**b**) of g-CN and g-CN-1 mg PTCDA. XRD patterns (**c**), and FT-IR spectra (**d**) of the samples of g-CN and g-CN-x mg PTCDA (x = 1, 10, 100, 1000).

XRD was used to study the crystal structure of the samples. As shown in Figure 2c, g-CN-x mg PTCDA (x = 1, 10, 100) showed two distinct diffraction peaks at around 13° and 27° which correspond to the (100) and (002) crystal planes as g-CN [31]. The (100) crystal plane is ascribed to the repeating motif of tri-s-triazine in-plane and the (002) plane is from the stacking of aromatic systems of the interlayer, which indicates that the crystal structures are well retained after the introduction of PTCDA. When the amount of PTCDA is raised to 1 g, the XRD pattern was clearly different from g-CN but similar to that of the PTCDA powder, and the diffraction peaks belong to the π–π stacking of excessive unreacted PTCDA.

FT-IR spectra were recorded to index the chemical structure and the covalent interactions between PTCDA and g-CN (Figure 2d). g-CN-x mg PTCDA (x = 1, 100) exhibited approximate characteristic absorption bands as g-CN, the typical peak at 810, 1200–1600, and 3000–3500 cm^{-1} are attributed to the bending vibration of the triazine ring, the stretching vibrations of the aromatic heptazine heterocycles, and the NH$_2$ groups located on the edges of the g-CN nanosheets, respectively. When the amount of PTCDA was raised up to 1 g, it can be clearly seen that the typical peak at 1742 cm^{-1} arising from the C=O in the anhydride groups of PTCDA disappeared, and the peak at 1683 cm^{-1} is newly generated, which can be ascribed to the stretching vibrations of the C=O in the diimide group. The result provides solid evidence that the DA structures successfully constructed between PTCDA and g-CN via covalent bonds.

The surface chemical states of C, N, O and the interfacial interaction were studied by XPS. Figure 3a is the XPS survey spectra of g-CN and g-CN-PTCDA, from which we can see that the peaks of O 1s emerged with the introduction of PTCDA and the intensity increased when the PTCDA content increased. A high resolution of the C 1s spectra is presented in Figure 3b. For pure g-CN, the binding energy of C 1s is dominantly situated at 287.8 eV, which corresponds to the sp^2-hybridized C from N-C=N. The C 1s spectra of g-CN-x mg PTCDA (x = 1, 10) are deconvoluted into three peaks, except for the sp^2-hybridized N-C=N, in which the peaks at 286.07 and 284.81 eV are attributed to sp^2-hybridized C=O and C=C coming from PTCDA and the peak areas increase along with the added PTCDA weight percentage [22,32]. Figure 3c presents the N 1s binding energies of g-CN. The divided three peaks at 398.59, 400.31, and 401.36 eV of g-CN belong to the sp^2-hybridized N involved in the triazine rings, the bridging sp^3-hybridized N in the melem motif center, and the amino group -NH_x, respectively [33–35]. The peaks of sp^2-hybridized N (C-N=C) in g-CN-x mg PTCDA (x = 1, 10) both shift about 0.27 eV to a higher binding energy (398.86 eV) in contrast with g-CN, illustrating when the strong electron-withdrawing group PTCDA is introduced, the electron cloud density in heptazine ring decreased, demonstrating the efficient DA structures established in the matrix of g-CN-PTCDA. In addition, the mass ratio of C/N obtained from XPS are showed in Table S1, and the value of the C/N ratio increased along with the content of PTCDA raised. The binding energy of O 1s that originated from O=C (Figure 3d) was 531.9 eV and the intensity of the peaks also increased with the improved content of PTCDA.

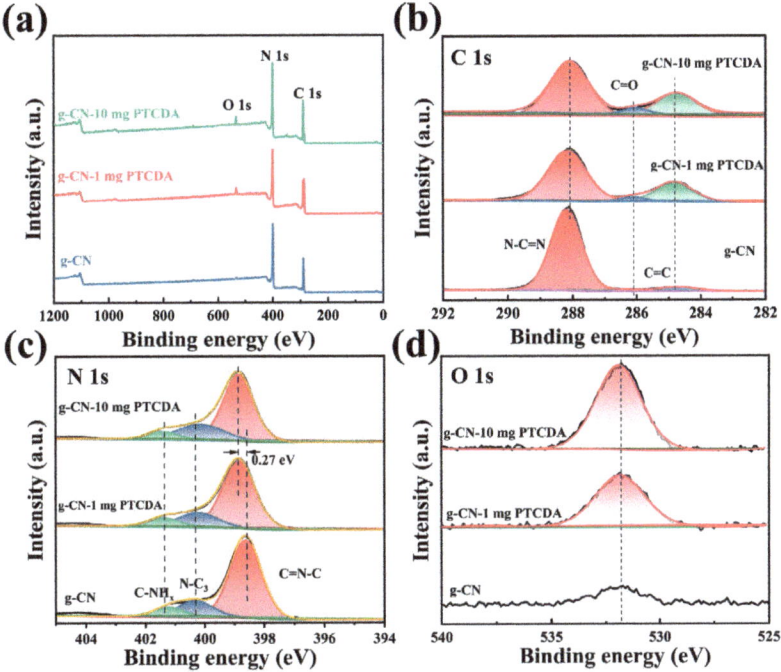

Figure 3. XPS-survey (**a**), high-resolution of C 1s (**b**), N 1s spectra (**c**), and O 1s spectra (**d**) of the samples of g-CN, g-CN-1 mg PTCDA, and g-CN-10 mg-PTCDA.

UV-vis DRS spectra were conducted to study the optical absorption properties of the g-CN and g-CN-x mg PTCDA. As shown in Figure 4a, the absorption band edges of g-CN-x mg PTCDA (x = 1, 10, 100, 1000) are gradually red-shifted, and the visible-light response range steadily expanded to the full spectra region accompanying the amount of PTCDA raised, fully indicating that PTCDA covalently connected with g-CN via imide bonds. The corresponding band gaps of g-CN-x mg PTCDA (x = 1, 10) are respectively calculated as 2.51, 2.44 eV based on the Tauc plots method (Figure 4b), and they are narrower than the 2.75 eV of g-CN, which can be explained by the π electrons being delocalized along the expanded conjugation system between PTCDA and g-CN.P

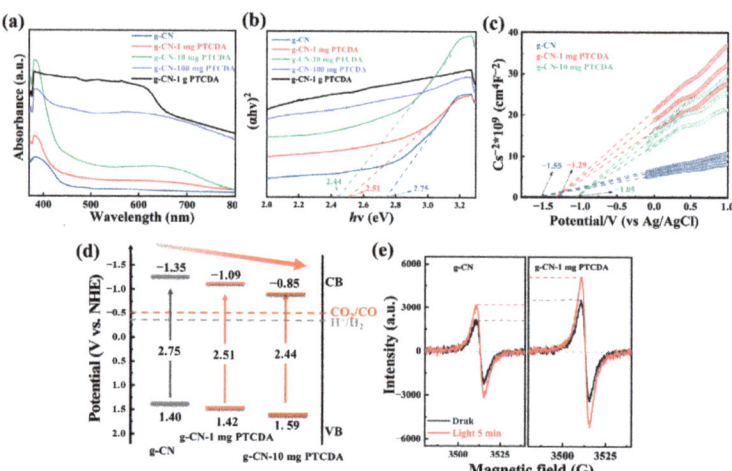

Figure 4. Optical absorption and band positions of g-CN, g-CN-1 mg PTCDA, and g-CN-10 mg PTCDA samples. UV-vis DRS spectra (**a**) and the corresponding Tauc plots (**b**), Mott–Schottky plots (**c**), energy diagrams of the band structure (**d**). EPR spectra of the g-CN and g-CN-1 mg PTCDA at room temperature (**e**).

The conduction band energy was obtained by testing the Mott–Schottky curve, in which the straight-line part of the curve is extrapolated to the abscissa axis, and the intersection point is the flat-band potential. Generally, for n-type semiconductors, the position of the conduction band bottom is consistent with the flat band potential, which can be considered as the position of the conduction band bottom. From Figure 4c, we can learn that g-CN-PTCDA is a typical n-type semiconductor due to the positive slope. The flat-band potential of g-CN, g-CN-1 mg PTCDA, and g-CN-10 mg PTCDA are located at −1.55, −1.29, and −1.05 eV (vs. Ag/AgCl), which could be converted to −1.35, −1.09, and −0.85 V (vs. NHE), respectively and are more negative than that of the required potential of reducing CO_2 to CO (−0.53 V). Regarding the combination with the band gap values determined in Figure 4b, the CB potentials of g-CN, g-CN-1 mg PTCDA, and g-CN-10 mg PTCDA are calculated to be 1.40, 1.42, and 1.59 V, respectively, as shown in Figure 4d.

The electronic structure of the g-CN and g-CN-PTCDA were explored by EPR measurements. As can be seen in Figure 4d, g-CN-1 mg PTCDA showed dramatically improved signals of the Lorentzian line with a g value of 2.0034 than that of g-CN, which originated from the unpaired electrons in the conduction bands of g-CN-1 mg PTCDA [36]. It means that a high concentration of unpaired electrons generated due to the DA structures. Besides, the charge separation efficiency of g-CN-1 mg PTCDA was more notable than that of g-CN under illumination conditions, which is conductive to promote the photocatalytic CO_2 reduction performance.

3. Photocatalytic Performance

The photocatalytic performances of the prepared g-CN-PTCDA were assessed by CO_2 reduction. Considering that the gas CO_2 has higher solubility in the organic solvent, and in order to suppress the proton reduction as a competition reaction, CH_3CN was chosen as the reaction solvent. TEOA was selected as the sacrificial agent to capture the unreacted photoinduced holes because it was extensively used and displayed the highest photocatalytic CO_2 reduction activity compared with other sacrificial agents [37–39], and the volume ratio of CH_3CN/TEOA was 4:1. The photocatalytic reaction was carried out under 5 W LED lamp (λ > 420 nm) irradiation and the reaction temperature was set at 6 °C with circulating condensed water. The control experiments showed that there was negligible CO or CH_4

detected as the reduction products without one of the factors including the irradiation, the photocatalyst, TEOA, or using N_2 instead of CO_2.

As presented in Figure 5a, the experiment results showed that CO was the main reduction product along with some amount of H_2 generated as the byproduct, and the H_2 was mainly from the partial degradation of TEOA. The CO evolution rate was 0.23 µmol g^{-1} h^{-1} when using unmodified g-CN as the photocatalyst, along with a H_2 generation rate of 0.43 µmol g^{-1} h^{-1} and the selectivity was 35% over proton reduction. When a certain amount of PTCDA was introduced into the skeleton of g-CN and the intramolecular DA system was constructed, the photocatalytic activity remarkably improved. The CO evolution rate can reach up to 5.25 µmol g^{-1} h^{-1} when the loading content of PTCDA was 1 mg, which is nearly 23-folds that of pristine g-CN. The H_2 generation rate was 1.14 µmol g^{-1} h^{-1} and the selectivity improved to 82%. The CO evolution rate began to descend with the continuous increment of PTCDA, as introducing too much PTCDA would destroy the conjugate structural integrity of g-CN and further weaken its light-harvesting ability.

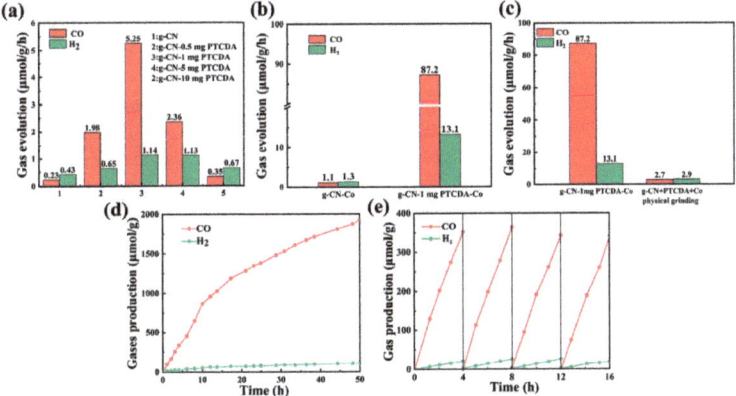

Figure 5. The effects of different PTCDA contents on the evolution of CO and H_2 (**a**), g-CN-Co and g-CN-1 mg PTCDA-Co gases production rate (**b**), comparison of CO and H_2 evolution rates of the covalent bond connection and physical mixing of g-CN and PTCDA (**c**), CO_2 reduction stability of g-CN-1 mg PTCDA-Co under 50 h light condition (**d**), the yields of CO and H_2 under the catalytic action of g-CN-1 mg PTCDA-Co over four continuous cycles and each cycle lasted 4 h (**e**).

Afterward, the photocatalyst g-CN-1 mg PTCDA-Co was fabricated by hydrothermal synthesis, where Co^{2+} was introduced, acting as the co-catalytic center, and the CO evolution rate dramatically increased to 87.2 µmol g^{-1} h^{-1}. P The H_2 generation rate was 13.1 µmol g^{-1} h^{-1} and the selectivity was 87%, which was more than 79-folds that of g-CN-Co (Figure 5b). This indicates that the electrons separated and accumulated on the PTCDA unit can quickly transfer to the catalytic active sites Co^{2+}, thus the catalytic performance dramatically enhanced. From Table S2, we can learn that the photocatalysts g-CN-1 mg PTCDA and g-CN-1 mg PTCDA-Co performed satisfactory photocatalytic CO_2 reduction performance.

The control experiment was also conducted to investigate the cause for the elevated catalytic activity (Figure 5c). The PTCDA compounded with g-CN through physical grinding and the CO evolution rate exhibited certain enhancement (2.7 µmol g^{-1} h^{-1}) with a selectivity of 48% over H_2, implying that the covalent connection between PTCDA and g-CN is indispensable.

The stability and reusability of the g-CN-1 mg PTCDA-Co were further evaluated. As shown in Figure 5d, the CO evolution rate maintained a near linear growth of up to 50 h with the average rate of 38.4 µmol g^{-1} h^{-1}, displaying the long-term robustness of the photocatalyst. As for the reusability, in the 16 h sequent reaction, that is, 4 consecutive cycles where each cycle lasts 4 h, no noticeable declination was observed (Figure 5e), signifying the excellent recyclability of g-CN-PTCDA-Co as a heterogeneous photocatalyst.

The properties of the recovered g-CN-1 mg PTCDA-Co after the long-term catalytic process were also studied by FT-IR, XRD, and TEM (Figure S3, Supplementary Materials). The characterization showed that the chemical and crystal structure of g-CN-PTCDA-Co remained unchanged and the morphology also remained intact, suggesting its fine and lasting photo stability.

4. Structure–Activity Relationship and Mechanism Discussion

A combination of the photocatalytic CO_2 reduction performance and the characterizations of g-CN-PTCDA, the enhanced photocatalytic activity, and the selectivity, can be ascribed to the following factors.

The construction of the intramolecular DA system plays a functional role [40–43]. As we know, the separation, migration, and transmission efficiency of photo-generated electron-holes are the key elements that affect the photocatalytic performance. The different electrons binding ability between PTDCA and g-CN boost the electrons transferred from g-CN to PTCDA and the separation efficiency of the charge carriers is greatly accelerated, which can be verified by the fluorescence and electrochemical tests.

Steady-state PL analysis is generally recognized as the technique used to characterize the recombination of photo-induced electron-holes, that is, the higher the peak intensity, the more recombinations of the photogenerated electrons and holes occur. As can be seen in Figure 6a, the PL intensity of g-CN-1 mg PTCDA decreased sharply in comparison with that of g-CN, which exhibited a strong PL emission peak excited at 365 nm at room temperature, fully illustrating that the recombination of the photogenerated electron-holes has been greatly inhibited. In addition, time resolved PL spectra were also performed to evaluate the dynamic photoinduced electron behaviors. In Figure 6b, the g-CN-1 mg PTCDA exhibits a shorter average fluorescence lifetime (4.29 ns) than that of g-CN (4.97 ns), indicating a more efficient non-radiative decay pathway between PTCDA and g-CN. The electron transfer rate (k_{ET}) calculated using the equation [44,45] was $3.2 \times 10^7 \text{ s}^{-1}$ (Table S3) between g-CN and PTCDA, illustrating the fast spatial charge separation efficiency through the intramolecular DA system.

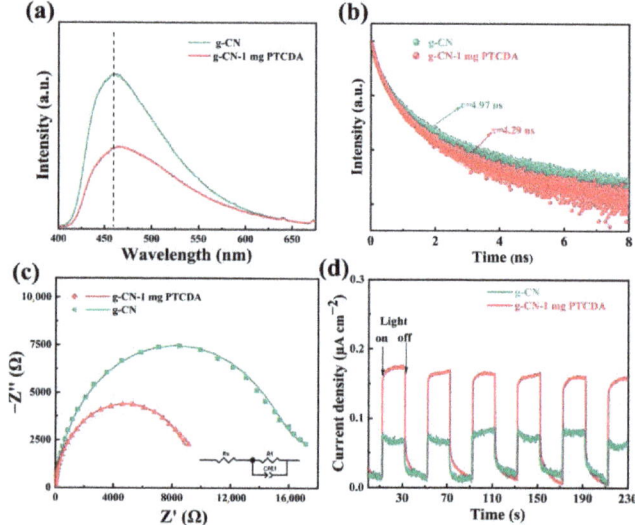

Figure 6. Steady-state PL spectra (**a**), time-resolution PL decay spectroscopy (**b**), Nyquist plots with the inset of the simulated circuit (**c**), photocurrent response (**d**) of the samples of g-CN and g-CN-1 mg PTCDA.

As for the EIS, a smaller arc radius commonly means lower electron-transfer resistance. In Figure 6c, the arc radius of g-CN-1 mg PTCDA is much smaller than that of g-CN and the fitting results are present in Table S4, implying the interfacial electrons' fast mobility of g-CN-1 mg PTCDA due to the lower resistance.

In addition, g-CN-1 mg PTCDA exhibited dramatically enhanced and steady photocurrent responses compared to g-CN, without significant decay after six repetitive cycles (Figure 6d), demonstrating that the DA system constructed between PTCDA and g-CN was more conductive to charge separation and transfer than g-CN, which is in accordance with the results of the PL and EIS analysis.

Furthermore, the electrostatic potential (ESP) surface distribution of the g-CN-PTCDA was built with a GaussView program. The method employed to optimize the structure model is B3LYP/6-31g. The computed result affords a feasible electron transfer route during this DA system. As can be seen in Figure 7, in this ESP map, the red color stands for the electrons affluent region while the blue color means the electron deficiency zone. As for the sample g-CN-PTCDA, the electrons migrate from g-CN to the oxygen atom of PTCDA, which can not only accelerate the separation of the electron-hole pairs, but PTCDA can also be regarded as an electron reservoir to drive the CO_2 reduction.

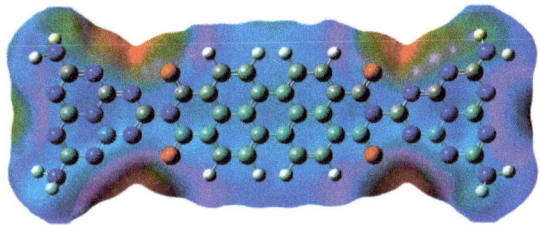

Figure 7. Optimized ESP surface distribution of the g-CN-PTCDA model.

Based on above analysis, a presumable mechanism using g-CN-1 mg PTCDA for CO_2 photoreduction is proposed (Figure 8). The photosensitizer g-CN absorbs visible light and the electrons excited from the valance band of g-CN to the LUMO level of the PTCDA, which is caused by the different electron affinities of PTCDA and g-CN. Then, electrons are accumulated on the PTCDA units while the holes are left on the g-CN parts. The intramolecular DA structures promote electrons migration and inhibit the electron-hole pairs recombination. Simultaneously, the enlarged specific surface area of g-CN-1 mg PTCDA facilitates more adsorption of CO_2 molecules. The process of the proton-coupled two electrons reduction is used to reduce CO_2 to CO (CO_2 + 2H$^+$ + 2e$^-$ →CO + H_2O, −0.53 V). Meanwhile, TEOA is added in the reaction system as a sacrificial electron donor to consume the unreacted holes to fasten the separation of electron-hole pairs and promote the CO_2 reduction performance.

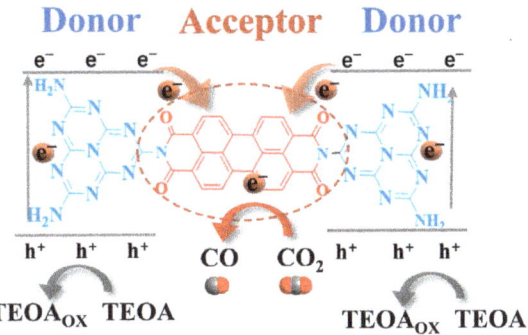

Figure 8. The mechanism of g-CN-1 mg PTCDA as a photocatalyst for CO_2 reduction is proposed.

5. Experimental Section

5.1. Reagents and Materials

Urea was obtained from J&K Scientific (Beijing, China). Acetonitrile (CH_3CN), triethanolamine (TEOA), and PTCDA were purchased from Aladdin (Shanghai, China). $CoCl_2 \cdot 6H_2O$ was provided by Maikelin (Fairfield, CT, USA). Na_2SO_4 was obtained from Kermel (Tianjin, China). All the chemical reagents were used as received without further treatment. Ultra-pure water used in the experiment was prepared by the instrument THM-50131954 (Thermo Scientific, Waltham, MA, USA).

5.2. Preparation of Catalysts

5.2.1. Preparation of g-CN

In detail, 20 g urea was placed in a crucible, covered with tin foil and heated in an ambient atmosphere to 550 °C in a muffle furnace at a heating rate of 3 °C min^{-1} and then kept at 550 °C for 2 h. The pale-yellow loose powder sample was obtained by natural cooling to room temperature.

5.2.2. Preparation of g-CN-x mg PTCDA

The samples with DA structures were synthesized by a facile one-pot thermal copolymerization method. In detail, 20 g urea and x mg PTCDA (x = 0.5, 1, 10, 100, 1000) were weighted and then fully grinded and mixed. The mixture was transferred to a crucible with a tin foil cover and heated to 550 °C for 2 h at a heating rate of 3 °C min^{-1} in a muffle furnace. When the reaction cooled to room temperature, the obtained samples were named g-CN-x mg PTCDA, where x stands for the initial mass of added PTCDA.

5.2.3. Preparation of g-CN-Co and g-CN-1 mg PTCDA-Co

The samples were prepared by a typical hydrothermal method. At first, g-CN (50 mg) or g-CN-1 mg PTCDA (50 mg) were uniformly dispersed in 10 mL deionized water and then 2.4 mg $CoCl_2 \cdot 6H_2O$ was added and stirred for 30 min to ensure sufficient dissolution. The mixture was then transferred to the autoclave for treatment at 120 °C for 12 h. When cooled to room temperature, the precipitate was centrifuged and washed successively with ultrapure water twice and EtOH once, and then dried at 80 °C for 12 h. The resultant samples were denoted as g-CN-Co or g-CN-1 mg PTCDA-Co.

5.3. Physicochemical Characterization

The morphologies of the samples were investigated by scanning electron microscopy (SEM, Nova NanoSEM 450, FEI, Hillsboro, OR, USA) and transmission electron microscope (TEM, JEM-2100F, Kitakyushu, Japan). Powder X-ray diffraction (XRD) was carried out on an X-ray diffractometer (Rigaku Dmax-2500, Tokyo, Japan) with a Cu K source. FT-IR spectra were recorded by VERTEX 70 spectrometer (Bruker, Billerica, MA, USA) in the range of 4000 to 400 cm^{-1}. The samples were prepared by mixing the sample with FT-IR-grade KBr and the KBr was used as the reference. X-ray photoelectron spectroscopy (XPS) measurements were conducted on the XPS instrument (ESCALAB 250XI, Thermo, Waltham, MA, USA), the excitation source for Al target Kα rays (1486.6 eV). Inductively coupled plasma atomic emission spectrometer (ICP) was performed with an Agilent 7800 ICP-MS. The UV-vis absorption was measured with UV-vis diffuse reflectance (UV-vis DRS, PerkinElmer Lambda 950, Waltham, MA, USA) using $BaSO_4$ as the reflectance standard reference. Steady and transient PL spectra were carried out using a fluorescence spectrometer (FLS 980, Edinburgh, UK) (exciting samples by 365 nm photons).

5.4. Photoelectrochemical Measurements

Photoelectrochemical measurements were monitored by a CHI 760E electrochemical workstation with a conventional three-electrode cell. The measurement processes were performed in a 0.2 M Na_2SO_4 aqueous solution. Ag/AgCl electrode and Pt plate were used as the reference electrode and counter-electrode. The working electrodes were fabricated by using an as-prepared catalyst coated on indium tin oxide (ITO) glass. The photocurrent as a function of time was recorded as the light was switched on and off every 20 s. The electrochemical impedance was also recorded in the range of 0.1 MHz–0.1 Hz. Mott-Schottky plots were recorded at frequencies of 1000, 1100, and 1200 Hz. N_2 was blown into the electrolyte for 20 min before any electrochemical measurement was made.

5.5. Photocatalytic CO_2 Reduction

In a typical photocatalytic reaction, 5 mg photocatalyst was dispersed in 4 mL CH_3CN in 50 mL quartz test bottle, along with 1 mL TEOA added as the electron sacrificial agent to trap the unreacted photo-induced holes. Before photocatalytic testing, the whole reaction system was evacuated and then bubbled with CO_2 (99%, Sigma-Aldrich, St. Louis, MO, USA) three times. The light source was a 5 W LED lamp (PCX-50C, Perfectlight, Beijing, China) with λ > 420 nm cutoff filter. The reaction temperature was set at 6 °C with circulating condensed water.

The gas products, including CO and H_2, were measured by gas chromatography (GC9790 II, Zhejiang Fuli, Wenling, China) equipped with a thermal conductivity detector (TCD) and a flame ionization detector (FID). In the cycle experiment, 5 mg g-CN-1 mg PTCDA-Co was used for photocatalysis, and the catalyst was recycled, washed, and dried in turn when each cycle ended.

6. Conclusions

The intramolecular DA system, g-CN-PTCDA, has been successfully constructed through copolymerization, employing PTCDA and urea as the starting material to elevate the photocatalytic performance towards CO_2 reduction. This strategy can not only enlarge the specific surface area of pristine g-CN and extend the visible light absorption range, but the constructed D-A structure can also boost the electrons transmission and suppress the recombination of the photo-induced electron-holes. The CO_2 reduction activity was 23-folds that of pristine g-CN and the CO evolution rate can reach up to 87.2 µmol g^{-1} h^{-1} when Co^{2+} is added as co-catalytic centers. The structure–activity relationship and the possible mechanism are also explored.

Supplementary Materials: The following supporting information can be downloaded at: https://www.mdpi.com/article/10.3390/catal13030600/s1, Figure S1. TG (a), and the FT-IR spectra (b) of PTCDA before and after calcination under copolymerization conditions, Figure S2. The photographic images of g-CN (a), g-CN-0.5 mg PTCDA (b), g-CN-1 mg PTCDA (c), g-CN-10 mg PTCDA (d), g-CN-100 mg PTCDA (e), g-CN-1 g PTCDA (f), and PTCDA (g), Figure S3. TEM (a), XRD pattern (b), and FT-IR spectra (c) of g-CN-1 mg PTCDA-Co sample before and after four cycles of catalytic reaction, Table S1. Surface elemental composition of g-CN, g-CN-1 mg PTCDA and g-CN-10 mg PTCDA based on XPS analysis, Table S2. The comparison with other DA structure photocatalysts towards CO_2 reduction under visible light, Table S3. The fitting parameters and average decay time of the time-resolved PL, rate constants of charge separation (kET) and quantum yield (ηET) for g-CN and g-CN-1 mg PTCDA, Table S4. The fitting values for the electrochemical impedance spectroscopy simulation fit for g-CN and g-CN-1 mg PTCDA. References [12,15,40,46–49] are cited in the supplementary materials.

Author Contributions: S.T. and F.B. conceived the idea. X.C. and J.W. performed the experiments. X.R. provided assistance in data analysis and characterization. All authors have read and agreed to the published version of the manuscript.

Funding: This work was financially supported by the National Natural Science Foundation of China (NSFC) U21A2085; Zhongyuan high level talents special support plan 204200510009; Scientific and Technological Innovation Team in University of Henan Province 20IRTSTHN001.

Conflicts of Interest: The authors declare no conflict of interest.

References

1. Xie, Y.; Ou, P.; Wang, X.; Xu, Z.; Li, Y.C.; Wang, Z.; Huang, J.E.; Wicks, J.; McCallum, C.; Wang, N.; et al. High carbon utilization in CO_2 reduction to multi-carbon products in acidic media. *Nat. Catal.* **2022**, *5*, 564–570. [CrossRef]
2. Xie, S.; Li, Y.; Sheng, B.; Zhang, W.; Wang, W.; Chen, C.; Li, J.; Sheng, H.; Zhao, J. Self-reconstruction of paddle-wheel copper-node to facilitate the photocatalytic CO_2 reduction to ethane. *Appl. Catal. B* **2022**, *310*, 121320. [CrossRef]
3. Yang, Z.; Yang, J.; Yang, K.; Zhu, X.; Zhong, K.; Zhang, M.; Ji, H.; He, M.; Li, H.; Xu, H. Synergistic Effect in Plasmonic CuAu Alloys as Co-Catalyst on $SnIn_4S_8$ for Boosted Solar-Driven CO_2 Reduction. *Catalysts* **2022**, *12*, 1588. [CrossRef]
4. Li, J.; Zhang, M.; Guan, Z.; Li, Q.; He, C.; Yang, J. Synergistic effect of surface and bulk single-electron-trapped oxygen vacancy of TiO_2 in the photocatalytic reduction of CO_2. *Appl. Catal. B* **2017**, *206*, 300–307. [CrossRef]
5. Cui, Y.; Ge, P.; Chen, M.; Xu, L. Research Progress in Semiconductor Materials with Application in the Photocatalytic Reduction of CO_2. *Catalysts* **2022**, *12*, 372. [CrossRef]
6. Wang, X.; Blechert, S.; Antonietti, M. Polymeric Graphitic Carbon Nitride for Heterogeneous Photocatalysis. *ACS Catal.* **2012**, *2*, 1596–1606. [CrossRef]
7. Zheng, J.Y.; Pawar, A.U.; Kang, Y.S. Preparation of C_3N_4 Thin Films for Photo-/Electrocatalytic CO_2 Reduction to Produce Liquid Hydrocarbons. *Catalysts* **2022**, *12*, 1399. [CrossRef]
8. Wang, K.; Li, Q.; Liu, B.; Cheng, B.; Ho, W.; Yu, J. Sulfur-doped g-C_3N_4 with enhanced photocatalytic CO_2-reduction performance. *Appl. Catal. B* **2015**, *176–177*, 44–52. [CrossRef]
9. Ong, W.-J.; Tan, L.-L.; Ng, Y.H.; Yong, S.-T.; Chai, S.-P. Graphitic Carbon Nitride (g-C_3N_4)-Based Photocatalysts for Artificial Photosynthesis and Environmental Remediation: Are We a Step Closer to Achieving Sustainability? *Chem. Rev.* **2016**, *116*, 7159–7329. [CrossRef]
10. Ding, J.; Tang, Q.; Fu, Y.; Zhang, Y.; Hu, J.; Li, T.; Zhong, Q.; Fan, M.; Kung, H.H. Core-Shell Covalently Linked Graphitic Carbon Nitride-Melamine-Resorcinol-Formaldehyde Microsphere Polymers for Efficient Photocatalytic CO_2 Reduction to Methanol. *J. Am. Chem. Soc.* **2022**, *144*, 9576–9585. [CrossRef] [PubMed]
11. Wu, X.; Li, D.; Huang, Y.; Chen, B.; Luo, B.; Shen, H.; Wang, M.; Yu, T.; Shi, W. Construction of binary donor-acceptor conjugated copolymer in g-C_3N_4 for enhanced visible light-induced hydrogen evolution. *Appl. Surf. Sci.* **2021**, *565*, 150012. [CrossRef]
12. Zhang, X.; Song, X.; Yan, Y.; Huo, P. Construction of carbon nitride based intramolecular D-A system for effective photocatalytic reduction of CO_2. *Catal. Lett.* **2022**, *152*, 559–569. [CrossRef]
13. Yu, F.; Wang, Z.; Zhang, S.; Wu, W.; Ye, H.; Ding, C.; Gong, X.; Hua, J. Construction of polymeric carbon nitride and dibenzothiophene dioxide-based intramolecular donor-acceptor conjugated copolymers for photocatalytic H_2 evolution. *Nanoscale Adv.* **2021**, *3*, 1699–1707. [CrossRef] [PubMed]
14. Zhang, J.-W.; Pan, L.; Zhang, X.; Shi, C.; Zou, J.-J. Donor-acceptor carbon nitride with electron-withdrawing chlorine group to promote exciton dissociation. *Chin. J. Catal.* **2021**, *42*, 1168–1175. [CrossRef]
15. Song, X.; Zhang, X.; Wang, M.; Li, X.; Zhu, Z.; Huo, P.; Yan, Y. Fabricating intramolecular donor-acceptor system via covalent bonding of carbazole to carbon nitride for excellent photocatalytic performance towards CO_2 conversion. *J. Colloid Interface Sci.* **2021**, *594*, 550–560. [CrossRef]
16. Zhu, C.; Wei, T.; Wei, Y.; Wang, L.; Lu, M.; Yuan, Y.; Yin, L.; Huang, L. Unravelling intramolecular charge transfer in donor-acceptor structured g-C_3N_4 for superior photocatalytic hydrogen evolution. *J. Mater. Chem. A* **2021**, *9*, 1207–1212. [CrossRef]
17. Song, J.; Chen, Y.; Sun, D.; Li, X. Perylenetetracarboxylic diimide modified $Zn_{0.7}Cd_{0.3}S$ hybrid photocatalyst for efficient hydrogen production from water under visible light irradiation. *Inorg. Chem. Commun.* **2018**, *92*, 27–34. [CrossRef]
18. Sun, T.; Song, J.; Jia, J.; Li, X.; Sun, X. Real roles of perylenetetracarboxylic diimide for enhancing photocatalytic H_2-production. *Nano Energy* **2016**, *26*, 83–89. [CrossRef]
19. Wei, W.; Zhu, Y. TiO_2@Perylene Diimide Full-Spectrum Photocatalysts via Semi-Core-Shell Structure. *Small* **2019**, *15*, 1903933. [CrossRef]
20. Yang, X.; Li, Y.; Chen, G.Y.; Liu, H.R.; Yuan, L.L.; Yang, L.; Liu, D. $ZnSnO_3$ Quantum Dot/Perylene Diimide Supramolecular Nanorod Heterojunction Photocatalyst for Efficient Phenol Degradation. *ACS Appl. Nano Mater.* **2022**, *5*, 9829–9839. [CrossRef]
21. Zhang, Z.; Wang, J.; Liu, D.; Luo, W.; Zhang, M.; Jiang, W.; Zhu, Y. Highly Efficient Organic Photocatalyst with Full Visible Light Spectrum through π-π Stacking of TCNQ-PTCDI. *ACS Appl. Mater. Interfaces* **2016**, *8*, 30225–30231. [CrossRef]
22. Wang, X.; Meng, J.; Yang, X.; Hu, A.; Yang, Y.; Guo, Y. Fabrication of a Perylene Tetracarboxylic Diimide-Graphitic Carbon Nitride Heterojunction Photocatalyst for Efficient Degradation of Aqueous Organic Pollutants. *ACS Appl. Mater. Interfaces* **2019**, *11*, 588–602. [CrossRef]
23. Zhang, J.; Zhao, X.; Wang, Y.; Gong, Y.; Cao, D.; Qiao, M. Peroxymonosulfate-enhanced visible light photocatalytic degradation of bisphenol A by perylene imide-modified g-C_3N_4. *Appl. Catal. B* **2018**, *237*, 976–985. [CrossRef]
24. Xing, C.; Yu, G.; Chen, T.; Liu, S.; Sun, Q.; Liu, Q.; Hu, Y.; Liu, H.; Li, X. Perylenetetracarboxylic diimide covalently bonded with mesoporous g-C_3N_4 to construct direct Z-scheme heterojunctions for efficient photocatalytic oxidative coupling of amines. *Appl. Catal. B* **2021**, *298*, 120534. [CrossRef]
25. Wang, Y.; Zhou, Y.; Hao, X.; Wang, Y.; Zou, Z. Z-scheme PTCDA/g-C_3N_4 photocatalyst based on interfacial strong interaction for efficient photooxidation of benzylamine. *Appl. Surf. Sci.* **2018**, *456*, 861–870. [CrossRef]
26. Yuan, Y.-J.; Shen, Z.-K.; Wang, P.; Li, Z.; Pei, L.; Zhong, J.; Ji, Z.; Yu, Z.-T.; Zou, Z. Metal-free broad-spectrum PTCDA/g-C_3N_4 Z-scheme photocatalysts for enhanced photocatalytic water oxidation. *Appl. Catal. B* **2020**, *260*, 118179. [CrossRef]

27. Wang, X.; Maeda, K.; Thomas, A.; Takanabe, K.; Xin, G.; Carlsson, J.M.; Domen, K.; Antonietti, M. A metal-free polymeric photocatalyst for hydrogen production from water under visible light. *Nat. Mater.* **2009**, *8*, 76–80. [CrossRef]
28. Tian, S.; Chen, S.; Ren, X.; Hu, Y.; Hu, H.; Sun, J.; Bai, F. An efficient visible-light photocatalyst for CO_2 reduction fabricated by cobalt porphyrin and graphitic carbon nitride via covalent bonding. *Nano Res.* **2020**, *13*, 2665–2672. [CrossRef]
29. Zong, X.; Niu, L.; Jiang, W.; Yu, Y.; An, L.; Qu, D.; Wang, X.; Sun, Z. Constructing creatinine-derived moiety as donor block for carbon nitride photocatalyst with extended absorption and spatial charge separation. *Appl. Catal. B* **2021**, *291*, 120099. [CrossRef]
30. Xie, Z.; Wang, W.; Ke, X.; Cai, X.; Chen, X.; Wang, S.; Lin, W.; Wang, X. A heptazine-based polymer photocatalyst with donor-acceptor configuration to promote exciton dissociation and charge separation. *Appl. Catal. B* **2023**, *325*, 122312. [CrossRef]
31. Wang, C.; Wan, Q.; Cheng, J.; Lin, S.; Savateev, A.; Antonietti, M.; Wang, X. Efficient aerobic oxidation of alcohols to esters by acidified carbon nitride photocatalysts. *J. Catal.* **2021**, *393*, 116–125. [CrossRef]
32. Yu, H.; Shi, R.; Zhao, Y.; Bian, T.; Zhao, Y.; Zhou, C.; Waterhouse, G.I.N.; Wu, L.-Z.; Tung, C.-H.; Zhang, T. Alkali-Assisted Synthesis of Nitrogen Deficient Graphitic Carbon Nitride with Tunable Band Structures for Efficient Visible-Light-Driven Hydrogen Evolution. *Adv. Mater.* **2017**, *29*, 1605148. [CrossRef] [PubMed]
33. Chen, Q.; Li, S.; Xu, H.; Wang, G.; Qu, Y.; Zhu, P.; Wang, D. Co-MOF as an electron donor for promoting visible-light photoactivities of g-C_3N_4 nanosheets for CO_2 reduction. *Chin. J. Catal.* **2020**, *41*, 514–523. [CrossRef]
34. Fu, J.; Zhu, B.; Jiang, C.; Cheng, B.; You, W.; Yu, J. Hierarchical Porous O-Doped g-C_3N_4 with Enhanced Photocatalytic CO_2 Reduction Activity. *Small* **2017**, *13*, 1603938. [CrossRef]
35. Li, Q.; Wang, S.; Sun, Z.; Tang, Q.; Liu, Y.; Wang, L.; Wang, H.; Wu, Z. Enhanced CH_4 selectivity in CO_2 photocatalytic reduction over carbon quantum dots decorated and oxygen doping g-C_3N_4. *Nano Res.* **2019**, *12*, 2749–2759. [CrossRef]
36. Qin, J.; Wang, S.; Ren, H.; Hou, Y.; Wang, X. Photocatalytic reduction of CO_2 by graphitic carbon nitride polymers derived from urea and barbituric acid. *Appl. Catal. B* **2015**, *179*, 1–8. [CrossRef]
37. Zhang, H.; Wei, J.; Dong, J.; Liu, G.; Shi, L.; An, P.; Zhao, G.; Kong, J.; Wang, X.; Meng, X.; et al. Efficient visible-light-driven carbon dioxide reduction by a single-atom implanted metal-organic framework. *Angew. Chem. Int. Ed.* **2016**, *55*, 14310–14314. [CrossRef]
38. Wang, S.; Yao, W.; Lin, J.; Ding, Z.; Wang, X. Cobalt Imidazolate Metal-Organic Frameworks Photosplit CO_2 under Mild Reaction Conditions. *Angew. Chem. Int. Ed.* **2014**, *53*, 1034–1038. [CrossRef]
39. Wang, S.; Lin, J.; Wang, X. Semiconductor-redox catalysis promoted by metal-organic frameworks for CO_2 reduction. *Phys. Chem. Chem. Phys.* **2014**, *16*, 14656–14660. [CrossRef]
40. Zhang, X.; Wang, M.; Song, X.; Yan, Y.; Huo, P.; Yan, Y.; Yang, B. Boosting charge carrier separation efficiency by constructing an intramolecular DA system towards efficient photoreduction of CO_2. *New J. Chem.* **2021**, *45*, 6042–6052. [CrossRef]
41. Fan, X.; Zhang, L.; Cheng, R.; Wang, M.; Li, M.; Zhou, Y.; Shi, J. Construction of Graphitic C_3N_4-Based Intramolecular Donor-Acceptor Conjugated Copolymers for Photocatalytic Hydrogen Evolution. *ACS Catal.* **2015**, *5*, 5008–5015. [CrossRef]
42. Song, X.; Li, X.; Zhang, X.; Wu, Y.; Ma, C.; Huo, P.; Yan, Y. Fabricating C and O co-doped carbon nitride with intramolecular donor-acceptor systems for efficient photoreduction of CO_2 to CO. *Appl. Catal. B* **2020**, *268*, 118736. [CrossRef]
43. Wei, Y.; Li, X.; Liu, Q.; Zhang, Y.; Zhang, K.; Huo, P.; Yan, Y. Leaf-Vein structure like g-C_3N_4/P-MWNTs donor-acceptor hybrid catalyst for efficient CO_2 photoreduction. *Carbon* **2022**, *188*, 59–69. [CrossRef]
44. Qu, D.; Liu, J.; Miao, X.; Han, M.; Zhang, H.; Cui, Z.; Sun, S.; Kang, Z.; Fan, H.; Sun, Z. Peering into water splitting mechanism of g-C_3N_4-carbon dots metal-free photocatalyst. *Appl. Catal. B* **2018**, *227*, 418–424. [CrossRef]
45. Jiang, W.; Zong, X.; An, L.; Hua, S.; Miao, X.; Luan, S.; Wen, Y.; Tao, F.F.; Sun, Z. Consciously Constructing Heterojunction or Direct Z-Scheme Photocatalysts by Regulating Electron Flow Direction. *ACS Catal.* **2018**, *8*, 2209–2217. [CrossRef]
46. Song, X.; Mao, W.; Wu, Y.; Wang, M.; Liu, X.; Zhou, W.; Huo, P. Fabricating Carbon Nitride-Based 3D/0D Intramolecular Donor-Acceptor Catalysts for Efficient Photoreduction of CO_2. *New J. Chem.* **2022**, *46*, 20225–20234. [CrossRef]
47. Hayat, A.; Rahman, M.U.; Khan, I.; Khan, J.; Sohail, M.; Yasmeen, H.; Liu, S.; Qi, K.; Lv, W. Conjugated Electron Donor-Acceptor Hybrid Polymeric Carbon Nitride as a Photocatalyst for CO_2 Reduction. *Molecules* **2019**, *24*, 1779. [CrossRef] [PubMed]
48. Zhong, H.; Hong, Z.; Yang, C.; Li, L.; Xu, Y.; Wang, X.; Wang, R. A Covalent Triazine-Based Framework Consisting of Donor-Acceptor Dyads for Visible-Light-Driven Photocatalytic CO_2 Reduction. *ChemSusChem* **2019**, *12*, 4493–4499. [CrossRef]
49. Xu, N.; Diao, Y.; Qin, X.; Xu, Z.; Ke, H.; Zhu, X. Donor-Acceptor Covalent Organic Frameworks of Nickel (II) Porphyrin for Selective and Efficient CO_2 Reduction into CO. *Dalton Trans.* **2020**, *49*, 15587–15591. [CrossRef]

Disclaimer/Publisher's Note: The statements, opinions and data contained in all publications are solely those of the individual author(s) and contributor(s) and not of MDPI and/or the editor(s). MDPI and/or the editor(s) disclaim responsibility for any injury to people or property resulting from any ideas, methods, instructions or products referred to in the content.

Article

Surface Plasmon Resonance Induced Photocatalysis in 2D/2D Graphene/g-C₃N₄ Heterostructure for Enhanced Degradation of Amine-Based Pharmaceuticals under Solar Light Illumination

Faisal Al Marzouqi [1] and Rengaraj Selvaraj [2,*]

[1] Department of Engineering, International Maritime College Oman, National University of Science and Technology, Falaj Al Qabail, P.O. Box 532, Suhar 322, Oman
[2] Department of Chemistry, College of Science, Sultan Qaboos University, Al-Khoudh, P.O. Box 36, Muscat 123, Oman
* Correspondence: rengaraj@squ.edu.om; Tel.: +968-2414-2436

Abstract: Pharmaceuticals, especially amine-based pharmaceuticals, such as nizatidine and ranitidine, contaminate water and resist water treatment. Here, different amounts of graphene sheets are coupled with g-C₃N₄ nanosheets (wt% ratio of 0.5, 1, 3 and 5 wt% of graphene) to verify the effect of surface plasmon resonance introduced to the g-C₃N₄ material. The synthesized materials were systematically examined by advanced analytical techniques. The prepared photocatalysts were used for the degradation of amine-based pharmaceuticals (nizatidine and ranitidine). The results show that by introducing only 3 wt% graphene to g-C₃N₄, the absorption ability in the visible and near-infrared regions dramatically enhanced. The absorption in the visible range was 50 times higher when compared to the pure sample. These absorption features suggest that the surfaces of the carbon nitride sheet are covered by the graphene nanosheet, which would effectively apply the LSPR properties for catalytic determinations. The enhancement in visible light absorption in the composite was confirmed by PL analysis, which showed greater inhibition of the electron-hole recombination process. The XRD showed a decrease in the (002) plan due to the presence of graphene, which prevents further stacking of carbon nitride layers. Accordingly, the Gr/g-C₃N₄ composite samples exhibited an enhancement in the photocatalytic performance, specifically for the 5% Gr/g-C₃N₄ sample, and close to 85% degradation was achieved within 20 min under solar irradiation. Therefore, applying the Gr/g-C₃N₄ for the degradation of a pharmaceutical can be taken into consideration as an alternative method for the removal of such pollutants during the water treatment process. This enhancement can be attributed to surface plasmon resonance-induced photocatalysis in a 2D/2D graphene/g-C₃N₄ heterostructure.

Keywords: amine-based pharmaceutical; graphene; g-C₃N₄; nizatidine; photocatalysts; solar irradiation

1. Introduction

Protecting the environment from pollution and delivering a sufficient amount of energy is vital to maintain our natural life on earth. Scientists are considering all the possible ways to keep the environment at a suitable level where the energy is supplied from a green sustainable source [1,2]. One of the possible solutions to maintain the green energy demand with the minimum environmental condition are the engineering and synthesizing of artificial photocatalysts with super photocatalytic activities. This field has attracted a huge scientific interest, resulting in many commercial products based on this technology. The current achievements in semiconductor materials are state-of-the-art technology, which has been applied in several necessary fields [1]. For example, water decontamination, environmental remediation, hydrogen production, photosensitive sensing, energy harvesting and energy-storing devices are intensively investigating the new implementation of semiconductor materials [2–6]. One of the goals is to fully utilize the abundantly free solar irradiation for driving significant chemical reactions for large-scale applications. There are numerous categories of semiconductor photocatalysts that have been investigated, such as oxynitrides, sulfides, oxides and metal-free semiconductors [7–9]. Free metal semiconductors are highly considered; graphitic carbon nitride (g-C_3N_4) has shown a new possible application in the photocatalytic field due to its compatible properties [10–13]. This property allows the synthesis of a visible active material with more accessible properties; based on that, a universal consideration has been provided to investigate this specific material.

Hence, the g-C_3N_4 materials demonstrated that they were suitable and compatible with energy and sustainability applications. A thorough review of the literature can give the readers a general idea about the conditions required to synthesize carbon nitride materials, which are relatively easy and faster compared to the synthesis of metal oxides and metal sulfides photocatalysts. In general, the carbon nitride samples can be obtained via the thermal polycondensation reaction (450–550 °C) of nitrogen-rich precursors in a semi-close system. So far, many precursors have been used as starting materials, such as cyanamide, dicyanamide, triazine, heptazine, melamine, urea, thiourea and so on. The most dominated one is melamine because the yield generated is much higher and the defect is less compared to other starting materials.

The research reported data present more and more understanding of the photocatalyst's material's methods of working. Therefore, the attention has been orientated to synthesizing narrow band gap semiconductors and efficient in visible light absorption, as opposed to what is commonly used, namely, materials with wide band gap photocatalysts, such as TiO_2 and ZnO [7,8]. The drawbacks of these traditional semiconductor-based photocatalysts are weak light absorption in the visible light region, toxicity, short-term stability and/or high material costs [9–12]. On the other hand, a polymeric semiconductor based on graphitic carbon nitride g-C_3N_4 has been investigated for water splitting under visible light irradiation. g-C_3N_4 has a band gap value of 2.7 eV, indicating its visible light response [14,15]. This semiconductor exhibits excellent needed properties, such as a 2D shape structure, suitable redox potential, low band gap, high thermal chemical stability and suitability for large-scale production from low-cost precursors, such as urea, thiourea and melamine [16–20].

The main limitation in using g-C_3N_4 material as a photocatalyst is the high recombination rate of photogenerated electrons and holes, which limits the photocatalytic efficiency. Thus, finding an appropriate way to overcome the stated problems is challenging. There are numerous methods used to enhance the photocatalytic activity. Coupling materials were used to approach the desired enhancement in photocatalytic performance, such as CdS/g-C_3N_4, TiO_2/g-C_3N_4 and ZnO/g-C_3N_4 [13–17]. On the other hand, an interesting approach is introducing a surface plasmon resonance (SPR) effect on the surface of the semiconductor material to enhance the electron-hole separation ability [21–30]. In a characteristic SPR phenomenon, the plasmonic electrons on the surface interact with the absorbed photons and oscillate on the surface of the metallic nanomaterials, resulting in an improvement of the local electromagnetic field and stimulating active electrons on the semiconductor, which leads to enhancement in the visible light response [18,20,21]. Primarily, noble metal nanoparticles like Au, Pt and Ag were used to introduce a localized surface plasmon resonance SPR to the surface of g-C_3N_4. However, noble metals are expensive and have a shallow surface interaction with the surface of the g-C_3N_4 material. On the other hand, graphene (Gr) is a more affordable non-metallic material and has a strong ability to introduce surface plasmon resonance similar to metallic particles [31–35].

Recent studies showed that 2D/2D stacking can provide a better interaction compared to 0D/2D and 1D/2D stacking structures. Graphene can provide strong 2D/2D stacking interaction with g-C_3N_4; therefore, Gr/g-C_3N_4 composite material is a very attractive catalyst from both experimental and theoretical aspects [24–26]. To date, only a few studies have demonstrated the synthesis of Gr/g-C_3N_4 composite material via different methods, such as hydrothermal, solvothermal, ionic-liquid and hydrolysis routes. Compared to these preparation methods, the thermal method dramatically reduces the experimental time and enhances the product purity [36–39]. The other methods are relatively long and require multiple steps to obtain the final product. The direct thermal method shows great potential in reducing the process time and steps. The advantage of combining carbon nitride with graphene is that both materials are chemically and thermally stable.

The reported Gr/g-C_3N_4 composite was mainly used for hydrogen evolution, lithium batteries and photocatalytic degradation of dyes. Currently, more research is oriented in the direction of wastewater and drinking water treatment from pharmaceuticals [40,41]. Amine-based pharmaceuticals, such as nizatidine and ranitidine, have garnered maximum researcher attention due to the ability of these compounds to generate toxic nitrogenous disinfection by-products, which are formed during the disinfection step. On the other hand, only a few studies have been performed in the area of photocatalytic degradation of amine-based pharmaceuticals as an alternative method for conventional wastewater treatment [42–45].

Herein, an easy direct thermal method for synthesizing 2D/2D Gr/g-C_3N_4 nanocomposite material is reported. The amount of Gr attached to the g-C_3N_4 sheets was investigated by varying the percentage of Gr from 0% up to 5% (pure g-C_3N_4, 0.5% Gr/g-C_3N_4, 1% Gr/g-C_3N_4, 3% Gr/g-C_3N_4 and 5% Gr/g-C_3N_4). Then, the effect of SPR on the g-C_3N_4 surface was examined. The photocatalytic performance of the photocatalysts was evaluated by the degradation of amine-based pharmaceuticals, nizatidine and ranitidine, under stable LED and direct solar light. The improvement in the visible light absorption in the combination was validated by optical and physical analyses, which showed superior inhibition of the electron-hole recombination process. Consequently, the Gr/g-C_3N_4 composite samples demonstrated a boost in the photocatalytic performance, specifically for the 5% Gr/g-C_3N_4 sample, and up to 85% degradation was achieved within 20 min under solar irradiation.

2. Results and Discussion

2.1. XRD Analysis

An X-ray diffraction analysis was performed to investigate the crystalline properties of the synthesized carbon nitride material. Figure 1 shows the XRD patterns of the Gr/g-C_3N_4 prepared via a direct thermal method. All the samples exhibited two main diffraction peaks. The first peak at around 27.90° corresponds to the (002) diffraction peaks characteristic of interlayer stacking of aromatic systems, and the second diffraction peak at around 13.05° is indexed to the (100) peak that represents inter-planar separation. These diffraction peaks are in good agreement with those reported for g-C_3N_4 and were retained during thermal oxidation [11,12], indicating the existence of the graphite-like structure of g-C_3N_4. The results show that increasing the amount of graphene amount reduces the diffraction peak 002. This kind of decrease in the 002 plan growth direction is expected. It can be attributed to the presence of graphene, which prevents further stacking of carbon nitride layers.

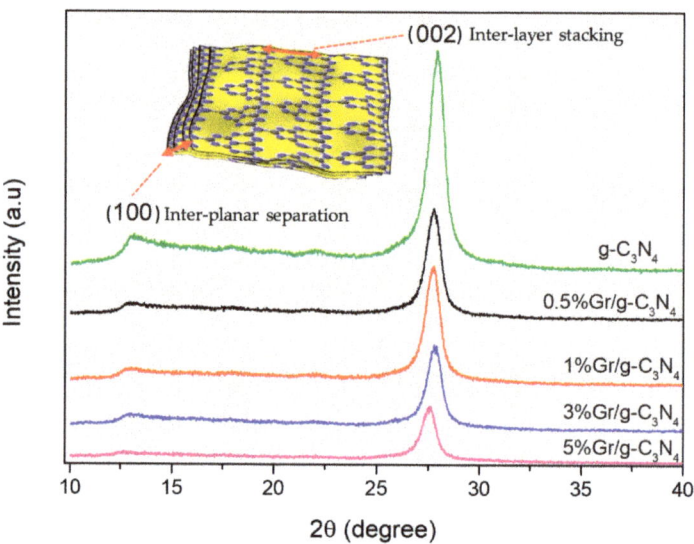

Figure 1. The XRD patterns of the Gr/g-C$_3$N$_4$ were prepared via the thermal method (inset schematic presentation of graphene stacking between g-C$_3$N$_4$ layers).

2.2. SEM, EDX and TEM Analysis

The morphological features of the Gr/g-C$_3$N$_4$ samples prepared via the direct heating method were determined by scanning electron microscopy (SEM). Figure 2 shows the SEM images of different percentages of the Gr/g-C$_3$N$_4$ samples. The g-C$_3$N$_4$ image depicted sheet-like microstructures. Moreover, the fabrication of g-C$_3$N$_4$ in the presence of graphene did not change the sheet structure of carbon nitride. The amount of graphene added to the surface of the g-C$_3$N$_4$ sheet was increased from 0.5% to 5%. However, the 2D/2D type of composite was expected to enhance the photocatalytic performance due to the higher area of interaction [24]. On the other hand, energy dispersive X-ray spectroscopy (EDXs) was also used to identify and confirm the elemental composition of the synthesized samples. The elemental composition of the prepared composite samples has been measured via EDXs analysis and the result is shown in Figure 3. It is seen that the sample was composed mainly of three main elements: carbon, nitrogen and oxygen. The atomic ratio of C:N was 56.9:38.5 wt%. These results further confirm the high purity of the produced Gr/g-C$_3$N$_4$ sample. To confirm the coupling of the graphene nanosheets and the carbon nitride nanosheets in a 2D/2D structure, a TEM analysis was carried out. Figure 4a shows the TEM image of the pure graphene sample. Figure 4b shows the TEM image of the Gr/g-C$_3$N$_4$ sample. High magnification on the selected area is presented in Figure 4c. The image clearly shows the presence of both sheets, indicating the formation of the Gr/g-C$_3$N$_4$ composite.

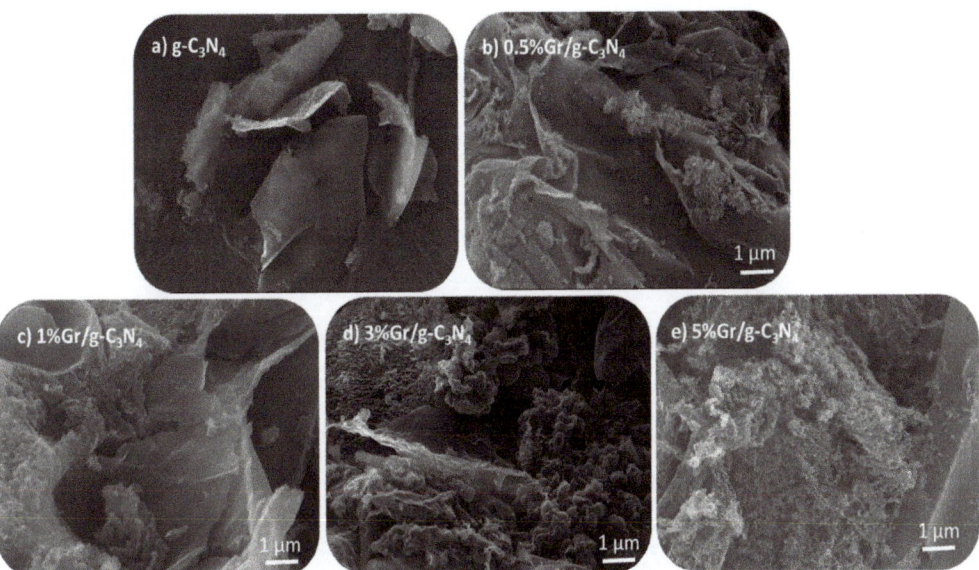

Figure 2. The SEM images of (**a**) g-C$_3$N$_4$, (**b**) 0.5%Gr/g-C$_3$N$_4$, (**c**) 1%Gr/g-C$_3$N$_4$, (**d**) 3%Gr/g-C$_3$N$_4$ and (**e**) 5%Gr/g-C$_3$N$_4$ samples.

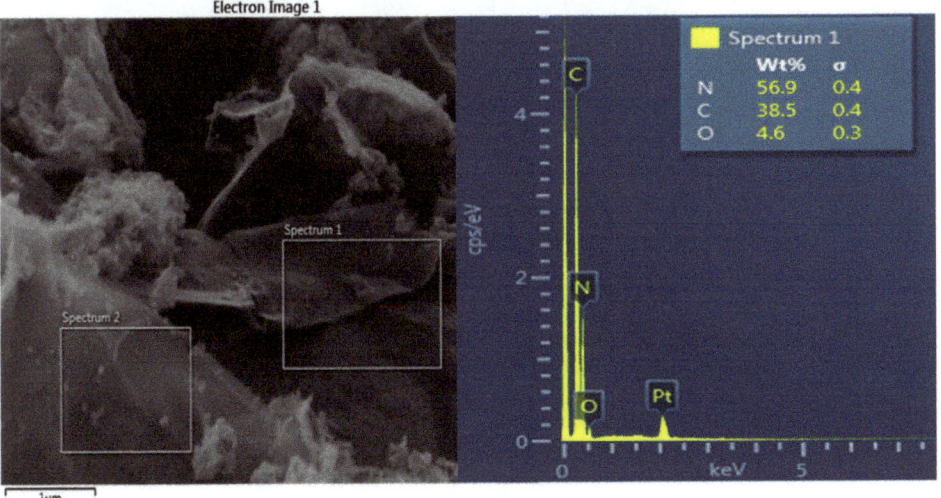

Figure 3. The element's presence in the Gr/g-C$_3$N$_4$ sample.

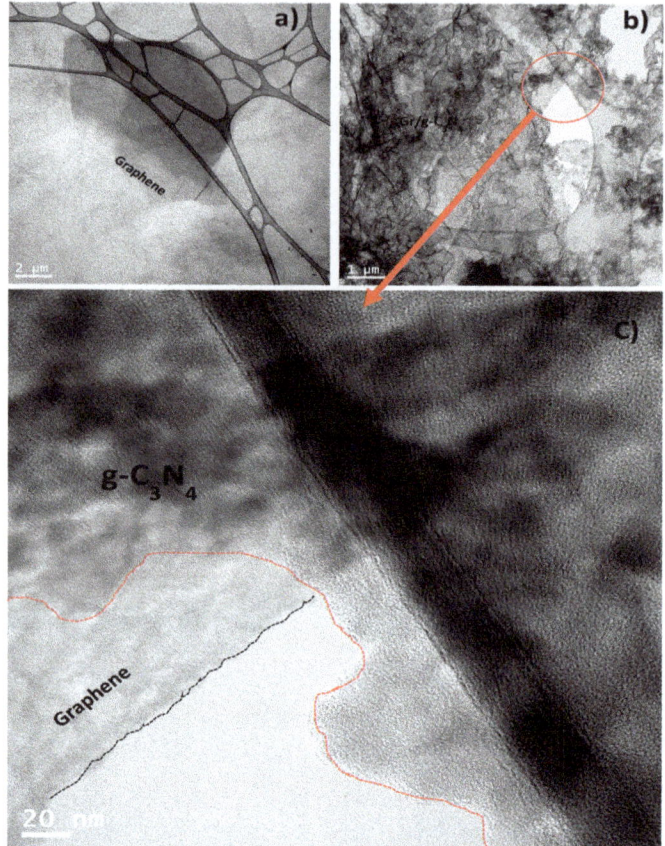

Figure 4. TEM images of (**a**) the graphene sheet, (**b**) the Gr/g-C₃N₄ composite samples and (**c**) high resolution of a selected area (red arrow) of the Gr/g-C₃N₄ composite.

2.3. UV-DRS Analysis

The UV-Vis diffuse reflectance spectra (UV-DRS) were used to investigate the optical properties of the as-prepared photocatalysts (Figure 5). In general, the absorbance spectra of direct inter-band transition energies of the prepared material are located at the edge of the visible region, which is compatible with the small band gap energy (2.7 eV). The UV diffuse reflectance spectrum of the synthesized samples is shown in Figure 5a. The fundamental absorption edge of the g-C₃N₄ material was about 450 nm, which is considered to be in the visible light range. Moreover, the coupling of g-C₃N₄ with graphene showed a small red shift of the band edge, which is expected to enhance the photocatalytic performance of the heterostructure. The optical band gap was calculated according to the following Tauc equation (Equation (1)):

$$\alpha h\nu = A(h\nu - E_g)^{n/2} \qquad (1)$$

where α, ν, A and E_g are the absorption coefficient, light frequency, proportionality constant and band gap, respectively. The band gap energy is obtained from the slope drawn near the band edge. The band gap values for g-C₃N₄, 0.5%Gr/g-C₃N₄, 1%Gr/g-C₃N₄, 3%Gr/g-C₃N₄ and 5%Gr/g-C₃N₄ samples were 2.76, 2.75, 2.72, 2.73 and 2.73 eV, respectively (Figure 5b–f). The small changes observed in the band gap were expected because the main role of graphene is to facilitate the separation of charge carriers. Moreover, increasing the amount of graphene ends up elevating the absorption tail due to the plasmonic effect of free electrons on the surface of the graphene [22,23]. The addition of 3% and 5% graphene on the g-C$_3$N$_4$ structure enhanced the absorption in the visible range by 50 times when compared to the pure sample. These absorption features suggest that the surfaces of the carbon nitride sheet are covered by the graphene nanosheet, which would effectively apply the LSPR properties for catalytic determinations.

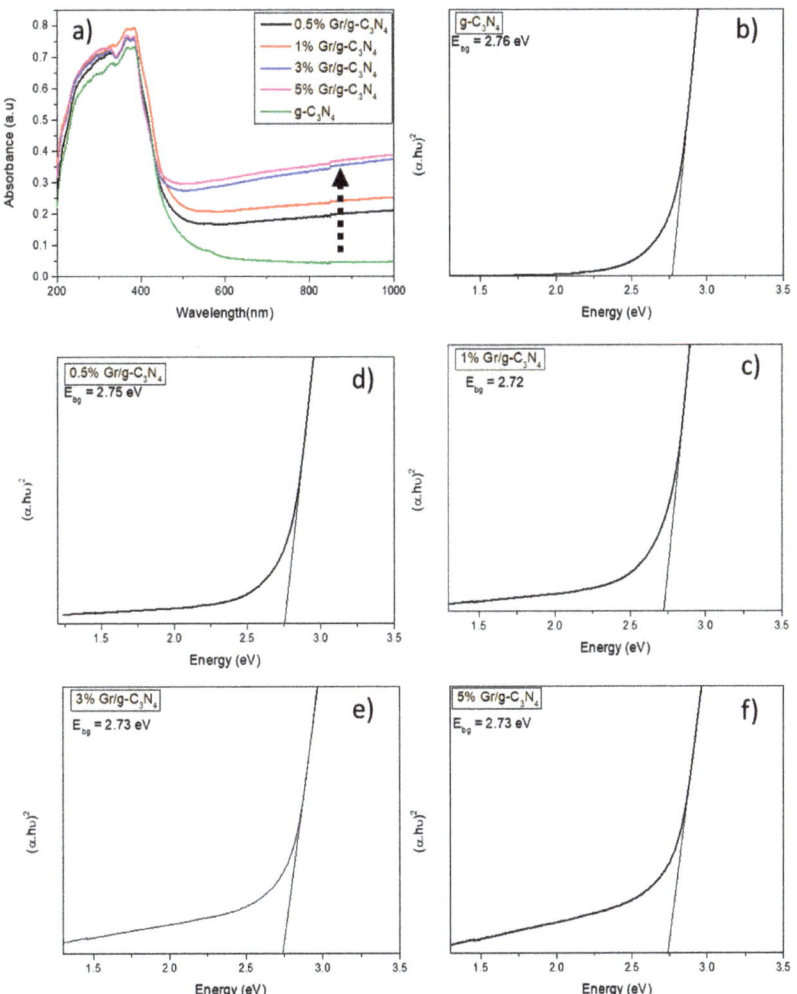

Figure 5. (a) UV diffuse reflectance spectra of the obtained samples; (b–f) the corresponding Tauc plot of the samples.

2.4. Photoluminescence Analysis (PL)

The photoluminescence emission peak is mainly considered a result of the recombination process of the photo-generated electrons and hole pairs. In general, the photoluminescence emission peak intensity is higher, indicating a higher recombination rate for photo-generated electrons and holes [34]. Figure 6a shows the PL emission spectra of the g-C$_3$N$_4$ and Gr/g-C$_3$N$_4$ composite samples. All the samples were exposed to an excitation process at a wavelength of 370 nm at room temperature and the main emission peak is observed at about 450 nm. The PL intensities of g-C$_3$N$_4$ reduced dramatically after coupling g-C$_3$N$_4$ with graphene, indicating the inhabitation of the recombination process of free charge carriers in the composite samples. Moreover, the 5% Gr/g-C$_3$N$_4$ has the lowest PL peak intensity compared to the other samples. The Gaussian fitting was used to convolute the photoluminescence peaks, as shown in Figure 6a, which helps us to obtain a clear understanding of excitons in the Gr/g-C$_3$N$_4$ samples and the origin of the emission concerning the initial precursors. All the samples showed three emission peaks. The carbon nitride materials are expected to have three states formed due to the presence of the sp3 C–Nσ band, sp2 C–Nπ band and the lone pair (LP) state of the bridge nitride atom [35,36]. To confirm the PL shift, the Commission Internationale de l'Eclairage (CIE) chromaticity diagram of the pure g-C$_3$N$_4$ sample and the Gr/g-C$_3$N$_4$ composite sample is presented in Figure 6b. The CIE (x, y) coordinate of the pure g-C$_3$N$_4$ samples located at (0.17, 0.18) were the CIE (x, y) coordinate of the Gr/g-C$_3$N$_4$ composite located at (0.19, 0.19), which further confirm that the PL emission is shifted toward light blue-violet. This further confirms that the PL emission is covering the blue-violet to close green light region. Additionally, the emission of carbon nitride products obtained from 5% graphene is located close to the edge of the blue-violet region, whereas the sample obtained from pure carbon nitride showed deep blue-violet. The differences in the colours further confirms the enhancement in visible light absorbance.

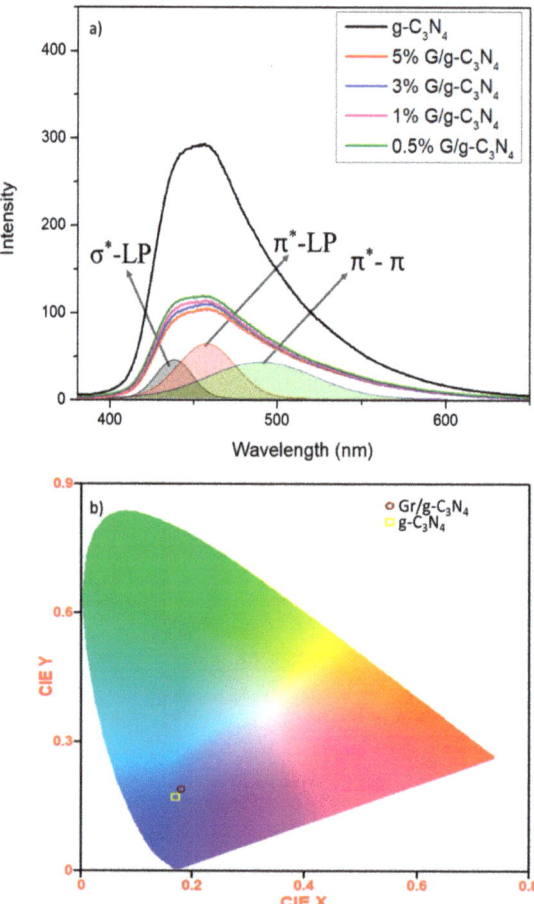

Figure 6. (a) The PL emission spectra of the prepared samples; (b) the (CIE) chromaticity diagram of the pure g-C$_3$N$_4$ sample and the Gr/g-C$_3$N$_4$ composite.

2.5. FTIR Analysis

The overlay FTIR spectra were measured to identify the characteristic peaks of the prepared samples. The response was recorded for the samples at the wavelength range of 600 to 4000 cm^{-1}. In general, all the samples demonstrate their graphitic structure, which can normally be shown as three main regions. Figure 7a shows the Fourier transform infrared (FTIR) spectrum of the as-prepared samples to identify the specific interaction of the functional groups. The result indicates the presence of the graphite-like structure of carbon nitride. The N-H stretching modes and the O-H from water absorbed on the surface are present in the broad peak observed in the range of 3000–3500 cm^{-1}. The bands around 1200–1600 cm^{-1} are characteristic of a typical stretching mode of CN heterocycles. In addition, the s-triazine ring mode was observed at 801 cm^{-1}. However, there was some broadening in the peak at 3000–3500 cm^{-1}, which is indexed to CO vibration. One of the interesting properties that carbon nitride has is a surface of multiple functional groups, as presented in Figure 7b. These groups influence the behaviors of the prepared materials. The most common functional groups that appear while the preparation of carbon nitride is primary are secondary amine groups (CNH$_2$ and C$_2$NH) due to a small amount of hydrogen remaining from the initial precursor. The presence of an amine group makes carbon nitride exhibit electron-rich properties with basic behaviors and the ability for H-bonding motifs formation [11,12]. This impurity makes carbon nitride more applicable as catalysis compared to perfect and defect-free g-C$_3$N$_4$. This amine group makes carbon nitride materials more suitable for the removal of acidic toxic compounds via chemical adsorption on the surface with the help of electrostatic interactions.

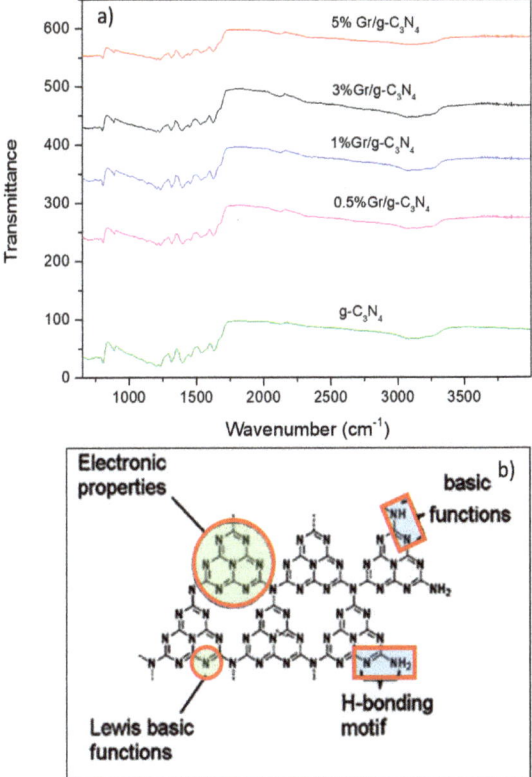

Figure 7. (a) FTIR spectrum of the g-C$_3$N$_4$ samples; (b) the functional groups on g-C$_3$N$_4$.

2.6. Photocatalytic Activity Test

The degradation of pharmaceuticals using the obtained Gr/g-C_3N_4 composite samples by the direct heating method was investigated under visible light and under direct solar light irradiation. The degradation of the selected pharmaceutical compounds was followed with the help of a UV spectrophotometer. The maximum absorbance peak for nizatidine and ranitidine was the same and was located at 312 nm (lambda max). Figure 8a represents the concentration changes of nizatidine starting from an initial concentration of 5 mg/L of the nizatidine aqueous solution at pH = 5.6 (with 5% Gr/g-C_3N_4). Figure 8b represents the concentration changes of ranitidine starting at the same concentration. Moreover, a control experiment was carried out to verify the effect of the visible light on the pure g-C_3N_4. The initial concentration remained almost the same for the pure sample, indicating that the g-C_3N_4 by itself is not very active under this condition. However, the degradation was dramatically enhanced after the addition of the prepared Gr/g-C_3N_4 catalysts. All the prepared composite samples showed higher degradation performance. As expected, the Gr/g-C_3N_4 composite samples exhibited superior photocatalytic performance compared to the pure sample. The 5% Gr/g-C_3N_4 showed the best performance among all the prepared samples (see Figure 8c,d). The enhancement noticed for the 5% Gr/g-C_3N_4 sample could be attributed to the coupling of two materials, which facilitates an effective separation of the charge carriers, as shown in the UVDRS and PL results. Finally, the best sample was chosen to be tested under direct solar light irradiation (Figure 9a). The sample showed more enhanced performance under solar irradiation due to more light intensity, more than 80% degradation achieved within 20 min. Figure 9b shows a schematic presentation of the charge carrier formation and the mechanism of the degradation.

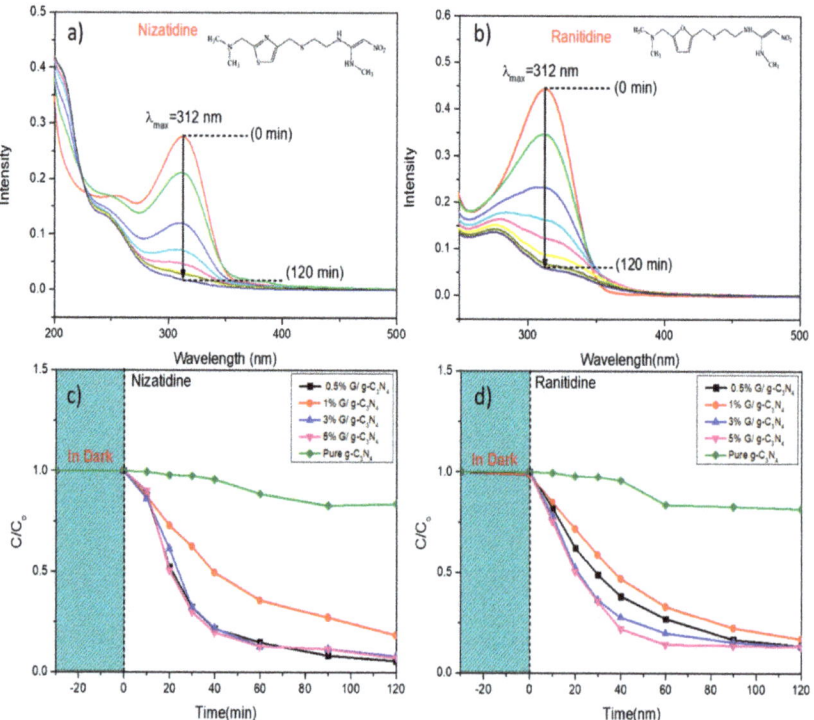

Figure 8. (**a**) The degradation rate of nizatidine and (**b**) ranitidine at an initial concentration of 5 mg/L and pH = 5.6 with 5% Gr/g-C_3N_4. (**c,d**) C/C_o Plots for nizatidine and ranitidine, respectively, under UV irradiation.

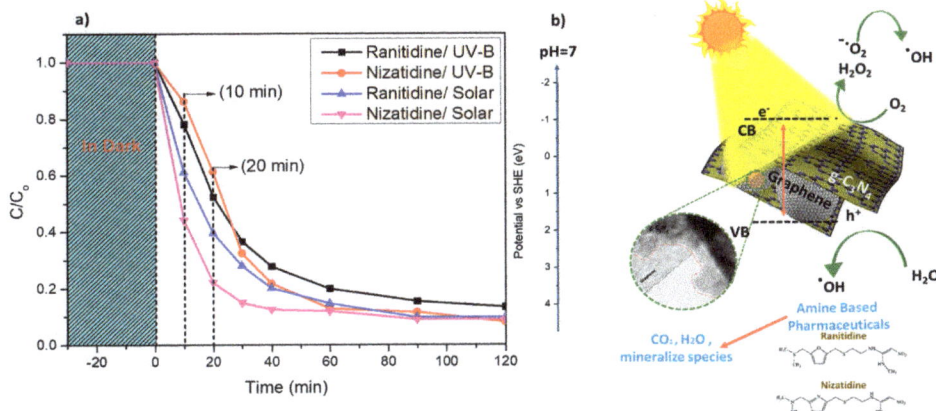

Figure 9. (a) C/C_0 Plots for nizatidine and ranitidine under UV and Solar irradiation. (b) the photocatalytic mechanism under solar light.

3. Materials and Methods

3.1. Materials

The melamine powder (M2659 Aldrich, St. Louis, MO, USA) and graphene oxide (763705-100 ML, St. Louis, MO, USA) were purchased from Sigma-Aldrich and were used without further purification.

3.2. Characterization

The prepared samples were examined by an X-ray diffraction (XRD, Malvern, UK) test using an XRD Panalytical X-pert Pro instrument equipped with graphite monochromatized Cu Kα radiation (λ = 1.540 A°). The detector was NaI (T1). The samples' morphologies were observed using a field emission scanning electron microscope (FESEM, JSM-7800F JOEL, Tokyo, Japan) with a maximum working voltage of 30 kV, a maximum resolution of 0.8 nm and a working distance of 10 mm used during measurements where the elements were present near the surface, and analyzed by energy dispersive X-ray spectroscopy (EDX, JOEL, Tokyo, Japan). The transmission electron microscope (TEM, JOEL, Tokyo, Japan) model JEM-1400-JEOL was used for high-resolution analysis. The UV-Vis diffuse reflection spectroscopy (UV-Vis DRS, Waltham, MA, USA) measurements were conducted using the Perkin Elmer Lambda 650S spectrometer. The photoluminescence (PL, Waltham, MA, USA) behavior was evaluated using a Perkin-Elmer LS 55 Luminescence Spectrometer. The degradation of amine-based pharmaceuticals (nizatidine and ranitidine) was analyzed by a Shimadzu UV-1800 UV/Visible Scanning Spectrophotometer (Shimadzu, Kyoto, Japan).

3.3. Synthesis of Gr/g-C_3N_4 Composite Materials

The graphene (Gr)-based carbon nitride materials (g-C_3N_4) were prepared by mixing a specific volume of a graphene oxide solution with melamine powder. One gram of melamine powder was mixed with the required wt% ratio of 0.5, 1, 3 and 5 wt% of graphene, then placed in an alumina crucible and covered with a lid. The mixture of the graphene oxide and melamine powder was left to dry overnight in an air oven at 80 °C. The dried sample was crushed to a fine powder to ensure a homogeneous distribution of graphene in the mixture. Then, direct thermal heating was applied up to 550 °C at a heating rate of 20 °C/min and stabilized at 550 °C for 3 h. After cooling, the product was collected for further analysis and the process was repeated to obtain the required amount. Simplified schematic steps of the synthesis are presented in Figure 10. There was no further washing or purification performed. A pure carbon nitride sample was obtained by the same method without graphene added to the melamine powder.

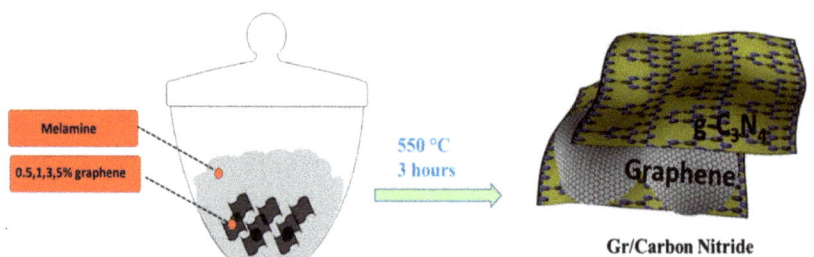

Figure 10. Schematic steps of the synthesis of Gr/g-C_3N_4.

3.4. Photocatalytic Test of (Gr/g-C_3N_4)

The photocatalytic activity performances of the as-prepared nanosheet were measured by following the degradation of two amine-based pharmaceutical compounds (nizatidine and ranitidine). The photoreaction analyses were conducted using a batch system reactor involving a cylindrical borosilicate glass vessel with an effective volume of 500 mL. All photoreaction was performed in an open atmosphere at a stable temperature (25 °C) with a cold water circulation system. The reactor was attached to an air diffuser machine to uniformly disperse the air into the solution. The suspensions (catalyst and pollutant solution) were prepared by adding 0.1 g of the prepared Gr/g-C_3N_4 powder into 250 mL of an aqueous solution of a nizatidine- and ranitidine-contaminated solution with an initial concentration of 5 mg/L (5 ppm). The system was chosen to be conducted in an open atmosphere with an air diffuser fixed at the reactor to mimic a real-life situation and uniformly disperse the air into the solution. The reaction suspensions were magnetically stirred for 30 min in the dark to ensure adsorption–desorption equilibrium between the photocatalyst and the pharmaceuticals. One sample was taken after this step to evaluate the adsorption amount. During illumination with light, about 6 mL of the suspension solution was taken from the reactor at a scheduled interval. The samples were filtered to remove the catalyst. For a stable UV source of light, a Mic-LED-365 (from Prizmatix 420 mW) was used to activate the photocatalytic reaction, then the experiment was repeated under direct solar irradiation with an average intensity in the range of 1100–1300 W/m^2.

4. Conclusions

In conclusion, the Gr/g-C_3N_4 composite photocatalytic materials were successfully synthesized by the direct thermal method and were demonstrated to be a highly competitive catalytic system with superior activity for visible-light-induced degradation of amine-based pharmaceuticals (nizatidine and ranitidine). The results from X-ray diffraction, the samples' morphologies and the surface analyzed by energy dispersive X-ray spectroscopy (EDX) and PL indicate that the selected method can simplify the preparation of Gr/g-C_3N_4 compared to other techniques. The photocatalytic tests of the prepared Gr/g-C_3N_4 samples showed higher efficiency for the degradation of amine-based pharmaceutical models under solar light irradiation. The sample with a 5% graphene to g-C_3N_4 ratio showed the highest photocatalytic activity compared to lower graphene percentages (0.5%, 1%, 2% and 3%). The degradation reached 85% within only 20 min. Therefore, applying Gr/g-C_3N_4 for the degradation of a pharmaceutical can be taken into consideration as an alternative method for the removal of such pollutants during the water treatment process. This enhancement can be attributed to surface plasmon resonance-induced photocatalysis in a 2D/2D graphene/g-C_3N_4 heterostructure. Thus, 2D/2D graphene/g-C_3N_4 heterostructures can be useful materials for the removal of pharmaceutical pollutants from water using solar energy.

Author Contributions: Methodology, R.S.; formal analysis, F.A.M.; investigation, F.A.M.; writing—original draft preparation, F.A.M.; writing—review and editing, R.S.; supervision, R.S.; project administration, R.S.; funding acquisition, R.S. All authors have read and agreed to the published version of the manuscript.

Funding: This research received no external funding.

Data Availability Statement: The data presented in this study are available on request from the corresponding author. The data are not publicly available due to the nature of the data.

Acknowledgments: Faisal Al Marzouqi wishes to thank the International Maritime College Oman, Sultanate of Oman for the support (2022/CRG 09). Rengaraj Selvaraj acknowledges the Surface Science Lab, Department of Physics, College of Science, Sultan Qaboos University and The Central Analytical and Applied Research Unit (CAARU) College of Science, Sultan Qaboos University, Oman.

Conflicts of Interest: The authors declare no conflict of interest.

References

1. Ahmad, R.; Ahmad, Z.; Khan, A.U.; Mastoi, N.R.; Aslam, M.; Kim, J. Photocatalytic systems as an advanced environmental remediation: Recent developments, limitations and new avenues for applications. *J. Environ. Chem. Eng.* **2016**, *4*, 4143–4164. [CrossRef]
2. Al Balushi, B.S.; Al Marzouqi, F.; Al Wahaibi, B.; Kuvarega, A.T.; Al Kindy, S.M.; Kim, Y.; Selvaraj, R. Hydrothermal synthesis of CdS sub-microspheres for photocatalytic degradation of pharmaceuticals. *Appl. Surf. Sci.* **2018**, *457*, 559–565. [CrossRef]
3. Chen, X.; Shen, S.; Guo, L.; Mao, S.S. Semiconductor-based photocatalytic hydrogen generation. *Chem. Rev.* **2010**, *110*, 6503–6570. [CrossRef] [PubMed]
4. Fresno, F.; Portela, R.; Suárez, S.; Coronado, J.M. Photocatalytic materials: Recent achievements and near future trends. *J. Mater. Chem. A* **2014**, *2*, 2863–2884. [CrossRef]
5. Liu, J.; Zhang, Y.; Lu, L.; Wu, G.; Chen, W. Self-regenerated solar-driven photocatalytic water-splitting by urea derived graphitic carbon nitride with platinum nanoparticles. *Chem. Commun.* **2012**, *48*, 8826–8828. [CrossRef]
6. Tauc, J.; Grigorovici, R.; Vancu, A. Optical properties and electronic structure of amorphous germanium. *Phys. Status Solidi B* **1966**, *15*, 627–637. [CrossRef]
7. Al Marzouqi, F.; Al-Balushi, N.A.; Kuvarega, A.T.; Karthikeyan, S.; Selvaraj, R. Thermal and hydrothermal synthesis of WO3 nanostructure and its optical and photocatalytic properties for the degradation of Cephalexin and Nizatidine in aqueous solution. *Mater. Sci. Eng. B* **2021**, *264*, 114991. [CrossRef]
8. Al Sarihi, F.T.; Al Marzouqi, F.; Kuvarega, A.T.; Karthikeyan, S.; Selvaraj, R. Easy conversion of BiOCl plates to flowers like structure to enhance the photocatalytic degradation of endocrine disrupting compounds. *Mater. Res. Express* **2020**, *6*, 125537. [CrossRef]
9. Meetani, M.A.; Alaidaros, A.; Hisaindee, S.; Alhamadat, A.; Selvaraj, R.; Al Marzouqi, F.; Rauf, M.A. Photocatalytic degradation of acetaminophen in aqueous solution by Zn-0.2 Cd0. 8S catalyst and visible radiation. *Desalination Water Treat.* **2019**, *138*, 270–279. [CrossRef]
10. Al Marzouqi, F.; Kim, Y.; Selvaraj, R. Shifting of the band edge and investigation of charge carrier pathways in the CdS/g-C3N4 heterostructure for enhanced photocatalytic degradation of levofloxacin. *New J. Chem.* **2019**, *43*, 9784–9792. [CrossRef]
11. Al Marzouqi, F.; Selvaraj, R.; Kim, Y. Thermal oxidation etching process of g-c3n4nanosheets from their bulk materials and its photocatalytic activity under solar light irradiation. *Desalination Water Treat.* **2018**, *116*, 267–276. [CrossRef]
12. Al Marzouqi, F.; Selvaraj, R.; Kim, Y. Rapid photocatalytic degradation of acetaminophen and levofloxacin using g-C3N4 nanosheets under solar light irradiation. *Mater. Res. Express* **2020**, *6*, 125538. [CrossRef]
13. Bai, Y.; Chen, T.; Wang, P.; Wang, L.; Ye, L.; Shi, X.; Bai, W. Size-dependent role of gold in g-C3N4/BiOBr/Au system for photocatalytic CO_2 reduction and dye degradation. *Sol. Energy Mater. Sol. Cells* **2016**, *157*, 406–414. [CrossRef]
14. Cheng, F.; Yin, H.; Xiang, Q. Low-temperature solid-state preparation of ternary CdS/g-C3N4/CuS nanocomposites for enhanced visible-light photocatalytic H2-production activity. *Appl. Surf. Sci.* **2017**, *391*, 432–439. [CrossRef]
15. He, Y.; Wang, Y.; Zhang, L.; Teng, B.; Fan, M. High-efficiency conversion of CO_2 to fuel over ZnO/g-C3N4 photocatalyst. *Appl. Catal. B Environ.* **2015**, *168*, 1–8. [CrossRef]
16. Li, K.; Gao, S.; Wang, Q.; Xu, H.; Wang, Z.; Huang, B.; Dai, Y.; Lu, J. In-situ-reduced synthesis of Ti^{3+} self-doped TiO_2/g-C3N4 heterojunctions with high photocatalytic performance under LED light irradiation. *ACS Appl. Mater. Interfaces* **2015**, *7*, 9023–9030. [CrossRef]
17. Al Marzouqi, F.; Al Farsi, B.; Kuvarega, A.T.; Al Lawati, H.A.; Al Kindy, S.M.; Kim, Y.; Selvaraj, R. Controlled Microwave-Assisted Synthesis of the 2D-BiOCl/2D-g-C3N4 Heterostructure for the Degradation of Amine-Based Pharmaceuticals under Solar Light Illumination. *ACS Omega* **2019**, *4*, 4671–4678. [CrossRef]
18. Li, H.; Jing, Y.; Ma, X.; Liu, T.; Yang, L.; Liu, B.; Yin, S.; Wei, Y.; Wang, Y. Construction of a well-dispersed Ag/graphene-like gC3N4 photocatalyst and enhanced visible light photocatalytic activity. *RSC Adv.* **2017**, *7*, 8688–8693. [CrossRef]

19. Peng, B.; Tang, L.; Zeng, G.; Fang, S.; Ouyang, X.; Long, B.; Zhou, Y.; Deng, Y.; Liu, Y.; Wang, J. Self-powered photoelectrochemical aptasensor based on phosphorus doped porous ultrathin g-C3N4 nanosheets enhanced by surface plasmon resonance effect. *Biosens. Bioelectron.* **2018**, *121*, 19–26. [CrossRef]
20. Tonda, S.; Kumar, S.; Shanker, V. Surface plasmon resonance-induced photocatalysis by Au nanoparticles decorated mesoporous g-C3N4 nanosheets under direct sunlight irradiation. *Mater. Res. Bull.* **2016**, *75*, 51–58. [CrossRef]
21. Wang, H.; Sun, T.; Chang, L.; Nie, P.; Zhang, X.; Zhao, C.; Xue, X. The g-C3N4 nanosheets decorated by plasmonic Au nanoparticles: A heterogeneous electrocatalyst for oxygen evolution reaction enhanced by sunlight illumination. *Electrochim. Acta* **2019**, *303*, 110–117. [CrossRef]
22. Farmani, A.; Mir, A. Graphene sensor based on surface plasmon resonance for optical scanning. *IEEE Photonics Technol. Lett.* **2019**, *31*, 643–646. [CrossRef]
23. Islam, M.; Sultana, J.; Biabanifard, M.; Vafapour, Z.; Nine, M.; Dinovitser, A.; Cordeiro, C.; Ng, B.-H.; Abbott, D. Tunable localized surface plasmon graphene metasurface for multiband superabsorption and terahertz sensing. *Carbon* **2020**, *158*, 559–567. [CrossRef]
24. Ji, M.; Di, J.; Ge, Y.; Xia, J.; Li, H. 2D-2D stacking of graphene-like g-C3N4/Ultrathin Bi4O5Br2 with matched energy band structure towards antibiotic removal. *Appl. Surf. Sci.* **2017**, *413*, 372–380. [CrossRef]
25. Liu, W.; Qiao, L.; Zhu, A.; Liu, Y.; Pan, J. Constructing 2D BiOCl/C3N4 layered composite with large contact surface for visible-light-driven photocatalytic degradation. *Appl. Surf. Sci.* **2017**, *426*, 897–905. [CrossRef]
26. Wang, Q.; Wang, W.; Zhong, L.; Liu, D.; Cao, X.; Cui, F. Oxygen vacancy-rich 2D/2D BiOCl-g-C3N4 ultrathin heterostructure nanosheets for enhanced visible-light-driven photocatalytic activity in environmental remediation. *Appl. Catal. B Environ.* **2018**, *220*, 290–302. [CrossRef]
27. Tompsett, G.A.; Conner, W.C.; Yngvesson, K.S. Microwave synthesis of nanoporous materials. *Chemphyschem* **2006**, *7*, 296–319. [CrossRef]
28. Pu, X.; Zhang, D.; Gao, Y.; Shao, X.; Ding, G.; Li, S.; Zhao, S. One-pot microwave-assisted combustion synthesis of graphene oxide–TiO2 hybrids for photodegradation of methyl orange. *J. Alloys Compd.* **2013**, *551*, 382–388. [CrossRef]
29. Cruz, M.; Gomez, C.; Duran-Valle, C.J.; Pastrana-Martínez, L.M.; Faria, J.L.; Silva, A.M.; Faraldos, M.; Bahamonde, A. Bare TiO2 and graphene oxide TiO2 photocatalysts on the degradation of selected pesticides and influence of the water matrix. *Appl. Surf. Sci.* **2017**, *416*, 1013–1021. [CrossRef]
30. Harish, S.; Archana, J.; Sabarinathan, M.; Navaneethan, M.; Nisha, K.; Ponnusamy, S.; Muthamizhchelvan, C.; Ikeda, H.; Aswal, D.; Hayakawa, Y. Controlled structural and compositional characteristic of visible light active ZnO/CuO photocatalyst for the degradation of organic pollutant. *Appl. Surf. Sci.* **2017**, *418*, 103–112. [CrossRef]
31. Bing, J.; Hu, C.; Zhang, L. Enhanced mineralization of pharmaceuticals by surface oxidation over mesoporous γ-Ti-Al2O3 suspension with ozone. *Appl. Catal. B Environ.* **2017**, *202*, 118–126. [CrossRef]
32. Liu, C.; Wang, J.; Chen, W.; Dong, C.; Li, C. The removal of DON derived from algae cells by Cu-doped TiO2 under sunlight irradiation. *Chem. Eng. J.* **2015**, *280*, 588–596. [CrossRef]
33. Sharma, A.; Ahmad, J.; Flora, S. Application of advanced oxidation processes and toxicity assessment of transformation products. *Environ. Res.* **2018**, *167*, 223–233. [CrossRef] [PubMed]
34. Wen, J.; Xie, J.; Chen, X.; Li, X. A review on g-C3N4-based photocatalysts. *Appl. Surf. Sci.* **2017**, *391*, 72–123. [CrossRef]
35. Dong, G.; Zhang, Y.; Pan, Q.; Qiu, J. A fantastic graphitic carbon nitride (g-C3N4) material: Electronic structure, photocatalytic and photoelectronic properties. *J. Photochem. Photobiol. C Photochem. Rev.* **2014**, *20*, 33–50. [CrossRef]
36. Xu, J.; Shalom, M.; Piersimoni, F.; Antonietti, M.; Neher, D.; Brenner, T.J. Color-Tunable Photoluminescence and NIR Electroluminescence in Carbon Nitride Thin Films and Light-Emitting Diodes. *Adv. Opt. Mater.* **2015**, *3*, 913–917. [CrossRef]
37. Liang, Z.; Xue, Y.; Wang, X.; Zhang, X.; Tian, J. Structure engineering of 1T/2H multiphase MoS2 via oxygen incorporation over 2D layered porous g-C3N4 for remarkably enhanced photocatalytic hydrogen evolu-tion. *Mater. Today Nano* **2022**, *18*, 100204. [CrossRef]
38. Fang, B.; Xing, Z.; Sun, D.; Li, Z.; Zhou, W. Hollow semiconductor photocatalysts for solar energy con-version. *Adv. Powder Mater.* **2022**, *1*, 100021. [CrossRef]
39. Liu, X.; Han, X.; Liang, Z.; Xue, Y.; Zhou, Y.; Zhang, X.; Cui, H.; Tian, J. Phosphorous-doped 1T-MoS2 decorated nitrogen-doped g-C3N4 nanosheets for enhanced photocatalytic nitrogen fixation. *J. Colloid Interface Sci.* **2022**, *605*, 320–329. [CrossRef]
40. Wang, C.; Liu, K.; Wang, D.; Wang, G.; Chu, P.K.; Meng, Z.; Wang, X. Hierarchical CuO–ZnO/SiO2 fibrous membranes for efficient removal of congo red and 4-nitrophenol from water. *Adv. Fiber Mater.* **2022**, *4*, 1069–1080. [CrossRef]
41. Wang, Y.; Li, Z.; Fu, W.; Sun, Y.; Dai, Y. Core–Sheath CeO2/SiO2 Nanofibers as Nanoreactors for Stabilizing Sinter-Resistant Pt, Enhanced Catalytic Oxidation and Water Remediation. *Adv. Fiber Mater.* **2022**, *4*, 1278–1289. [CrossRef]
42. Li, M.; Chen, X.; Li, X.; Dong, J.; Zhao, X.; Zhang, Q. Controllable strong and ultralight aramid nanofiber-based aerogel fibers for thermal insulation applications. *Adv. Fiber Mater.* **2022**, *4*, 1267–1277. [CrossRef]
43. Li, S.; Wang, C.; Liu, Y.; Liu, Y.; Cai, M.; Zhao, W.; Duan, X. S-scheme MIL-101 (Fe) octahedrons modified Bi2WO6 microspheres for photocatalytic decontamination of Cr (VI) and tetracycline hydrochloride: Synergistic insights, reaction pathways, and toxicity analysis. *Chem. Eng. J.* **2023**, *455*, 140943. [CrossRef]

44. Cai, M.; Liu, Y.; Wang, C.; Lin, W.; Li, S. Novel $Cd_{0.5}Zn_{0.5}S/Bi_2MoO_6$ S-scheme heterojunction for boosting the photodegradation of antibiotic enrofloxacin: Degradation pathway, mechanism and toxicity assessment. *Sep. Purif. Technol.* **2023**, *304*, 122401. [CrossRef]
45. Andreou, E.K.; Koutsouroubi, E.D.; Vamvasakis, I.; Armatas, G.S. Ni2P-Modified P-Doped Graphitic Carbon Nitride Hetero-Nanostructures for Efficient Photocatalytic Aqueous Cr (VI) Reduction. *Catalysts* **2023**, *13*, 437. [CrossRef]

Disclaimer/Publisher's Note: The statements, opinions and data contained in all publications are solely those of the individual author(s) and contributor(s) and not of MDPI and/or the editor(s). MDPI and/or the editor(s) disclaim responsibility for any injury to people or property resulting from any ideas, methods, instructions or products referred to in the content.

MDPI
St. Alban-Anlage 66
4052 Basel
Switzerland
www.mdpi.com

Catalysts Editorial Office
E-mail: catalysts@mdpi.com
www.mdpi.com/journal/catalysts

Disclaimer/Publisher's Note: The statements, opinions and data contained in all publications are solely those of the individual author(s) and contributor(s) and not of MDPI and/or the editor(s). MDPI and/or the editor(s) disclaim responsibility for any injury to people or property resulting from any ideas, methods, instructions or products referred to in the content.